THE OBSERVATION AND ANALYSIS
OF STELLAR PHOTOSPHERES

THE OBSERVATION AND ANALYSIS OF STELLAR PHOTOSPHERES

DAVID F. GRAY

University of Western Ontario
London, Ontario

A WILEY-INTERSCIENCE PUBLICATION

JOHN WILEY AND SONS, New York · London · Sydney · Toronto

Copyright © 1976 by John Wiley & Sons, Inc.

All rights reserved. Published simultaneously in Canada.

No part of this book may be reproduced by any means,
nor transmitted, nor translated into a machine language
without the written permission of the publisher.

Library of Congress Cataloging in Publication Data:

Gray, David F   1938–
  The observation and analysis of stellar photospheres.

  "A Wiley-Interscience publication."
  Includes bibliographies and index.
  1. Stellar photospheres.   I. Title.

QB809.G67   523.8'2   75-19229
ISBN 0-471-32380-2

Printed in the United States of America

10 9 8 7 6 5 4 3 2 1

PREFACE

The remarkable nature of stars is transmitted to us by the light they send. The light escapes from the outer layers of the star—called, by definition, the atmosphere. The complete atmosphere of a star can be viewed comprehensively as a transition from the stellar interior to the interstellar medium. And yet almost the whole visible stellar spectrum comes from a relatively thin part called the photosphere. Obviously we cannot disconnect the photosphere from the adjacent portions of the atmosphere, but in actual fact it is the only region we can study extensively for most stars. It is for this reason that the photosphere has taken its place as the central theme of this book.

Several books have appeared during the last decade dealing with the *theory* of stellar atmospheres. These works are for the most part excellent. It is to the material largely omitted by these books that the present treatise is directed. My students and I have felt for some time the need of a book that presents the basics of the field through the eyes of an observer and analyzer of stellar atmospheres.

An introduction to a subject, in my opinion, should be presented in a way that can be understood by a reader who has not studied the topic before. It follows that the material should be presented in as simple and straightforward a manner as possible. The Fourier transform (as covered in Chapter 2) is a unifying theme helping to accomplish this aim. Transforms come naturally into the material on data collection, optical instruments, the instrumental profile, line absorption coefficients, velocity fields, and spectral line analysis. In addition, I have selected and developed topics that I consider to be important to those of us who look at stars and attempt to understand what we see. At the same time, I have tried to present the material in the least complicated manner. The word "complicated" is affixed to things that are difficult to understand. Complicated things consequently are often

v

unsuitable topics for the novice. We should seek not the most general case conceivable, but the least complicated case that is serviceable (a version of the principle of minimum assumption).

The development of each of the main topics starts at an elementary level, proceeds with discussion of the topic, and ends by pointing the direction to the more advanced literature. It should be easy to expand from this book into the areas holding attraction for you. Realizing that astrophysics is a very dynamic field, I have documented (or otherwise made clear) the source of the material used in examples. When no source is indicated, the material is from observations or calculations of my own. The references in general have been selected because they are good illustrations of the material being discussed or because they have a basic lasting approach to the subject. I have also biased the referencing toward good starting points in the literature and toward review articles and journals to which the student is likely to have access. The references are listed at the end of each chapter and ordered according to the author's name and date of publication.

The first two chapters contain preparatory material. The main theme starts in Chapter 3 with a discussion of spectroscopic tools. Generally the continuous spectrum topics are developed first, followed by the somewhat more involved subject of the line spectrum. From Chapter 14 (Chemical Analysis) through Chapter 18, the material is oriented completely toward analysis and deduction. These later chapters are closely interlinked with the preceding chapters.

The book is suitable as a text for a one year course and as a reference to the more advanced reader.

DAVID F. GRAY

London, Ontario
April 1975

ACKNOWLEDGMENTS

I would like to thank the many colleagues, students, and other friends who have encouraged me in this work and given me information, instruction, and guidance. The funding of my research by the National Research Council of Canada over the past eight years has allowed me to do much of the research and studies out of which this book has grown. I am grateful for this material assistance. Thanks are due to Margot Neil for clerical help and to Gary Roberts for technical assistance. To my wife and children I express appreciation for their sacrifices and patience in hearing me talk through my work day in and day out. Special thanks go to Madeline Wallace and Mira Rasche for their kind and faithful services, encouragement, and dedication.

D. F. G.

CONTENTS

Chapter 1   Background   **1**

What is a Stellar Atmosphere? 1
The Spectral Type, 3
Magnitudes and Color Indices, 7
The Gas Law, 9
The Velocity Distributions, 10
Atomic Excitation and Ionization in
Thermodynamic Equilibrium, 11
Stellar Catalogs, Tables, and Atlases, 14
References, 16

Chapter 2   Fourier Transforms   **18**

The Definition, 18
Some Common Transforms, 22
Data Sampling and Data Windows, 28
The Convolution, 29
Convolution with a δ Function, 31
Convolutions of Gaussian and Dispersion Profiles, 32
Several Useful Theorems, 32
The Resolution Theorem, 34
Sampling and Aliasing, 36
Numerical Calculation of Transforms, 38
Noise Transfer between the Fourier Domains, 40
References, 42

Chapter 3   Spectroscopic Tools   **43**

Aspects of Telescopes, 43
Spectrographs—Some General Relations, 46
Diffraction Gratings, 48
The Blazed Reflection Grating, 55
The Wavelength of the True Blaze, 60
Shadowing, 62
Grating Ghosts, Satellites, and Anomalies, 63
Spectrograph Slit Magnification and Spectral
Purity, 68
Interferometers, 69
References, 73

Chapter 4   Detectors   **75**

What is a Photographic Plate? 75
Photomultiplier Tubes, 77
Quantum Efficiency and Spectral Response, 78
Detector Dark Output, 82
Linearity and the Photographic Plate, 85
Linearity and Other Photomultiplier
Characteristics, 88
Spacial Resolution, 91
Noise, 94
Other Detectors, 97
References, 97

Chapter 5   Radiation Terms and Definitions   **100**

Specific Intensity, 100
Flux, 102
The K Integral and Radiation Pressure, 104
The Absorption Coefficient and Optical Depth, 106
The Emission Coefficient and the Source Function, 106
Pure Isotropic Scattering, 107
Pure Absorption, 108
The Einstein Coefficients, 109
References, 110

Chapter 6   The Black Body and Its Radiation   111

Observed Relations, 112
Planck's Radiation Law, 114
Calculation of Black Body Radiation, 117
The Black Body as a Radiation Standard, 118
References, 120

Chapter 7   Radiative and Convective Energy Transport   121

The Transfer Equation and Its Formal Solution, 121
The Transfer Equation for Different Geometries, 123
The Flux Integral, 126
The Mean Intensity and the K Integrals, 128
Exponential Integrals, 129
Radiative Equilibrium, 130
The Grey Case, 132
Convective Transport, 136
Condition for Convective Flow, 137
References, 139

Chapter 8   The Continuous Absorption Coefficient   140

Origins of Continuous Absorption, 140
The Stimulated Emission Factor, 141
Neutral Hydrogen, 142
The Negative Hydrogen Ion, 147
Other Hydrogen Continuous Absorbers, 150
The Negative Helium Ion, 150
The Metals, 152
Scattering, 153
Other Sources of Opacity, 154
The Total Absorption Coefficient, 155
References, 155

Chapter 9   The Model Photosphere   158

The Hydrostatic Equation, 159
The Temperature Distribution in the Sun, 162
Temperature Distributions in Other Stars, 166

The P_g–P_e–T Relation, 171
Completion of the Model, 174
The Geometrical Depth Scale, 174
Computation of the Spectrum, 175
Properties of Models: Pressure Relations, 178
The Effect of Chemical Composition, 182
Changes with Temperature, 185
Model Tabulations, 186
Reflection, 187
References, 190

Chapter 10   The Measurement and Behavior of Stellar Continua   192

Scanners and Polychromators, 193
The Observation, 198
Absolute Calibration, 200
Photometric Standard Stars, 204
Measured Continua, 204
Continua from Photospheric Models, 206
Line Absorption, 208
A Comparison of Model to Stellar Continua, 213
Bolometric Flux, 215
References, 218

Chapter 11   The Line Absorption Coefficient   221

The Natural Atomic Absorption, 222
The Damping Constant for Natural Broadening, 227
Pressure Broadening, 229
The Impact Approximation, 231
Theoretical Evaluation of γ_n, 234
Numerical Calculation of γ_n, 237
Limitation of the Impact Model, 240
The Ionic Broadening of Hydrogen Lines, 241
The Addition of Electron Broadening in
Hydrogen Lines, 247
Thermal Broadening, 248
Combining Absorption Coefficients, 249
The Mass Absorption Coefficient for Lines, 254
Other Broadening Mechanisms, 256
References, 256

Chapter 12   The Measurement of Spectral Lines   **259**

The Coudé Grating Spectrometer, 260
Photographic Cameras, 263
Photoelectric Line Measurement, 265
The Instrumental Profile, 267
The Reconstruction Process, 269
Noise and Its Complications, 271
Fourier Noise Filters, 274
The Discrete Fourier Transform, 276
Other Techniques, 279
The Instrumental Profile of an Interferometer, 279
δ Function Spectra, 279
Measurement of the Instrumental Profile, 282
Scattered Light, 283
Measurement of Scattered Light, 286
Corrections for Scattered Light, 288
Line Measurements with Low Resolution, 290
Choice of Fourier Domains for Analysis, 293
Spectrophotometric Standard Stars, 294
References, 296

Chapter 13   The Behavior of Spectral Lines   **298**

The Line Transfer Equation, 298
The Line Source Function, 300
The Level Populations, 302
Other Formalisms for S_ν, 303
Computations of a Line Profile in LTE, 305
Contribution Functions and the Depth of
Formation of Spectral Lines, 307
The Behavior of the Line Strength, 312
The Temperature Dependence, 312
The Pressure Dependence, 318
The Abundance Dependence, 323
A Comparison of Theory and Observation, 329
References, 333

Chapter 14   Chemical Analysis   **335**

What Can We Determine? 336
The Curve of Growth: Scaling Relations, 337

The Curve of Growth: Other Characteristics, 340
Curves of Growth for Analytical Models:
A Historical Note, 345
Derivation of Abundances via the Curve of
Growth, 346
The Differential Analysis, 348
The Synthesis Method, 349
The Solar Chemical Composition, 351
Stellar Abundances, 353
Comments, 354
References, 355

Chapter 15 The Measurement of Stellar Temperatures and Radii   359

Speckle Photometry, 360
The Interferometers, 362
Lunar Occultations, 363
Eclipsing Binaries, 363
The Surface Brightness Method, 364
The Bolometric Flux Method, 365
Radii from Absolute Flux, 366
Effective Temperatures from Absolute Flux, 367
Effective Temperatures from Model Photospheres, 370
The Paschen Continuum, 371
The Balmer Jump, 373
The Hydrogen Lines, 374
Metal Lines as Temperature Indicators, 374
References, 375

Chapter 16 The Measurement of Photospheric Pressure   378

The Continuum as a Pressure Indicator, 379
The Hydrogen Lines, 380
Other Strong Lines, 381
The Weaker Lines, 384
The Gravity–Temperature Diagram, 385
The Helium Abundance, 386
Surface Gravity from Visual Binaries, 387
The Eclipsing Binary Data, 387
The Variation of Surface Gravity along the Main
Sequence, 389
References, 390

Chapter 17   Stellar Rotation   **392**

The Rotation Profile, 394
Measurement of Rotation, 400
Fourier Analysis of the Rotation Profile, 402
Statistical Correction for Axial Projection, 405
Observed Rotation of Main Sequence Stars, 407
Observed Rotation above the Main Sequence, 408
Rotation in Open Star Clusters, 409
Rapid Rotation, 410
Rotation in Binary Stars, 411
Conclusions, 413
References, 413

Chapter 18   Turbulence in Stellar Photospheres   **416**

From Velocity to Spectrum, 418
Microturbulence in Line Computations, 419
Macroturbulence in Line Computations, 421
The Affect of Turbulence on Line Profiles, 423
Techniques for Measuring Turbulence, 427
Fourier Analysis for Macroturbulence, 431
Fourier Analysis for Microturbulence, 436
The Apparent Behavior of ξ in the HR Diagram, 438
The Depth Dependence and Anisotropy of
Microturbulence, 439
Observations of Macroturbulence, 441
Comments, 443
References, 443

Epilog   **447**

Appendix A.   A Table of Useful Constants   **449**
Appendix B.   Table of sinc $\pi\sigma$ **451**
Appendix C.   Fast Fourier Transform FORTRAN Program   **453**
Appendix D.   Ionization Potentials and Partition Functions   **455**
                      References, 459
Appendix E.   The Strongest Lines in the Solar Spectrum   **460**
Appendix F.   Computation of Random Errors   **462**

Index   **465**

CHAPTER ONE

BACKGROUND

We study stellar spectra, including both lines and continua, because we are interested in the nature of the star's atmosphere. At the same time the behavior of the atmosphere is controlled by the density of the gases in it and the energy escaping through it. These in turn depend on the mass and age of the star and to a lesser extent on chemical composition and angular momentum. Stellar atmospheres are the connecting links between the observations and the rest of stellar astrophysics. In this way two philosophies arise. One is the study of the atmosphere for its own sake and the other is the use of the atmosphere as a tool to connect our observations to other parameters of interest. This book should be useful for students of both philosophies.

The topics brought together to form this chapter are background material which the reader will need to facilitate understanding of the presentation. The more advanced reader can profitably skim through to Chapter 2. The true beginner in astronomy can find additional background material in several general astronomy texts such as Smith and Jacobs (1973), Abell (1969), Menzel et al. (1970), and McLaughlin (1961).

WHAT IS A STELLAR ATMOSPHERE?

The broad concept of a stellar atmosphere is, as mentioned in the preface, a transition region from the stellar interior to the interstellar medium. We can quantify this description by discussing the change in kinetic temperature with height as observed in the sun. Figure 1-1 shows the solar temperature profile based on Vernazza et al. (1973). The subphotosphere, photosphere,

1

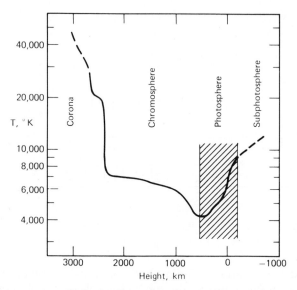

Fig. 1-1 The temperature distribution in the outer layers of the sun is shown as a function of geometrical depth in kilometers. The center of the sun is toward the right. The extent of the photosphere is indicated by the shaded region. (Adapted from Vernazza et al. 1973.)

chromosphere, and corona are indicated in the figure. For an *observer* of stellar atmospheres, one concept is very important: *the major portion of the visible stellar spectrum originates in the region marked "photosphere."* Relatively speaking, very little is known about stellar atmospheres above or below the photosphere because of this fact.

In the sun this layer is of the order of 10^3 km thick as illustrated in Fig. 1-1. The geometrical extent of the photosphere in other stars differs inversely with the surface gravity defined by

$$g = g_\odot \frac{M}{R^2} \tag{1-1}$$

where $g_\odot$ is the surface gravity of the sun (2.738×10^4 cm/sec^2) and the mass, M, and radius, R, of the star are in solar units. The thickness also depends upon the opacity of the gases comprising the photosphere.

The second physical variable that most strongly affects the nature of the atmosphere is its "average" temperature. Typically the temperature drops by somewhat more than a factor of 2 from the bottom to the top of the photosphere and instead of choosing a temperature at some depth to characterize this "average" temperature parameter, it is customary to use the effective temperature. The effective temperature of a star is defined in terms of the total power per square centimeter radiated by the star,*

$$\int_0^\infty \mathscr{F}_v \, dv \equiv \sigma T_{\text{eff}}^4 \qquad (1\text{-}2)$$

where the total radiant power per square centimeter is given by the integral and $\sigma = 5.67 \times 10^{-5}$ erg/sec cm^2 deg^4. $\mathscr{F}_v$ is the flux leaving the stellar surface and is defined formally in Chapter 6.

Traditionally and of historical necessity, the more empirical parameters of spectral type, magnitude, and color index have been used to describe stellar surface gravity and temperature.

THE SPECTRAL TYPE

Photographic records of stellar line spectra are used to classify a star according to temperature and photospheric pressure. Spectrograph dispersions of ~ 125 Å/mm giving ~ 2 Å resolution are found best for classification. The spectrograms are centered in the blue region of the spectrum where photographic emulsions are most sensitive and line strengths and ratios are estimated visually by comparing an untyped spectrogram with standards. Figure 1-2 shows a set of standard spectra in order of decreasing effective temperature. The effective temperature associated with each spectral type is discussed in Chapter 15. Those spectra toward the cool end of the sequence are referred to as "late-type" spectra and those on the hot end as "early-type" spectra. In principle the spectral classification separates the observed (normal) stars into about 60 temperature categories.

The pressure sensitive features in the spectrum show markedly less variation and only five or six classes are recognized. It is an observed fact that luminosity and pressure are strongly correlated and the pressure classification is universally referred to as the luminosity classification. Luminosity effects are illustrated in Fig. 1-3 and discussed in detail in Chapters 13 and 16.

Before classification can begin, the observer must acquire spectrograms of the standard stars using the same equipment and technique he intends to use in his investigation of other stars. The first step in classification is to identify the lines whose strengths serve as the criterion for classification. Needed lines can be identified by their position relative to the comparison spectrum lines or to the few prominent features of the spectrum itself.

---

* Equation 1-2 has the form of the Stefan–Boltzmann law, eq. 6-3, making T_{eff} the temperature of a black body having the same power per square centimeter output as the star. But the distribution of power across the spectrum may differ dramatically from a black body at the same effective temperature.

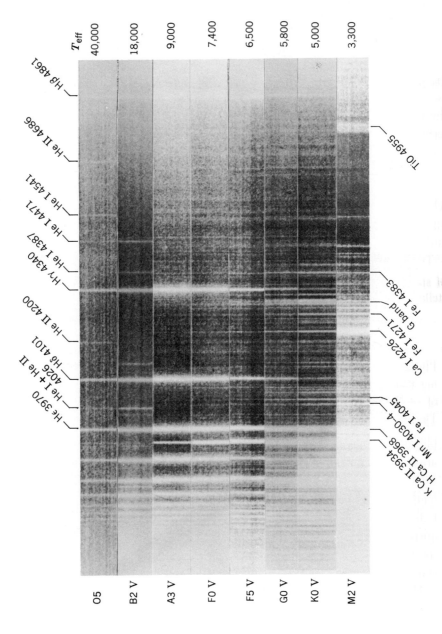

Fig. 1-2 The temperature classification is illustrated by selected samples of standard spectra from Abt et al. (1968).

4

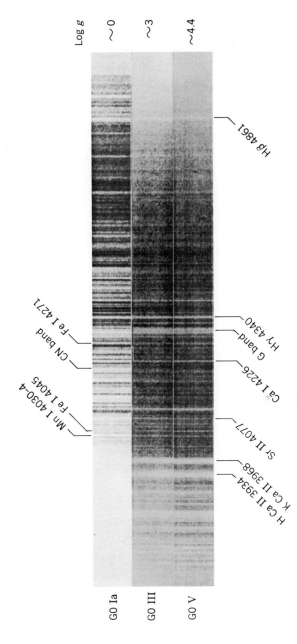

Fig. 1-3 The luminosity classification is illustrated by selected samples of standard spectra from Abt et al. (1968).

Iron and titanium are commonly used comparison spectra. In the stellar spectrum we easily recognize the hydrogen lines from class O to F5 and the less conspicuous lines can be referred to them. H_β remains distinguishable as the most intense line at the long wavelength end of the spectrogram (O-type emulsion) as late as K0. From F5 to M, the G band near the center of the spectrum and the H and K lines on the short wavelength end can be used.

The spectrum is first classed roughly. Does it have few lines, many lines, or does it show molecular bands? This rough classification is refined in successive steps. It is generally considered unwise to attempt an immediate accurate classification. Table 1-1 gives a summary of the classification criteria used in the atlas of Abt et al. (1968). Although these special features are singled out, it is the appearance of the whole blue region spectrum that is ultimately compared to the standards. The atlas by Morgan et al. (1943) is similar (see also Johnson and Morgan 1953). Most modern spectral types

Table 1-1   AMMT Classification

Temperature Criteria	
O4–B0	He I 4471/He II 4541 increasing with type
	He I + He II 4026/He II 4200 increasing with type
B0–A0	H lines increase with type
	He I lines reach max at B2
	Ca K line becomes visible at B8
A0–F5	Ca II K/Hδ increasing with type
	Neutral metals become stronger
	G band visible starting at F2
F5–K2	H lines decrease
	Neutral metals increase
	G band strengthens
K2–M5	G band changes appearance
	Ca I 4226 increases rapidly
	TiO starts near M0 in dwarfs, K5 in giants

Luminosity Criteria	
O9–A5	H and He lines weaker with increasing luminosity
	Fe II becomes prominent in A0–A5 supergiants
F0–K0	Blend 4172–9 in early F strengthens with luminosity
	Sr II 4077 increases with luminosity
	CN 4200 increases with luminosity
K0–M6	CN 4200 increases with luminosity
	Sr II 4077 increases with luminosity

are on the Morgan–Keenan system (MK) or the Abt et al. system (AMMT). A review of spectral classification is given by Morgan and Keenan (1973) along with revised standards for the MK system.

MAGNITUDES AND COLOR INDICES

The brightness of a star is often expressed in magnitude units defined by

$$m = -2.5 \log \int_0^\infty F_v W(v)\, dv + \text{const} \qquad (1\text{-}3)$$

in which the flux of the star, F_v, is recorded in the spectral interval specified by $W(v)$. The constant is arbitrary and usually adjusted to suit an adopted set of standard stars. Vega, for example, has a visual magnitude of 0.00 according to Iriarte et al. (1965).

Standard magnitude systems have been set up. The most common are the *RGU* system of Becker (1948, 1954), the *UBV* system of Johnson and Morgan (1953), the *uvby* system of Strömgren (1966) and the *DDO* system of McClure and van den Bergh (1968). The standard magnitude transmission bands are defined in terms of filters and detectors. The *UBV* response functions are shown in Fig. 1-4.

By comparison of the magnitudes for the same star measured in different parts of the spectrum, a measurement is made of the shape of the stellar energy

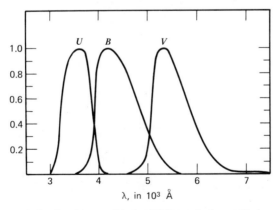

Fig. 1-4 The transmission band passes for the *UBV* standard magnitude system are shown (from Johnson 1965). The *U* band is obtained with Schott UG2 and Corning 7910 glass filters each 2 mm thick. The *B* band requires 1 mm of Schott BG12 plus 2 mm Schott GG13 filters. The *V* band needs only Corning 3384 4 mm thick. An S-4 photocathode is used.

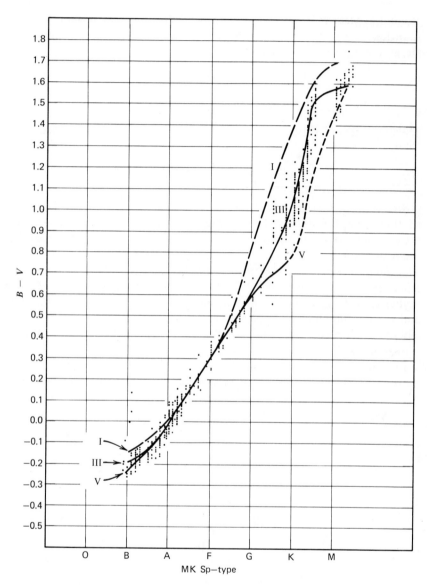

Fig. 1-5 Mean relations between the $B - V$ color index and the MK spectral type for bright stars are based on the catalog of Iriarte et al. (1965). Individual points for luminosity classes III and V only are shown. Almost all the points from B to G spectral type are dwarfs (V), while almost all the points from G to M spectral type are giants (III).

distribution. The $B - V$ color index is related to the flux and the transmission bands by

$$B - V = -2.5 \log \left(\frac{\int_0^\infty F_v W_B(v) \, dv}{\int_0^\infty F_v W_V(v) \, dv} \right) + \text{const} \qquad (1\text{-}4)$$

and a similar expression is written for the $U - B$ color index. The constants by definition are zero for an average A0V star. The $B - V$ color index is a relative measure of the temperature of the star through the slope of the Paschen continuum. With 0.01 magnitude precision in $B - V$, the temperature sequence can be divided into about 200 intervals. The $U - B$ color index is a measure of the Balmer discontinuity in early-type stars.

Figure 1-5 relates $B - V$ and spectral type empirically. We can use this relation to estimate the $B - V$ of a star from its spectral type or vice versa.

Interstellar extinction alters the continuous energy distribution of a star and consequently its color indices. Stars affected in this way deviate from the relation shown in Fig. 1-5. The spectral lines are unaffected by the interstellar extinction because the wavelength dependence of the extinction is negligible over the extent of an absorption line. There are, however, interstellar absorption lines such as the calcium HK lines and the sodium D lines arising from the gaseous component of the interstellar material.

Interstellar extinction is a serious problem when the continuous spectrum of a star is to be analyzed because there is no really satisfactory way of correcting for the effect with the precision needed in stellar atmosphere analyses. For stars near enough to have distances determined geometrically, no detectable reddening is expected ($r \lesssim 100$ pc). Inside this relatively small volume, the shape of the continuum and hence color indices such as $B - V$ are independent of distance.

THE GAS LAW

In all the stars we consider, the perfect gas law can be used to describe the relation among temperature, density, and pressure. This is true because in the photospheric layers the gas pressure is $\lesssim 10^5$ dynes/cm^2 (i.e., small) and it is well known that the perfect gas law holds exactly in the limit of low pressures. We can write this law as

$$PV = nRT \qquad (1\text{-}5)$$

where P is the gas pressure of the volume of gas V at a temperature T in degrees Kelvin. The number of moles of gas is n and the gas constant R has a value of 8.314×10^7 erg/mole deg. More frequently we use the gas law in a slightly different form given by dividing eq. 1-5 by V as follows. Let the number of particles/per mole (2.02204×10^{23}) be denoted by N_0. Then

$$P = \frac{n N_0}{V} \frac{R}{N_0} T = NkT \tag{1-6}$$

where N is the number of particles/per cubic centimeter. Boltzmann's constant, k, has a value of 1.38066×10^{-16} erg/deg. In fact the partial pressures of each of the species in the atmosphere when summed together give the total gas pressure

$$P_g = kT \sum_j N_j. \tag{1-7}$$

One of these partial pressures is due to the free electrons in the ionized gas of the photosphere,

$$P_e = N_e kT \tag{1-8}$$

where N_e is the number of electrons/per cubic centimeter.

THE VELOCITY DISTRIBUTIONS

The velocities of particles in a hot gas are distributed in a Gaussian or Maxwellian distribution. The orientation of the coordinate system is arbitrary since this thermal velocity distribution is isotropic. In rectangular coordinates the fraction of particles in the velocity intervals v_x to $v_x + dv_x$, v_y to $v_y + dv_y$, and v_z to $v_z + dv_z$ is

$$\frac{dN(v_x, v_y, v_z)}{N_{\text{total}}} = \left(\frac{m}{2\pi kT}\right)^{3/2} e^{-mv^2/2kT} \, dv_x \, dv_y \, dv_z \tag{1-9}$$

where we have used

$$v^2 = v_x^2 + v_y^2 + v_z^2.$$

The temperature of the gas is T and the particles have a mass m. Because they produce Doppler shifts, the line of sight velocities have a distribution that is an important special case for spectroscopy

$$\frac{dN(v_r)}{N_{\text{total}}} = \left(\frac{m}{2\pi kT}\right)^{3/2} e^{-mv_r^2/2kT} \, dv_r$$

where v_r is the radial (line of sight) velocity component.

In situations where energy aspects of the gas are being considered, the speed distribution, also called the Maxwell–Boltzmann distribution, may be more appropriate. It is

$$\frac{dN\,(v)}{N_{\text{total}}} = \left(\frac{2}{\pi}\right)^{1/2}\left(\frac{m}{kT}\right)^{3/2} v^2 e^{-mv^2/2kT}\,dv. \tag{1-10}$$

That is, the fraction of particles in the speed interval $(v, v + dv)$ is dN/N_{total}. The maximum of the speed distribution occurs at

$$v_1 = \left(\frac{2kT}{m}\right)^{1/2} = \text{most probable velocity.} \tag{1-11}$$

The average velocity, v_2, is

$$v_2 = \left(\frac{8\,kT}{\pi\,m}\right)^{1/2} = 1.128v_1 \tag{1-12}$$

and the root mean square velocity v_3 is

$$v_3 = \left(\frac{3kT}{m}\right)^{1/2} = 1.225v_1 \tag{1-13}$$

These are shown for clarity in Fig. 1-6.

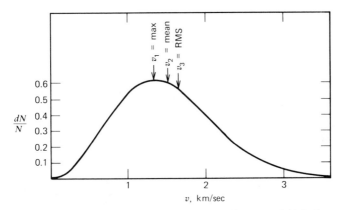

Fig. 1-6 The Maxwell–Boltzmann speed distribution according to eq. 1-10 for iron atoms at a temperature of 6000°K.

ATOMIC EXCITATION AND IONIZATION IN THERMODYNAMIC EQUILIBRIUM

In Fig. 1-7 a schematic energy level diagram shows the nth level among others and the continuum at the top. The energy above the ground level is called the excitation potential, χ. The energy difference between the ground level

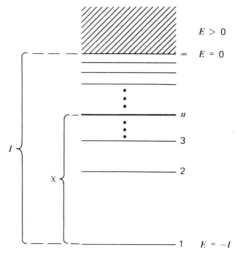

Fig. 1-7 A qualitative schematic energy level diagram showing the excitation potential for an arbitrary level n, the ionization potential I, and on the right, the energy of the electron in the continuum $E > 0$, at the ionization limit $E = 0$, and in the ground level $E = -I$.

and the continuum is called the ionization potential, I. Although all of the bound levels have negative energy, the ground level is most negative so that χ and I are positive numbers.

The relative population of each level depends in a detailed way on the mechanisms for populating and depopulating the levels. The mechanisms of interest in a stellar photosphere are radiative, collisional, and spontaneous transitions. As we shall see in these pages, in some cases the collisional interaction is dominant over the radiative and in other cases the reverse is true. When collisions do dominate we can calculate the populations from the equations that hold in thermodynamic equilibrium as shown below.

The fraction of atoms (or ions) excited to the nth level is proportional to a statistical weight g_n and the so-called Boltzmann factor,

$$N_n = \text{const } g_n e^{-\chi_n/kT}.$$

The statistical weight is $2J + 1$ where J is the inner quantum number (obtainable from Moore 1945). For hydrogen, $g_n = 2n^2$. The temperature T is in degrees Kelvin as usual.

The ratio of populations in two levels m and n is then

$$\frac{N_n}{N_m} = \frac{g_n}{g_m} e^{-\Delta\chi/kT} \tag{1-14}$$

with $\Delta\chi = \chi_n - \chi_m$. Similarly, the number of atoms/per cubic centimeter in a level n expressed as a fraction of all atoms of the same species is

$$\frac{N_n}{N} = \frac{g_n e^{-\chi_n/kT}}{g_1 + g_2 e^{-\chi_2/kT} + g_3 e^{-\chi_3/kT} + \cdots}$$

$$= \frac{g_n}{u(T)} e^{-\chi_n/kT}. \tag{1-15}$$

The partition function has been defined by $u(T) \equiv \sum g_i e^{-\chi_i/kT}$. Computation of the partition functions has been done for most elements of interest. Details can be found in Appendix D. Equation 1-15 is often expressed as a power of 10,

$$\frac{N_n}{N} = \frac{g_n}{u(T)} 10^{-\theta\chi_n} \tag{1-16}$$

with $\theta = \log e/kT = 5039.77/T$ (since k must be in units of electron volts/per degree to be compatible with χ in electron volts). Equation 1-15 or 1-16 is called the excitation or Boltzmann equation.

The ionization for the collision dominated gas can be calculated using

$$\frac{N_1}{N_0} P_e = \frac{(2\pi m)^{3/2}(kT)^{5/2}}{h^3} \frac{2u_1(T)}{u_0(T)} e^{-I/kT}$$

where N_1/N_0 is the ratio of ions to neutrals, u_1/u_0 is the ratio of ionic to neutral partition functions, m is the electron mass, and h is Planck's constant. This equation was first derived by M. N. Saha and sometimes bears his name.* Numerically the equation is

$$\log \frac{N_1}{N_0} P_e = \frac{-5040}{T} I + 2.5 \log T + \log \frac{u_1}{u_0} - 0.1762$$

or

$$\frac{N_1}{N_0} = \frac{\Phi(T)}{P_e} \tag{1-17}$$

where

$$\Phi(T) = 0.6665 \frac{u_1}{u_0} T^{5/2} 10^{-(5040I/T)}$$

$$= 1.2020 \times 10^9 \frac{u_1}{u_0} \theta^{-5/2} 10^{-\theta I}. \tag{1-18}$$

---

* A brief historical account of the astronomical side of Saha's work is given by Struve and Zebergs (1962).

The ratio of doubly ionized to singly ionized particles is calculated in the same way: $N_2/N_1 = \Phi(T)/P_e$ where the I in eq 1-18 now stands for the ionization potential of the ion and the partition function ratio is $u_2(T)/u_1(T)$. Derivations of these relations can be found in Aller (1963) or statistical mechanics expositions such as that by Tolman (1938).

In most calculations eqs. 1-16 and 1-17 are combined to find the number of atoms in a given level as a fraction of the total number of the element. For example, if we let N_n be the number in the nth level of the singly ionized atom, then

$$\frac{N_n}{N_{\text{total}}} = \frac{N_n}{N_0 + N_1 + N_2 + \cdots}$$

$$= \frac{N_n/N_1}{N_0/N_1 + 1.0 + N_2/N_1 + \cdots}.$$

The ratio N_n/N_1 comes from eq. 1-16 and each of the ratios $N_0/N_1, N_2/N_1 \cdots$ comes from eq. 1-17. Virtually all of an element is accounted for by including three stages of ionization. These relations are particularly useful in Chapter 10 where model photospheres are calculated.

Let us close this chapter with a list of sources of useful numerical data.

STELLAR CATALOGS, TABLES, AND ATLASES

A large body of useful material has been compiled over the years. Access to many of the following sources of data is necessary for successful observation and analysis.

Catalogue of Bright Stars (by D. Hoffleit, 3rd ed., Yale University Press, New Haven, Conn., 1964).

Lists 9091 stars by HR = BS number. Gives HD and ADS numbers, equatorial coordinates for 1900 and 2000, precession rates, visual magnitude, color index, spectral type, proper motion, parallax, radial velocity, and useful notes.

General Catalogue of Stellar Radial Velocities (by R. E. Wilson, Carnegie Institute of Washington Publication 601, Washington, D.C., 1953).

Radial velocity data for 15,107 stars.

Aller, L. H. 1963. *The Atmospheres of the Sun and Stars*, Ronald, New York, p. 99.

Becker, W. 1948. *Astrophys. J.* **107**, 278.

Becker, W. 1954. *Z. Astrophys.* **34**, 1.

Iriarte, B., H. L. Johnson, R. I. Mitchell, and W. K. Wisniewski 1965. *Sky and Telescope* **30**, 21.

Johnson, H. L. 1965. *Astrophys. J.* **141**, 940.

Johnson, H. L. and W. W. Morgan 1953. *Astrophys. J.* **117**, 313.

McClure, R. D. and S. van den Bergh, 1968. *Astron. J.* **73**, 313.

McLaughlin, D. B. 1961. *Introduction to Astronomy*, Houghton Mifflin, Boston.

Menzel, D. H., F. L. Whipple, and G. de Vaucouleurs 1970. *Survey of the Universe*, Prentice-Hall, Englewood Cliffs, N.J.

Moore, C. E. 1945. *A Multiplet Table of Astrophysical Interest*, National Bureau of Standards Technical Note 36, United States Department of Commerce Office of Technical Services, Washington.

Morgan, W. W. and P. C. Keenan 1973. *Annu. Rev. Astron. Astrophys.* **11**, 29.

Morgan, W. W., P. C. Keenan, and E. Kellman 1943. *An Atlas of Stellar Spectra*, University of Chicago Press, Chicago.

Smith, E. and K. Jacobs 1973. *Introductory Astronomy and Astrophysics*, Saunders, Philadelphia.

Strömgren, B. 1966. *Annu. Rev. Astron. Astrophys.* **4**, 433.

Struve, O. and V. Zebergs 1962. *Astronomy of the 20th Century*, Macmillan, New York.

Tolman, R. C. 1938. *The Principles of Statistical Mechanics*, Oxford University Press, London.

Vernazza, J. E., E. H. Avrett, and R. Loeser 1973. *Astrophys. J.* **184**, 605.

CHAPTER TWO

FOURIER TRANSFORMS

There is hardly a phase of modern astrophysics to which Fourier techniques do not lend some insight or practical advantage. Fourier pairs of variables come up in the context of absorption coefficients, spectral resolution, profile analysis, antenna patterns, and the study of noise. In these and other applications, convolutions appear in the physics of the situation. Usually it is so much easier to visualize a product in place of the convolution that again the Fourier transform via the convolution theorem can be applied to advantage.

This chapter forms an introduction to Fourier transforms for those unfamiliar with them and a useful refresher for those who have studied them in past years. The treatment is necessarily highly abbreviated. Nevertheless it does cover all the concepts used in the remainder of the book. Those wishing to learn the material in a more rigorous and extensive way are referred to the excellent book by Bracewell (1965). Practical technique and computation are nicely covered by Bell (1972).

THE DEFINITION

A function's Fourier transform is a specification of the amplitudes and phases of sinusoidals which, when added together, reproduce the function. Only one dimensional functions are treated here. Expansion to two or higher dimensions can be done by the reader with modest effort.

We take a function $F(x)$ and write

$$f(\sigma) = \int_{-\infty}^{\infty} F(x)e^{2\pi i x\sigma} \, dx. \tag{2-1}$$

This new function, $f(\sigma)$, is the Fourier transform of $F(x)$. The variables x and σ are called a Fourier pair. The inverse transform is

$$F(x) = \int_{-\infty}^{\infty} f(\sigma) e^{-2\pi i x \sigma} \, d\sigma. \tag{2-2}$$

Upper and lower case symbols denote, respectively, the functions and their transforms. Thus $f(\sigma)$ and $F(x)$ are transforms of each other. Notice the change in sign of the exponent between eqs. 2-1 and 2-2.

Not all functions have Fourier transforms. But we shall not expound on these details, we simply assume the transform exists for the functions we use and refer the reader to Bracewell's book for more detail. A derivation of the relations 2-1 and 2-2 can be found in a book by Karkevich (1960).

Suppose we expand eq. 2-1 using Euler's formula, that is, $e^{ix} = \cos x + i \sin x$. We obtain four integrals provided we allow $F(x)$ to be a complex function, $F(x) = F_R(x) + iF_I(x)$, consisting of a real, $F_R(x)$, and imaginary, $F_I(x)$ component. Then

$$f(\sigma) = \int_{-\infty}^{\infty} F_R(x) \cos 2\pi x \sigma \, dx + i \int_{-\infty}^{\infty} F_I(x) \cos 2\pi x \sigma \, dx$$

$$+ i \int_{-\infty}^{\infty} F_R(x) \sin 2\pi x \sigma \, dx - \int_{-\infty}^{\infty} F_I(x) \sin 2\pi x \sigma \, dx. \tag{2-3}$$

This equation is useful for visualizing the Fourier transform process. Consider the third integral as an example. It says we multiply the function $F_R(x)$ by a sine with periodicity of $1/\sigma$, σ being a constant for this integration. The sum of the area resulting from this product is the value needed. Figure 2-1 shows this process. Similar visualizations can be done with sines and cosines to obtain the other three integrals in eq. 2-3. This gives $f(\sigma)$ at one value of σ. Changing the frequency of the sines and cosines and repeating the process gives $f(\sigma)$ at a second point and so on. This process is the essence of the numerical evaluation of Fourier transforms which is discussed later in the chapter.

If we further write in eq. 2-3, $f(\sigma) = f_R(\sigma) + if_I(\sigma)$, we see that f_R is composed of the first and last integrals of eq. 2-3 while f_I is given by the center two integrals. So in the general case both $F(x)$ and $f(\sigma)$ are complex functions.

Under some physical circumstances $F(x)$ may be a real function. One such case is when $F(x)$ is the flux spectrum of starlight. Since $F_I(x) = 0$, $f(\sigma)$ is given by

$$f(\sigma) = \int_{-\infty}^{\infty} F_R(x) \cos 2\pi x \sigma \, dx + i \int_{-\infty}^{\infty} F_R \sin 2\pi x \sigma \, dx.$$

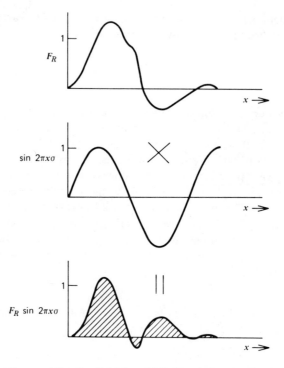

Fig. 2-1 The arbitrary real function $F_R(x)$ is multiplied by $\sin 2\pi x\sigma$ to give the bottom curve. The area is the value of the transform $f(\sigma)$ for one particular σ.

We see that $f(\sigma)$ is still a complex function. Only if $F_R(x)$ is an even function $[F_R(x) = F_R(-x)]$ as well as real do we find

$$f(\sigma) = \int_{-\infty}^{\infty} F_R(x) \cos 2\pi x\sigma \, dx.$$

Here the evenness of $F_R(x)$ combined with the oddness of $\sin 2\pi x\sigma$ gives exactly as much negative as positive area, causing the imaginary term to cancel. You can also see from eq. 2-3 that when $F(x)$ is an even function or an odd function that $f(\sigma)$ is even.

All complex numbers can be specified in rectangular or polar coordinates. In rectangular form we have written $f(\sigma) = f_R(\sigma) + if_I(\sigma)$. We can convert this to polar form in the usual way using an amplitude and phase.

$$f(\sigma) = |f(\sigma)|e^{i\psi} \qquad (2\text{-}4)$$

where the amplitude is

$$|f(\sigma)| = [f_R^2(\sigma) + f_I^2(\sigma)]^{1/2}$$

and

$$\tan \psi = \frac{f_I(\sigma)}{f_R(\sigma)}$$

where ψ is the phase.

We conclude this section by pointing out that the zero ordinate of the transform is the area under the original function. And because the original function is the transform of the transform, its zero ordinate specifies the area under the transform. Figure 2-2 illustrates the relation. This conclusion can

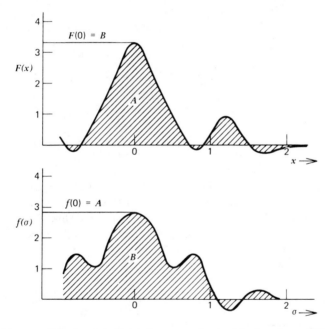

Fig. 2-2 The area of the function $F(x)$ is the zero ordinate of the transform, $f(0) = A$. Similarly $F(0) = B$, the area of the transform.

be reached trivially from the definitions, eqs. 2-1 and 2-2, by setting $\sigma = 0$, $f(0) = \int_{-\infty}^{\infty} F(x)\, dx$ and $F(0) = \int_{-\infty}^{\infty} f(\sigma)\, d\sigma$. These relations are useful, for example, when the Fourier transform of a spectral line is being investigated. The $\sigma = 0$ value is the total absorption caused by the line.

SOME COMMON TRANSFORMS

We now consider several simple functions and their transforms to illustrate some of the basic behavior. First consider the box shown in Fig. 2-3. It might represent graphically the light distribution in the plane of an opaque screen with a long slit of width W in it. We can denote the function by

$$B(x) = 0 \quad \text{for} \quad \frac{W}{2} < x < -\frac{W}{2}$$

$$B(x) = 1 \quad \text{for} \quad -\frac{W}{2} \leq x \leq \frac{W}{2}.$$

And immediately the Fourier transform can be written

$$b(\sigma) = \int_{-\infty}^{\infty} B(x)e^{2\pi ix\sigma} \, dx$$

$$= \int_{-W/2}^{W/2} e^{2\pi ix\sigma} \, dx = \left[\frac{e^{2\pi ix\sigma}}{2\pi i\sigma}\right]_{-W/2}^{W/2}$$

$$= \frac{1}{2\pi i\sigma} \left[e^{2\pi i(W/2)\sigma} - e^{-2\pi i(W/2)\sigma}\right]$$

$$= W \frac{\sin \pi W\sigma}{\pi W\sigma} \equiv W \text{ sinc } \pi W\sigma. \tag{2-5}$$

This function is shown in Fig. 2-3 on the right (it is also tabulated in Appendix B). The reciprocal behavior in the widths of these functions is particularly important. When W is large the transform is narrow, and vice versa. This illustrates a general behavior between all functions and their transforms. The first zero of the transform is at $\sigma = 1/W$. The other zeroes are equally spaced at $\Delta\sigma = 1/W$ along the σ axis.

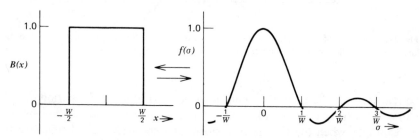

Fig. 2-3 The box of width W and its transform W sinc $\pi W\sigma$ of characteristic width $1/W$.

As a second function let us take the Gaussian given by*

$$G(x) = \frac{1}{\beta\sqrt{\pi}} e^{-x^2/\beta^2}.$$

Again we find the transform from the definition.

$$g(\sigma) = \int_{-\infty}^{\infty} G(x) e^{2\pi i x \sigma} \, dx$$

$$= \int_{-\infty}^{\infty} \frac{1}{\beta\sqrt{\pi}} e^{-x^2/\beta^2} e^{2\pi i x \sigma} \, dx.$$

But this function is real and symmetric so we can use the expansion as in eq. 2-3 to get

$$g(\sigma) = \frac{1}{\beta\sqrt{\pi}} \int_{-\infty}^{\infty} e^{-x^2/\beta^2} \cos 2\pi x \sigma \, dx.$$

Standard integral tables supply the necessary integration with the result that

$$g(\sigma) = \frac{1}{\beta\sqrt{\pi}} \frac{\sqrt{\pi}}{1/\beta} e^{-(2\pi\sigma)^2\beta^2/4}$$

$$= e^{-\pi^2\beta^2\sigma^2}. \tag{2-6}$$

In short, a Gaussian transforms into a Gaussian. If the dispersion of the original Gaussian is β, the dispersion of the transform is $(\pi\beta)^{-1}$, again an inverse relation.

Another function that comes up, particularly in the study of spectral line profiles, is the dispersion profile defined by

$$F(x) = \frac{1}{\pi} \frac{\beta}{x^2 + \beta^2}. \tag{2-7}$$

β is the half half width of this function which is normalized to unit area. The transform is

$$f(\sigma) = \frac{1}{\pi} \int_{-\infty}^{\infty} \frac{\beta}{x^2 + \beta^2} e^{2\pi i x \sigma} \, dx$$

$$= \frac{2}{\pi} \int_{0}^{\infty} \frac{\beta}{x^2 + \beta^2} \cos 2\pi x \sigma \, dx$$

* In statistics the notation is slightly different, with $\beta/\sqrt{2}$ rather than β being called the dispersion.

since this function is real and symmetric (see eq. 2-3). This integral is of the form

$$\int_0^\infty \frac{\cos mu}{1 + u^2} \, du = \frac{\pi}{2} e^{-m}.$$

We find that

$$f(\sigma) = e^{-2\pi\beta\sigma} \qquad \text{for} \quad \sigma > 0$$
$$= e^{2\pi\beta\sigma} \qquad \text{for} \quad \sigma < 0$$

or

$$f(\sigma) = e^{-2\pi\beta|\sigma|}. \tag{2-8}$$

Now let us turn to a very useful but slightly more complicated function, the delta function. Physically the δ-function represents an impulse. This is equivalent to saying that the physical process takes place so rapidly that it is beyond our power to resolve it. Therefore the integrated property of the impulse is what we observe, which for the δ function is defined as unity,

$$\int_{-\infty}^\infty \delta(x) \, dx = 1. \tag{2-9}$$

Although we cannot resolve the impulse we can specify its position or time of occurrence. Symbolically we write

$$\delta(x) = 0 \qquad \text{for} \quad x \neq 0 \tag{2-10}$$

which implies $\delta(x)$ occurs at $x = 0$. If we write $\delta(x - x_1)$, then we mean the impulse occurs at $x = x_1$.

Mathematically it is convenient to think of some unit area function that can be made infinitesimal in width. This process is necessary to provide a rigorous proof of many of the properties of a δ function. In keeping with the level of treatment in this chapter, we omit these proofs and so shall not pursue this avenue of thought. It should be understood, however, that the very large amplitude implied in making a unit area function go to infinitesimal width is not a practical problem because either the δ function is used in conjunction with another function or only its integral property is of concern. As an example, consider the product of the function $F(x)$ with the delta function,

$$\delta(x - x_1)F(x) = \delta(x - x_1)F(x_1) \tag{2-11}$$

since $\delta(x - x_1) = 0$ except at $x = x_1$. In other words, multiplication by the δ function retains only one point in the original function.

Further, if we integrate eq. 2-11 we have

$$\int_{-\infty}^{\infty} \delta(x - x_1)F(x)\, dx = F(x_1) \int_{-\infty}^{\infty} \delta(x - x_1)\, dx = F(x_1). \qquad (2\text{-}12)$$

This interesting property proves useful when the δ function is applied in analysis. Notice also that eq. 2-12 can be rearranged slightly and written as

$$\int_{-\infty}^{\infty} \delta(x_1 - x)F(x)\, dx = F(x_1). \qquad (2\text{-}13)$$

This expression has the form of a convolution which is defined and discussed later in this chapter.

The transform of $\delta(x - x_1)$ is given by

$$f(\sigma) = \int_{-\infty}^{\infty} \delta(x - x_1)e^{2\pi i x \sigma}\, dx.$$

But since $\delta(x - x_1) = 0$ except when $x = x_1$ we can write

$$f(\sigma) = e^{2\pi i x_1 \sigma} \int_{-\infty}^{\infty} \delta(x - x_1)\, dx = e^{2\pi i x_1 \sigma} \qquad (2\text{-}14)$$

and so the amplitude of $f(\sigma)$ is unity independent of σ, but there is a linear phase term, $2\pi x_1 \sigma$. These properties are shown in Fig. 2-4. This function

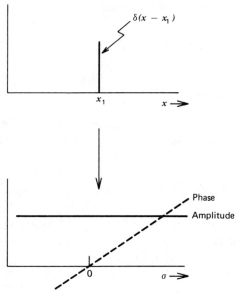

Fig. 2-4 A δ function at $x = x_1$ transforms into the constant amplitude function with linear phase $2\pi x_1 \sigma$.

represents the ultimate in the inverse dimension behavior: an infinitesimal width for the function gives an infinite extension in the transform. The linear phase term is shown as a dashed line with a slope of $2\pi x_1$. The relation between the x_1 shift in $\delta(x - x_1)$ and the phase term is discussed in more detail in the theorem's section below (theorem 2).

We can obtain the transforms of sines and cosines easily using delta functions and their transforms (eq. 2-14). Suppose, for example, we have two δ functions symmetrically placed about $x = 0$ as shown in Fig. 2-5. That is,

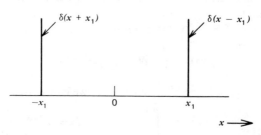

Fig. 2-5 The symmetrically placed δ function pair is the Fourier transform of a cosine.

$F(x) = \delta(x - x_1) + \delta(x + x_1)$, and because Fourier transforms are linear (theorem 1 below) we can write the transform as the sum of the individual transforms,

$$f(\sigma) = e^{2\pi i x_1 \sigma} + e^{-2\pi i x_1 \sigma}$$
$$= 2 \cos 2\pi x_1 \sigma.$$

In a similar way $F(x) = \delta(x - x_1) - \delta(x + x_1)$ transforms to $f(\sigma) = 2i \sin 2\pi x_1 \sigma$.

Instead of two δ functions, let us suppose we have an infinite array of them equally spaced at intervals of Δx. Then denoting this array by* $\mathrm{III}(x)$, we can write

$$\mathrm{III}(x) = \sum_{n=-\infty}^{\infty} \delta(x - n\, \Delta x), \qquad n = \text{integer.}$$

The transform of $\mathrm{III}(x)$ is another infinite set of δ functions, that is, $\mathrm{III}(\sigma) = \sum \delta(\sigma - n/\Delta x)$ and (as always) the spacing between peaks in $\mathrm{III}(\sigma)$ is inversely proportional to the spacing in $\mathrm{III}(x)$.

Figure 2-6 summarizes these examples along with the triangle and its transform.

* $\mathrm{III}(x)$ is the Shah function.

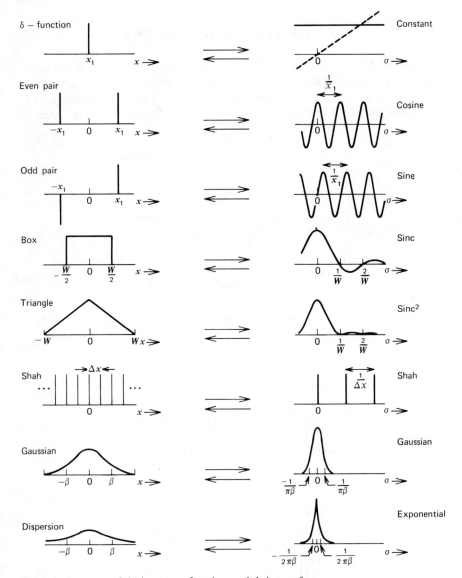

Fig. 2-6 Summary of the important functions and their transforms.

DATA SAMPLING AND DATA WINDOWS

One important application of $III(x)$ is to multiply it with a continuous function. The result is the same as if the function were sampled at the points $x = n\,\Delta x$. If we measure a spectral line profile by taking data points spaced in wavelength by $\Delta\lambda$ and call these measurements $D(\lambda)$, then we can write

$$D(\lambda) = III(\lambda)F(\lambda) \tag{2-15}$$

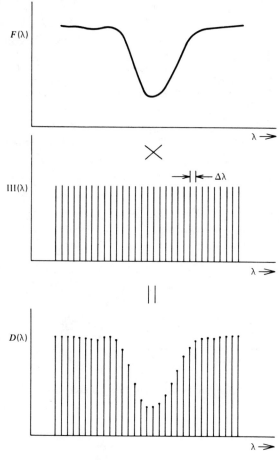

Fig. 2-7 The line profile (top) is sampled at a spacing of $\Delta\lambda$ to give the data shown at the bottom.

where $F(\lambda)$ is the real continuous spectrum. We are quite accustomed to thinking of $D(\lambda)$ as the measured line profile. We can extend this concept and imagine any function to be synthesized from δ functions. These δ functions can be as closely packed as necessary for the job at hand and will have amplitudes modulated by the true function in a relation like eq. 2-15. Figure 2-7 shows this process.

In addition to this data sampling, we record data in a window. When we measure a stellar spectrum, our data extends from λ_1 to λ_2 and these are the ends of our data window. Or when we measure a variable star from a time t_1 to a time t_2 this time interval forms a window. The windows can be expressed in our data string by multiplication with a box, $B(x)$ (or a series of boxes), having a width (or widths) equal to the window limits of our observation. With this in mind our spectral data shown in eq. 2-15 become

$$D(\lambda) = B(\lambda)\text{III}(\lambda)F(\lambda). \tag{2-16}$$

It is this data $D(\lambda)$ with which we deal and which we reduce and compare to theory. Chapter 12 contains a prime example of one such application.

THE CONVOLUTION

Multiplication of functions, as in eq. 2-16, clearly needs to be done. If we ask now, what is the Fourier transform of a product of functions, say $F(x)G(x)$, we find that it is a new process called a convolution and is defined by

$$K(\sigma) \equiv \int_{-\infty}^{\infty} f(\sigma_1)g(\sigma - \sigma_1)\, d\sigma_1 \tag{2-17}$$

where $f(\sigma)$ and $g(\sigma)$ are the transforms of $F(x)$ and $G(x)$. This integral is abbreviated by writing

$$K(\sigma) = f(\sigma) * g(\sigma).$$

The correspondence of multiplication in one Fourier domain with convolution in the other is called the convolution theorem.

It is relatively straightforward to look at eq. 2-17 and understand what convolution means. $K(\sigma)$ is to be evaluated point by point, the integration being done over σ_1 with σ being constant. The function $g(\sigma - \sigma_1)$ is translated by an amount σ relative to $f(\sigma_1)$ but because of the negative sign of σ_1 in the argument, $g(\sigma - \sigma_1)$ is reversed right for left. The product of these

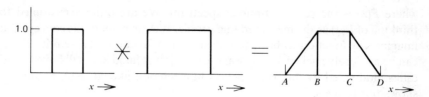

Fig. 2-8 The convolution of boxes results in a trapezoid. The dimensions AB and CD are equal to the extent of the smaller box. The dimension BC is the difference in width between the wider and the narrower box.

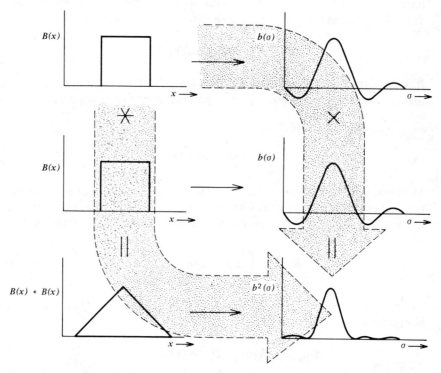

Fig. 2-9 In the x domain on the left, the convolution of a box with itself gives the triangle at the bottom. The transform of the triangle is $b^2(\sigma) = \mathrm{sinc}^2\, \pi\sigma$. An alternative route is to take the transform of the box $B(x) \rightarrow b(\sigma) = \mathrm{sinc}\, \pi\sigma$ which is then multiplied by itself, according to the convolution theorem, to give the $\mathrm{sinc}^2\, \pi\sigma$ transform at the bottom right.

30

functions is formed and the resulting net area is $K(\sigma)$ for that σ. Changing σ amounts to placing $g(\sigma - \sigma_1)$ at a new position. Tabulating $K(\sigma)$ then means sliding the reversed $g(\sigma)$ over $f(\sigma)$ and at each step recording the area under the product of the curves.

If we instead move a reversed $f(\sigma)$ over $g(\sigma)$, we obtain exactly the same result for $K(\sigma)$. The order of convolution does not matter,

$$f(\sigma) * g(\sigma) = g(\sigma) * f(\sigma). \tag{2-18}$$

Let us consider the simple case of the convolution of one unit amplitude box with another as shown in Fig. 2-8. This is a convenient example because the ordinates of each function are constant when they are not zero. When the boxes are separated the product is zero. When the boxes overlap the product is unity over the length of overlap. The length of overlap is proportional to the translation (A to B) until one of the boxes is contained within the other (B to C). Then the reduction of common area occurs as the boxes begin to separate (C to D) becoming zero beyond point D. A special case of interest is when the points B and C are the same. This corresponds to the convolution of two identical boxes or the convolution of the box with itself. We deduce that if the unit width box $B(x)$ transforms into sinc $\pi\sigma$ then the triangle $B(x) * B(x)$ transforms into sinc^2 $\pi\sigma$. Figure 2-9 illustrates this general pattern of analysis for the specific case just cited. There are two paths by which one can get from the function to the transform. One path is usually easier to visualize or to compute than the other.

With the exception of the δ function discussed below, all convolutions span a wider range of the variable than either of the individual functions. The convolution is a blurring or averaging process.

CONVOLUTION WITH A δ FUNCTION

The δ function has a transform that is unity for all values of σ. Convolution of a function $F(x)$ with $\delta(x)$ implies a multiplication of their transforms, which gives $f(\sigma)$ times unity for all σ. Clearly the convolution of $F(x)$ with $\delta(x)$ results in $F(x)$ again.

However, if we instead have $\delta(x - x_1)$ convolved with $F(x)$ we find $F(x - x_1)$. We refer to eq. 2-14 where we see the transform of a delta function that is not at the origin results in a phase term $e^{2\pi i x_1 \sigma}$. The product of transforms is then $f(\sigma)e^{2\pi i x\sigma_1}$, corresponding to a translation in $F(x)$ to the position of the δ function. (This is a special case of theorem 2 below.) We conclude that convolution with a δ function gives the function back again, but displaced to the position of the δ function.

CONVOLUTIONS OF GAUSSIAN AND DISPERSION PROFILES

Equation 2-6 can be written for two Gaussians and multiplied together

$$g_a(\sigma)g_b(\sigma) = e^{-\pi^2\beta_a^2\sigma^2}e^{-\pi_b^2\sigma^2}$$
$$= e^{-\pi^2\beta_c^2\sigma^2}$$

where $\beta_c^2 = \beta_a^2 + \beta_b^2$. The above transform is seen in this way to correspond to the Gaussian

$$G(x) = \frac{1}{\beta_c\sqrt{\pi}}e^{-x^2/\beta_c^2}$$

$$= G_a(x) * G_b(x). \qquad (2\text{-}19)$$

The convolution of two Gaussians is a third Gaussian.

In an identical way (using eq. 2-8) one can show that the convolution of two dispersion profiles is a new dispersion profile with a half half width given by $\beta_c = \beta_a + \beta_b$.

The convolution between a Gaussian and a dispersion function is called a Voigt function. If we let β_1 and β_2 represent the width parameters for Gaussian and dispersion profiles, then the transform of the Voigt function is

$$v(\sigma) = e^{-\pi^2\beta_1^2\sigma^2}e^{-2\pi\beta_2|\sigma|}. \qquad (2\text{-}20)$$

The convolution of two Voigt functions leads to a product of transforms of this form. Since the exponents can be grouped together, it is trivial to show that the result is a new Voigt function with $\beta_1^2 = (\beta_1^a)^2 + (\beta_1^b)^2$ and $\beta_2 = \beta_2^a + \beta_2^b$. These functions are considered in greater detail in Chapter 11.

SEVERAL USEFUL THEOREMS

Physical insight as well as computational saving can be gained by application of theorems such as the following. Most of these can be proved by reverting to the defining integrals, eqs. 2-1 and 2-2. In all cases $f(\sigma)$ is the transform of $F(x)$ and the arrow means "transforms to."

Theorem 1. Fourier transforms are linear. If

$$F(x) = F_1(x) + F_2(x) + \cdots$$

then

$$f(\sigma) = f_1(\sigma) + f_2(\sigma) + \cdots \qquad (2\text{-}21)$$

where $F_1(x) \rightarrow f_1(\sigma)$ and so on.

Theorem 2. Translation in one domain gives a phase term in the opposite domain.

$$F(x - x_1) \rightarrow e^{2\pi i x_1 \sigma} f(\sigma). \tag{2-22}$$

Theorem 3. Coordinate scaling in one domain leads to inverse coordinate scaling *and* inverse amplitude scaling in the opposite domain.

$$F(ax) \rightarrow \left| \frac{1}{a} \right| f\left(\frac{\sigma}{a} \right). \tag{2-23}$$

Theorem 4. Differentiation in one domain is equivalent to multiplication by $2\pi i\sigma$ in the other domain.

$$\frac{dF(x)}{dx} \rightarrow 2\pi i\sigma f(\sigma). \tag{2-24}$$

Integration is obviously the inverse of this.

Theorem 5. Convolution derivatives involve differentiation of only one or the other of the convolved functions but not both. If

$$K(x) = F(x) * G(x),$$

then

$$\frac{dK(x)}{dx} = \frac{dF(x)}{dx} * G(x) = F(x) * \frac{dG(x)}{dx}. \tag{2-25}$$

Theorem 6. If $G(x)$ is normalized to unit area, then $K(x) = F(x) * G(x)$ has the same area as $F(x)$, that is,

$$\int_{-\infty}^{\infty} K(x) \, dx = \int_{-\infty}^{\infty} F(x) \, dx.$$

Theorem 7. If we define the center of a function as $\bar{x}$ where

$$\bar{x} = \frac{\int_{-\infty}^{\infty} x F(x) \, dx}{\int_{-\infty}^{\infty} F(x) \, dx},$$

then the center of a convolution $K(x) = F(x) * G(x)$ is the sum of the centers of the two functions being convolved, that is,

$$\bar{x}_K = \bar{x}_F + \bar{x}_G. \tag{2-26}$$

Theorem 8. Rayleigh's theorem relates the amplitudes of the function and its transform.

$$\int_{-\infty}^{\infty} |F(x)|^2 \, dx = \int_{-\infty}^{\infty} |f(\sigma)|^2 \, d\sigma. \tag{2-27}$$

An almost identical theorem called the power theorem is

$$\int_{-\infty}^{\infty} F(x)G^*(x)\, dx = \int_{-\infty}^{\infty} f(\sigma)g^*(\sigma)\, d\sigma. \tag{2-28}$$

Where the asterisk denotes complex conjugate. However, in several common physical applications F and G are proportional to each other so that eq. 2-28 reverts to eq. 2-27 (examples of this are electric and magnetic fields in radiation; voltage and current in resistive dissipation).

THE RESOLUTION THEOREM

All of our instruments are finite. They measure for a finite time, over a finite bandwidth, and so forth. So necessarily all of our data can be written in the form of eq. 2-16. This has fundamental implications to which we now turn our attention. Let us first write the data $D(x)$ as if it were a continuous measure of the true function $F(x)$ over the region, W, defined by the nonzero extent of the unit height box $B(x)$. Then

$$D(x) = B(x)F(x).$$

The transform of this equation is

$$d(\sigma) = b(\sigma) * f(\sigma) \tag{2-29}$$

and further

$$b(\sigma) = \frac{W \sin \pi\sigma W}{\pi\sigma W}.$$

Let us suppose, for example, that $f(\sigma)$ showed some sharp spectral features as in Fig. 2-10. Compared to the extent of $b(\sigma)$ these features are impulses. The above convolution then gives $b(\sigma)$ centered on the positions of the impulses, and the sharpness of the features has been destroyed. The only way in which a sharp feature in $f(\sigma)$ can be resolved is to extend the bandwidth W until $1/W$ is less than the width of the feature.

We conclude that due to the finite span of all real measurements, the high frequency components of the transform are missing. Only those frequencies comprising $d(\sigma)$ with $x < W$ have been recorded. Therefore real structure in the transform is limited to $\Delta\sigma \lesssim 1/W$.

The most common astronomical examples we deal with are telescope resolution where W is the aperture of the objective (remember, we are treating only the one dimensional case here), spectral windows where W is the length

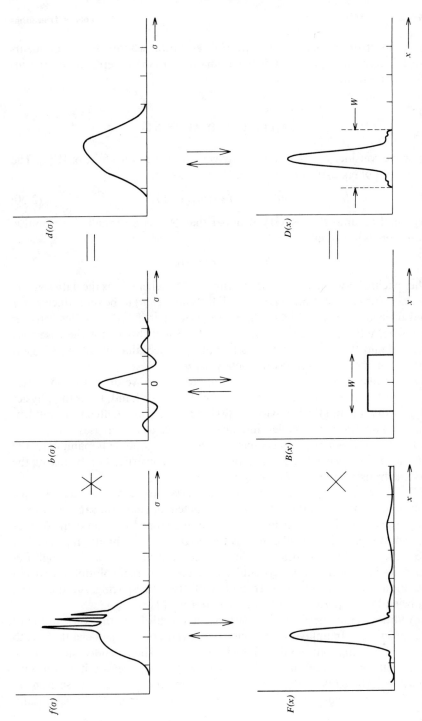

Fig. 2-10 The resolution theorem is illustrated for a finite observation of extent W. The data $D(x)$ is limited to the nonzero extent of $B(x)$. The data, $d(\sigma)$, in the transform domain is devoid of all sinusoidals with frequency greater than $1/W$.

of the spectrum measured, and spectral resolution where the measurements are taken in the opposite Fourier domain and $b(\sigma)$ is replaced by the instrumental profile.

SAMPLING AND ALIASING

Now we consider eq. 2-16 specifically with regard to the effect of $III(x)$. The transform of eq. 2-16 is of the form

$$d(\sigma) = b(\sigma) * III(\sigma) * f(\sigma). \tag{2-30}$$

We shall assume $B(x)$ is very wide so that $b(\sigma)$ is essentially an impulse. Equation 2-30 is then

$$d(\sigma) = III(\sigma) * f(\sigma).$$

The spacing between δ functions in $III(\sigma)$ is $1/\Delta x$ where Δx is the data spacing in $D(x)$. The convolution of $f(\sigma)$ with $III(\sigma)$ causes $f(\sigma)$ to be reproduced over and over as shown in Fig. 2-11. We see now that if $f(\sigma)$ becomes zero for $\sigma < 0.5/\Delta x$ these replications are separate. But if this is not the case they overlap and there is trouble. This trouble is called aliasing. $\sigma_N \equiv 0.5/\Delta x$ is called the cut-off or Nyquist frequency.

Generally we measure $D(x)$ and compute $d(\sigma)$ so we do not know if $f(\sigma)$ has significant amplitudes at $\sigma \gtrsim \sigma_N$. On the other hand, from the physics of the situation and the behavior of $f(\sigma)$ for $\sigma < \sigma_N$ we can often feel confident that $f(\sigma)$ decreases to smaller and smaller values as σ increases.

For the case where $f(\sigma)$ becomes zero for $\sigma \gtrsim \sigma_N$, we can manipulate the transform $d(\sigma)$ as needed and unambiguously reconstruct $F(x)$ by taking the inverse transform of $d(\sigma)$ for $-\sigma_N \le \sigma \le \sigma_N$.

For the case where successive patterns overlap, there is no longer a unique solution. Many different $F(x)$'s can be concocted that have the same transform $d(\sigma)$. These functions differ in the way the overlapped portion is divided, as shown in Fig. 2-12. This of course is the origin of the term aliasing. The low frequencies of the first peak or alias to the right are disguised as high frequencies in the central pattern and so on. The only real solution for such a situation is to redo the measurements with the data spacing, Δx, decreased sufficiently to separate the successive patterns of $f(\sigma)$.

Additional mixing of the $f(\sigma)$ patterns can result from the convolution of $b(\sigma)$ in eq. 2-30. In unfavorable circumstances $b(\sigma)$ may be sufficiently wide to result in aliasing that would otherwise not exist. The extended sidelobes of $b(\sigma) = \text{sinc } \sigma$ can accentuate this problem and for this reason it is necessary to smooth the ends of $B(x)$, bringing them smoothly to zero (as seen in eq. 12-16 and the related discussion). The smooth wings may still be significant,

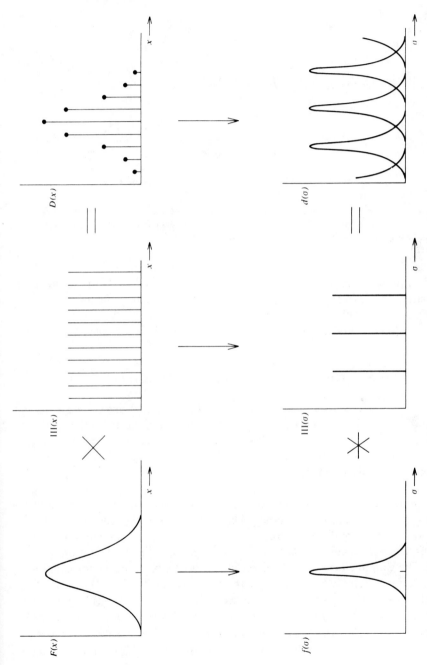

Fig. 2-11 Graphical illustration of the sampling of $F(x)$ and the multiplicity of $f(\sigma)$ it causes. The data transform, $d(\sigma)$, shows undesirable overlapping in the multiple structure (aliasing).

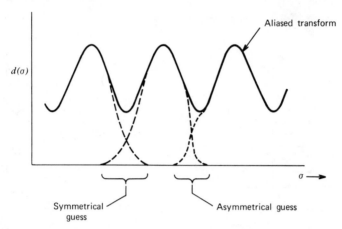

Fig. 2-12 In the aliased situation we obtain in $d(\sigma)$ only the vector sum of the overlapping functions (——). Without additional information, it is impossible to separate the overlap in a unique way. Two arbitrary examples of separation are shown.

and so to make certain $b(\sigma) * f(\sigma)$ is no larger than needed, there should be no large signals in $f(\sigma)$ that are unnecessary. Subtraction of the mean value of $F(x)$, for example, will often remove a large spike at $\sigma = 0$.

A detailed application of these principles is made in Chapter 12.

NUMERICAL CALCULATION OF TRANSFORMS

We have indicated above how some simple analytic functions can be transformed. In this section we are concerned with numerical computation of the transforms, particularly for data. It is possible to use quadrature formulae on eqs. 2-1 and 2-2 directly and such a procedure may be useful in some applications, especially when the data points are spaced nonuniformly along the abscissa.

A mild revolution has occurred with the invention of the fast Fourier transform. (Also called the Cooley–Tukey algorithm). This method of calculating transforms depends on a clever method of nesting sums and avoiding redundancy in calculating sines and cosines that the direct integration does not. The gain in efficiency over direct integration is phenomenal for lengthy calculations. Bell (1972) shows that the computation time for the classical method varies as $2N^2$ while the fast Fourier transform equivalent is $3N \log_2 N$. The ratio is $0.46N/\ln N$ where N is the number of points in the function to be transformed. Table 2-1 gives examples of this gain. The N's listed are integral powers of 2 as required by the Fourier transform. This requirement is

Table 2-1   A Comparison of the Efficiency of
Computation

N	Old*	FFT*	Ratio
128	32,768	2,689	12
256	131,072	6,147	21
512	524,288	13,830	38
1,024	2,097,152	30,734	68
2,048	8,388,608	67,614	124
4,096	33,554,432	147,521	227
8,192	134,217,728	319,629	420
16,384	536,870,912	688,432	780
32,768	2,147,483,648	1,475,212	1,456

* The number of computations needed with the
old and the fast Fourier transform methods.

easy to meet for any string of data simply by extension of the array to the
next larger power of 2 using zeroes.

The second requirement (or assumption) is that all data points be equally
spaced along the abscissa. The nesting procedure does not work without this
requirement. Many types of data can easily be taken in equal steps so for
these cases no hardship is introduced. In other cases treating the data as
if it were equally spaced results in negligible error.

We can write our data, bearing in mind eq. 2-16, as

$$D(x) = D(j\,\Delta x) = D(j)$$

in which Δx is the step and j an integer index running from 0 to $N - 1$ for
a total of N data points. The Fourier integral can then be written as the sum

$$d(\sigma) = \sum_{-\infty}^{\infty} D(x_j)e^{2\pi i x_j \sigma}\,\Delta x$$

$$= \sum_{0}^{N-1} D(j)e^{2\pi i(j\Delta x)\sigma}\,\Delta x.$$

We can also write $\sigma = k\Delta\sigma$ where $\Delta\sigma$ is the spacing of points in $d(\sigma)$ and k
is an integer. Then

$$d(\sigma) = d(k) = \sum_{0}^{N-1} D(j)e^{2\pi i jk\Delta x \Delta\sigma}\,\Delta x. \qquad (2\text{-}31)$$

Further, the highest frequency we can hope to obtain information at is

$$\sigma_N = \frac{1}{2 \, \Delta x}.$$

It is usually assumed that if there are N points defining $D(x)$, then N values of $d(\sigma)$ will be computed. If, in addition, $d(\sigma)$ extends equally far in negative as in positive σ, then there are $N/2$ points away from zero. The step in σ must then be

$$\Delta\sigma = \frac{\sigma_N}{N/2} = \frac{1}{\Delta x \, N}$$

or

$$\Delta\sigma \, \Delta x = \frac{1}{N}. \tag{2-32}$$

with this substitution, eq. 2-31 gives

$$d(k) = \sum_{0}^{N-1} D(j)e^{2\pi ijk/N} \, \Delta x. \tag{2-33}$$

It is a summation of this form that is evaluated in the fast Fourier transform technique and most versions also assume $\Delta x = 1$. The inverse transform corresponding to eq. 2-33 is

$$D(j) = \sum_{0}^{N-1} d(k)e^{-2\pi ijk/N} \, \Delta\sigma. \tag{2-34}$$

The same algorithm can be used to evaluate this sum but $\Delta\sigma$ must be explicitly taken into account using eq. 2-32. This is equivalent to renormalization,

$$D(j) = \frac{1}{\Delta x \, N} \sum_{0}^{N-1} d(k)e^{-2\pi ijk/N}. \tag{2-35}$$

A FORTRAN version of the fast Fourier transform is given in Appendix C.

NOISE TRANSFER BETWEEN THE FOURIER DOMAINS

Suppose we measure a function, $F(x)$, and take its transform. The measurements will contain noise and we ask how the noise appears in the transform. Let $E(x)$ be the error or noise in these measurements such that we can write

$$F(x) = F_0(x) + E(x)$$

where $F_0(x)$ is the true, error free function. Because of the linearity (theorem 1) we can write

$$f(\sigma) = f_0(\sigma) + e(\sigma)$$

where $e(\sigma)$ is the transform of $E(x)$.

Theorem 8 tells us that

$$\int_{-\infty}^{\infty} E^2(x)\, dx = \int_{-\infty}^{\infty} e^2(\sigma)\, d\sigma$$

but our measurements extend over a finite range in x, call it L, and our transform is unique only over a finite range in σ, say $\pm 1/2$, due to the data sampling spacing and the attendant aliases. We can rewrite theorem 8 as

$$\int_{x_0}^{x_0+L} E^2(x)\, dx = \int_{-l/2}^{l/2} e^2(\sigma)\, d\sigma \tag{2-36}$$

where x_0 is the starting value. If we denote the average by a vinculum, the relation can also be written

$$\overline{E^2}L = \overline{e^2}l. \tag{2-37}$$

This is the fundamental relation that says the mean square error times the length of the data is the same in both domains.

More specifically, the data we transform are usually an array of numbers. Let us also assume the fast Fourier transform is used. Then with N equally spaced data points Δx apart, eq. 2-36 takes the form

$$\sum_{0}^{N-1} E^2(x_j)\, \Delta x = \sum_{0}^{N-1} e^2(\sigma_j)\, \Delta\sigma$$

with

$$L = \Delta x\, N$$

$$\frac{l}{2} = \sigma_N = \frac{0.5}{\Delta x}.$$

In place of the mean square errors, we define the standard deviations

$$S_x = \left[\sum_{0}^{N-1} \frac{E^2(x_j)}{(N-1)} \right]^{1/2}$$

$$S_\sigma = \left[\sum_{0}^{N-1} \frac{e^2(\sigma_j)}{(N-1)} \right]^{1/2}.$$

The corresponding form for eq. 2-37 is

$$S_\sigma^2 l = S_x^2 L, \qquad (2\text{-}38)$$

or

$$
\begin{aligned}
S_\sigma &= S_x \sqrt{L/l} \\
&= S_x \sqrt{\Delta x\, L} \qquad (2\text{-}39)\\
&= S_x\, \Delta x\, \sqrt{N}.
\end{aligned}
$$

As might be expected the noise in the two domains is proportional. When the measured span, L, is more finely divided by making the sampling interval, Δx, small, S_σ is reduced because the noise is spread over a larger span in σ.

The common practice of extending the true data array with zeroes to reach an even power of 2 alters eq. 2-39 because the zeroes increase L or N but carry no noise. Consequently L or N in eq. 2-39 refers to the true data set before any modification for the fast Fourier transform calculation.

When the noise $E(x)$ is "white," meaning independent of frequency, the noise $e(\sigma)$ is statistically constant with σ. We have seen how it is important to choose Δx small enough so that all the signal transform is within $\sigma_N = 0.5/\Delta x$ (no aliasing). If σ_N is made still larger, the only nonzero amplitudes at the high frequencies are noise. In other words, it is possible to see the noise level in the σ domain (see Fig. 12-8 for example) and in some cases it is advantageous to calculate the noise in the original data by working back through eq. 2-39 to get S_x from S_σ.

We do, perhaps only semiconsciously, a similar thing in the x domain. When Δx is small, that is, well below the dimension associated with signal variations, adjacent measured points should be strongly correlated. Differences in value are then due to noise. We assume white noise and estimate the noise level from the "grass," which is the high frequency noise.

REFERENCES

Bell, R. J. 1972. *Introductory Fourier Transform Spectroscopy*, Academic, New York.

Bracewell, R. 1965. *The Fourier Transform and Its Applications*, McGraw-Hill, New York.

Karkevich, A. A. 1960. *Spectra and Analysis*, Consultants Bureau, New York.

CHAPTER THREE

SPECTROSCOPIC TOOLS

Instruments used for stellar photosphere observations are light collecting and analyzing devices. The measurements can be grouped into two categories: (1) the continuous spectrum and (2) the line spectrum. Depending on the type of measurement, the instrument can be large or small, simple or complex. But in each case there is a light collector, such as a telescope, followed by an analyzer, such as a spectrometer, and finally a light detector–recorder device or system. Salient aspects of telescopes lead the discussion. General features of analyzers are considered, with emphasis being placed on the diffraction grating and the general features of a spectrograph. Special features of spectrophotometric equipment designed for measuring continua and lines are discussed separately in Chapters 10 and 12, respectively. Chapter 4 is about detectors.

ASPECTS OF TELESCOPES

The telescope's main function for studies of stellar photospheres is to collect light for spectral analysis. The size and design of a telescope must be compatible with the spectroscopic equipment coupled to it. The focal ratio of the telescope, for instance, must match the focal ratio of the collimator in a spectrograph or interferometer to avoid light loss. In a Cassegrain optical system the telescope focal ratio can be controlled by the figure on the secondary mirror. Often telescopes are constructed with interchangeable secondary mirrors to accommodate several types of Cassegrain focus instruments and to give the long focal length needed for coudé operation (see Fig. 3-1). The f ratios of 7–20 are used at the Cassegrain focus; values of

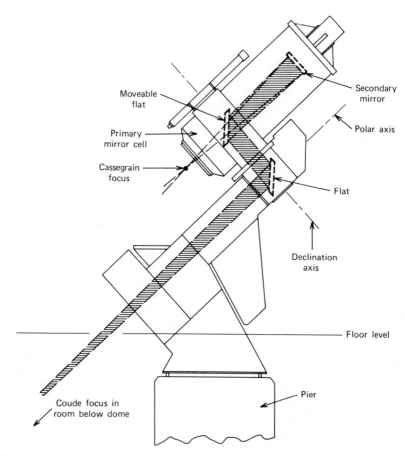

Fig. 3-1 A schematic diagram of a telescope. When the Cassegrain focus is used, the flat mirror in the telescope tube is moved out of the position shown and into a storage position out of the optical beam. The coudé mode of operation uses the two flat mirrors to bring the converging beam along first the declination axis and then the polar axis to the coudé focus on the floor below. In this way the coudé focus remains fixed and independent of the position of the star in the sky.

20–40 at the coudé focus. The *effective* focal length of a Cassegrain telescope is defined to be the focal ratio (angular convergence of the beam at the focus) multiplied by the aperture of the telescope, and it may be much larger than the physical distance from the primary mirror to the focus. The scale in the focal plane is given by

$$S = \frac{206{,}265}{F_{\text{eff}}} \; ''/\text{mm}$$

where the effective focal length, F_{eff}, is given in millimeters. Typical Cassegrain scales are 10–30″/mm, while at the coudé values of 2–6″/mm are common. A stellar seeing disc of 2″ would then be ~ 0.5 mm at a coudé focus. The actual distribution of light in the stellar seeing disc is sharply peaked toward the center and often shows a large but faint outer extension reaching to several seconds of arc. Seeing disc structure is discussed by Piccirillo (1973) and Young (1974). The speckle structure of the seeing disc is discussed in Chapter 15.

The light increasing power of the telescope is proportional to the square of the aperture size (with due allowance for the shadowing of the secondary mirror assembly) only if the whole of the stellar seeing disc is admitted into the entrance aperture of the analyzing apparatus. This is the case for an interferometer or a Cassegrain monochromer, but for a slit spectrograph the telescope's light increasing power varies linearly with aperture. The reason for this can be seen as follows. The linear diameter of the seeing disc is proportional to the effective focal length of the telescope. The effective focal length increases linearly with the aperture because the f ratio must stay fixed to match the spectrograph. Therefore the linear size of the seeing disc increases in both dimensions: along the entrance slit and across the entrance slit, in proportion to the aperture. The increased dimension along the slit lets more light through in proportion to the increase in length. The increased dimension across the slit accomplishes nothing since the light is stopped by the slit jaws. The slit width cannot be increased without increasing the collimator focal length or degrading the resolution (see eq. 3-21), options that are available even with the smaller aperture telescope. Image slicers can be used to reclaim a significant fraction of the light lost at the conventional spectrograph's entrance slit (see Fig. 12-3 and discussion). With an image slicer the efficiency increases nearly with the square of the telescope aperture again because most of the seeing disc is used.

Polarization introduced by the telescope is small at the Cassegrain focus but large at the coudé focus. The oblique reflections needed to bring the light to the coudé focus are the source of the large polarization. In the simple case shown in Fig. 3-1, two 45° reflections occur, one at the intersection of the primary mirror axis and the declination axis and the other at the intersection of the polar and declination axes. Each of these reflections produces about 6 or 7 % linear polarization. How these contributions combine depends on the declination to which the telescope is pointed. Additional oblique reflections are incorporated in many coudé arrangements. In general the state of the instrumentally introduced polarization is dependent on both the hour angle and the declination.

SPECTROGRAPHS—SOME GENERAL RELATIONS

The spectrograph is comprised of an entrance slit, dispersing element, light detector, and usually additional optical imaging components. The basic example of such a system is the Czerny–Turner system shown in Fig. 3-2. The imaging components in this case are the collimator and the camera mirrors. The dispersing element is a plane diffraction grating. The detector is a photomultiplier. The optical arrangement is particularly simple because the grating is illuminated in collimated light and the diffracted light of any given wavelength leaves the grating as collimated light. The entrance slit is at the focus of the collimator and the spectrum is in the focal plane of the camera.

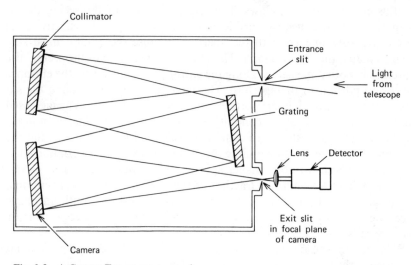

Fig. 3-2   A Czerny–Turner spectrograph.

In monochromatic light the image of the entrance slit comprises the spectrum when the slit is wide. The image is inverted and the magnification along the direction of the slit is the ratio of camera to collimator focal length. The magnification of the slit width depends on the orientation of the grating and is discussed later in this chapter (eq. 3-18). For very narrow slit settings, the structure of the grating diffraction pattern determines the width of a truly monochromatic image. When polychromatic white light is analyzed with the same instrument, the spectrum is composed of a continuum of monochromatic images spread out across the focal plane of the camera.

The chromatic purity of the light at a given position in the spectrum depends on the width, $\Delta\lambda$, of the monochromatic images. The term "chromatic resolution," or simply "resolution," is used to describe how well the color separation is made. Generally the higher the resolution, the happier the astronomer, but as nature would have it, high resolution is not easy to attain and often requires sacrificing a fraction of the precious stellar photons by narrowing the entrance slit. More specifically, we need $\Delta\lambda$ smaller than the width of the structure in the spectrum we wish to observe. A quantitative discussion requires a measurement of the monochromatic image produced by the spectrograph. The image is called the instrumental profile. A portion of Chapter 12 deals with measuring and allowing for the instrumental profile.

The angular dispersion given by the dispersing element and the focal length of the camera determine how spread out the spectrum will be in the focal plane of the camera. If two collimated beams having wavelengths λ and $\lambda + d\lambda$ leave the grating at angles differing by $d\beta$, then, as shown in Fig. 3-3, the linear separation in the spectral image is

$$dx = f_{\text{cam}} \, d\beta$$

and it follows immediately that the linear dispersion in angstroms per millimeter is given by

$$\frac{d\lambda}{dx} = \frac{d\lambda}{d\beta}\frac{d\beta}{dx} = \frac{1}{f_{\text{cam}}}\frac{1}{(d\beta/d\lambda)} \qquad (3\text{-}1)$$

where $d\beta/d\lambda$ is the angular dispersion given by the grating.

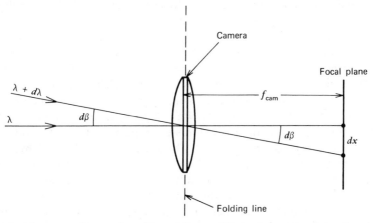

Fig. 3-3   The camera focuses the collimated beams differing in wavelength by $d\lambda$ to images differing in position by dx. If the camera is a mirror, the same diagram holds true if it is folded about a line through the lens.

Linear dispersion and resolution are often connected (as we shall see in Chapter 12) but not always. In astronomical work low dispersion, say 30 Å/mm and less,* is usually done at the Cassegrain focus while the high dispersion work is done at the coudé focus.

Now let us turn our attention to the plane diffraction grating of the type used in modern astronomical spectrographs.

DIFFRACTION GRATINGS

The plane diffraction grating has proved to be the most useful dispersing element. Its luminous efficiency is higher than concave gratings or prisms of similar angular dispersion. The linear dispersion can be changed by merely switching cameras, whereas concave gratings have a radius of curvature that fixes the dispersion. In addition, the plane grating can be used over a wide range of angles of incidence, which makes it possible to direct the dispersed light into any of several useful directions by rotating the grating.

Plane reflection gratings consist of a large number of parallel grooves ruled with a diamond tool into an aluminum or gold coating on a flat glass substrate. The grooves should all have the same shape, be smooth, and be equally spaced. A microscopic view of rulings is seen in Fig. 3-4. The best results so far have been obtained on ruling engines that are servo-controlled by interferometer monitors to maintain the strict positional requirement (Fig. 3-5). In high resolution applications, large gratings are needed. Wear on the diamond ruling tool limits the maximum ruled area. The largest gratings now available are 350 mm × 450 mm. Refer to Harrison (1973) for a discussion of ruling engines and grating quality.

The commonly encountered gratings are replica gratings constructed by using the ruled grating master as a mold. The replica can also be used as a mold to produce a second generation replica. A replica is far less expensive than an original master grating and usually has higher reflection efficiency.

Let us consider the simple transmission grating illuminated in collimated monochromatic light as in Fig. 3-6. (We will presently convert our results to the reflection gratings actually used in stellar spectrographs.) The slit width is denoted by b, the slit spacing by d. The angle of incidence, α, is measured from the grating normal. The diffraction angle, β, we choose to have the same sign as α when it is on the same side of the grating normal. α and β are of the same sign in Fig. 3-6. Because the rulings are so long compared to their width we can justifiably consider the grating as a one dimensional object. Further, let us restrict the angles α and β to the plane of the paper in Fig. 3-6.

---

* Note that low dispersion is a large number of angstroms per millimeter and vice versa.

Fig. 3-4 A microscopic photograph of the rulings on a plane reflection grating having 1180 lines/mm. Magnification is 18,000 × . (Courtesy of Jarrell–Ash.)

Fig. 3-5 A ruling engine for the manufacture of diffraction gratings. The grating blank is in the center of the figure with the ruling diamond poised above it. Part of the interferometer control can be seen at the lower left. (Courtesy of Bausch and Lomb, Analytical Systems Division, Rochester, New York.)

The incident plane scalar wave (a more rigorous vector wave treatment can be done, see Stroke 1967) can be written.

$$F(x, t) = F_0(t)e^{2\pi i x \sin \alpha/\lambda}. \tag{3-2}$$

Let us describe the grating transmission with the function $G(x)$,

$$G(x) = 1 \text{ for } x \text{ values corresponding to slits}$$
$$= 0 \text{ for all other } x \text{ values.}$$

It is the product of $F(x, t)$ and $G(x)$ that emerges from the back of the grating. The resultant wave going in any arbitrary direction β will be the sum of contributions, with their proper phase shift, from along the x coordinate. This can be written as

$$g(\beta) = \int_{-\infty}^{\infty} F(x, t)G(x)e^{2\pi i x \sin \beta/\lambda} \, dx$$

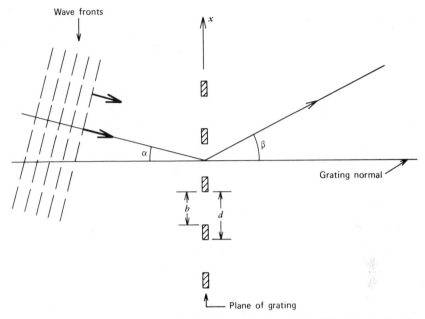

Fig. 3-6 The plane transmission grating of slit spacing d and slit width b lies in the plane perpendicular to the paper. The linear coordinate x runs across the rulings as shown. Light is incident at an angle α. The diffraction angle is β. Both of these angles have the same sign as drawn.

where $2\pi(x \sin \beta)/\lambda$ is the phase difference along the grating. Putting in eq. 3-2 we have

$$g(\beta) = F_0 \int_{-\infty}^{\infty} G(x)e^{2\pi i(x \sin \alpha + x \sin \beta)/\lambda}\, dx$$

and if we choose to measure x in units of λ while denoting $\sin \alpha + \sin \beta$ by θ, we have

$$g(\theta) = F_0 \int_{-\infty}^{\infty} G(x)e^{2\pi i x\theta}\, dx. \qquad (3\text{-}3)$$

Thus we see that the resultant waves behind the grating can be written as the Fourier transform of the grating transmission function, $G(x)$.

The description† of $G(x)$ in terms of the Shah function and boxes is

$$G(x) = B_1(x) * \text{III}(x)B_2(x). \qquad (3\text{-}4)$$

† Notice that we take $G(x)$ to be purely real, that is, an amplitude function. A more general case allows $G(x)$ to be complex. Gratings that have $G(x)$ purely imaginary are termed phase gratings.

The box $B_1(x)$ represents the transmission through a single slit (see Fig. 3-7) and so has a width b. The box $B_2(x)$ matches the total width of the grating, which we take to be W. The spacing of the δ functions in $III(x)$ is d. The grating diffraction pattern is shown on the upper right of Fig. 3-7. It is immediately clear that the different orders, n, arise from the many rulings, and the individual slit width, b, sets the diffraction envelope while the width of the grating, W, sets the width of the individual interference maxima. The amplitude diffraction pattern of the grating follows immediately from eqs. 3-3 and 3-4,

$$g(\theta) = III(\theta) * \frac{W \sin \pi\theta W}{\pi\theta W} b \frac{\sin \pi\theta b}{\pi\theta b}$$

where $III(\theta) = \sum_n \delta(\theta - n/d)$. The convolution with a δ function produces a translation to the position of the δ function and so

$$g(\theta) = \sum_n \frac{W \sin \pi(\theta - n/d)W}{\pi(\theta - n/d)W} \frac{b \sin \pi\theta b}{\pi\theta b}. \qquad (3\text{-}5)$$

The square of this wave amplitude function is proportional to the light intensity in the usual way.

In most applications we need the grating pattern as a function of β, but since $\sin \alpha + \sin \beta = \theta$ and α is fixed in any given case,

$$\beta = \sin^{-1}(\theta - \sin \alpha).$$

Further

$$d\theta = \cos \beta \, d\beta.$$

Often β is small in real instruments making $d\theta$ and $d\beta$ the same. Let us consider the width of the interference maxima. The first zero differs from the center of the maximum by

$$\Delta\theta = \frac{\lambda}{W}$$

or

$$\cos \beta \, \Delta\beta = \frac{\lambda}{W}. \qquad (3\text{-}6)$$

The grating resolution then is proportional to its overall width.*

---

* But the grating resolution does not limit the resolution in a stellar spectrograph except in the very high resolution applications; the slit widths do.

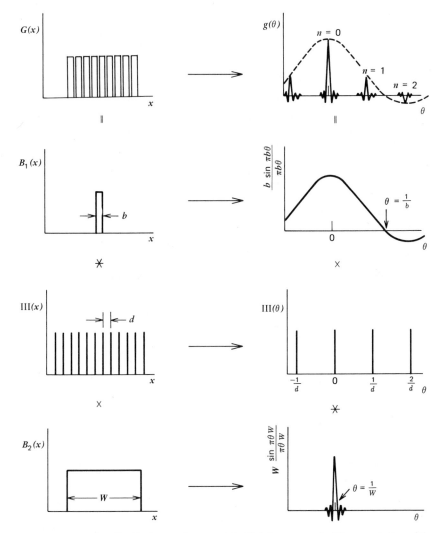

Fig. 3-7 The grating distribution on the upper left, $G(x)$, is composed of the convolution of the box $B_1(x)$ of width b with $III(x)$ of spacing d and multiplied by $B_2(x)$ representing the total width of the grating W. This is shown on the left side of the diagram. The right side shows the transforms. The (amplitude) diffraction pattern, $g(\theta)$, is seen to be the convolution of sinc $\pi\theta W$ with $III(\theta)$ which is multiplied by sinc $\pi\theta b$. The distribution of light in the diffraction pattern of the grating is proportional to $g^2(\theta)$.

The large maxima in $g(\theta)$ occur when the $\theta - n/d$ become zero. Each such maximum corresponds to a value of the order integer n. This condition for maxima is then

$$\frac{n\lambda}{d} = \theta = \sin \alpha + \sin \beta \qquad (3\text{-}7)$$

where the units of d are no longer λ units but d and λ must be in the same units. This relation is so fundamental in working with gratings that it is called *the* grating equation.

For gratings illuminated in polychromatic light, the collection of maxima for different wavelengths having one and the same n comprises the nth order spectrum. The zero order spectrum is the white light or undispersed order and falls in the $\beta = -\alpha$ direction. The larger the value of n, the larger β becomes for a given λ. The polychromatic light is spread out in θ in proportion to n (eq. 3-7) and any band of colors is seen successively farther from the zero order with increasing order. The position and orientation of visible wavelengths is shown in Fig. 3-8. The overlap of wavelengths at a given θ occur for all combinations of $n\lambda$ having the same value—a result following

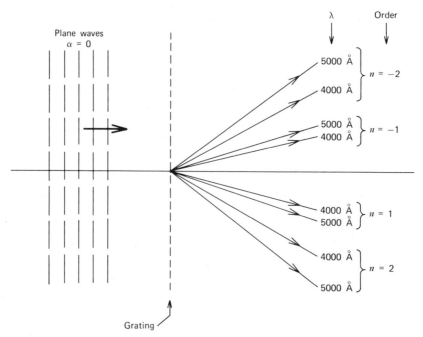

Fig. 3-8 The position and orientation of the first two orders on each side of the grating normal are illustrated for 4000 and 5000 Å light. The angles are calculated for a 600 lines/mm grating.

trivially from the grating equation. It is often necessary to place a filter in the light beam to remove light from unwanted orders.

The angular dispersion of the grating follows from eq. 3-7,

$$\frac{d\beta}{d\lambda} = \frac{n}{d \cos \beta}. \qquad (3\text{-}8)$$

Many spectrographs are designed so the grating normal points toward the camera. Then $\beta \sim 0$ and the angular dispersion is essentially constant with wavelength. Equation 3-8 shows us that the angular dispersion can be increased by observation in a higher order or by selection of a grating having smaller spacing between the rulings. The high density rulings are about 1200 lines/mm and values down to 300 lines/mm are typical of grating used in visible light spectrographs.

If we now combine the angular dispersion given by eq. 3-8 with the angular resolution from eq. 3-6 we find the chromatic resolution

$$\Delta\lambda = \Delta\beta \frac{d \cos \beta}{n} = \frac{\lambda}{W} \frac{1}{\cos \beta} \frac{d \cos \beta}{n}$$

or

$$\Delta\lambda = \frac{\lambda}{W} \frac{d}{n}. \qquad (3\text{-}9)$$

This is the wavelength resolution for the grating of width W, ruling spacing d, working in the nth order. Sometimes the ratio $\lambda/\Delta\lambda$ is used to describe the resolving power of the grating and, of course, $\lambda/\Delta\lambda$ is independent of wavelength. Resolving powers, of $\sim 10^6$ have been realized.

THE BLAZED REFLECTION GRATING

One of the objectionable features of the simple transmission gratings is that the diffraction envelope has its largest maximum at $\theta = 0$ where the chromatic dispersion is zero. A large part of the light is lost to the zero order. Should we desire to make use of the high orders to gain resolution we would find the light intensity reduced to uncomfortably low levels. To solve this problem it is clear that we wish to shift the diffraction envelope relative to the interference pattern to those θ values in which we wish to work. The Fourier shift theorem tells us that we can accomplish such a shift by introducing a phase term in $B_1(x)$ of eq. 3-4. When this is done, the grating is said to be blazed.

Conceptually we could blaze a transmission grating by placing a small prism over each slit as shown in Fig. 3-9. The beam would be deviated through

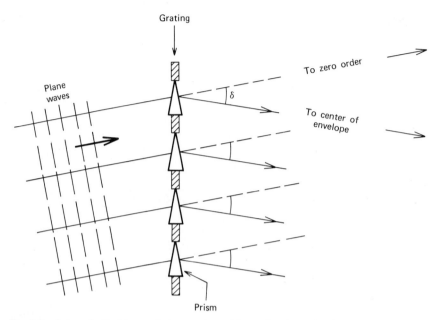

Fig. 3-9 A hypothetical transmission grating with a prism in each slit. A monochromatic incident wave experiences an angular deviation δ. The zero order image is no longer at the center of the diffraction envelope.

the angle δ thus shifting the diffraction envelope, but of course the interference pattern is not changed because the slits would still have the same separation and would still be oriented along the same plane. This system would not be very successful for polychromatic light because of the dispersive property of the prism.

The reflecting surface can also give the needed phase shift and is non-dispersive. The slits and prisms of the transmission grating are replaced by tilted "mirrors" (i.e., the rulings) the tilt giving the phase shift. Another unsought advantage also occurs at this point, namely, that the interslit or mirror spacing is no longer needed to distinguish one ruling from the next and can therefore be reduced to a minimum. This increases the efficiency of the grating. The grating cross section then makes the conceptual transition shown in Fig. 3-10. The interruling space is sufficiently small in real reflection gratings that we can take the slit width and the slit separation to be the same,

$$b = d.$$

The individual reflecting rulings are called facets. The facet angle, ϕ, is set by the manufacturer when he grinds and then sets the orientation of the ruling diamond in the ruling engine.

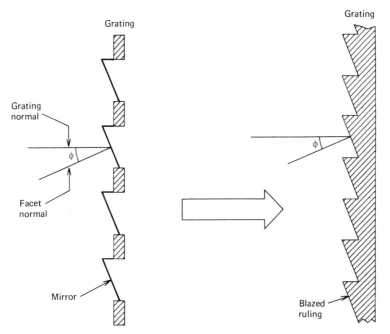

Fig. 3-10 On the left is a hypothetical grating with slits replaced by mirrors. The interslit spacing is conceptually eliminated to give the more realistic grating on the right. The facet normal makes an angle ϕ with the grating normal. The direction of positive angles is defined so that ϕ is positive.

The equations derived for the transmission grating hold for the reflection grating. The angles α and β are both positive when measured on the same side of the grating normal as the facet normal.

The distribution of light in the spectrum produced by a blazed grating for a white light source can be derived in the following way. Referring back to eq. 3-5 we see the diffraction envelope of $g(\theta)$ is given by

$$b \frac{\sin \pi\theta b}{\pi\theta b} = b \frac{\sin \pi b(\sin \alpha + \sin \beta)}{\pi b(\sin \alpha + \sin \beta)}.$$

Tilting the reflection facet through the angle ϕ changes the phase of the reflected wave such that α is replaced by $\alpha - \phi$ and β by $\beta - \phi$. Then the envelope light distribution is proportional to

$$I(\beta) = \left\{ \frac{\sin (\pi b/\lambda)[\sin (\alpha - \phi) + \sin (\beta - \phi)]}{(\pi b/\lambda)[\sin (\alpha - \phi) + \sin (\beta - \phi)]} \right\}^2 \tag{3-10}$$

where we have explicitly converted to b/λ in place of b measured in λ units. Again we have squared the amplitude pattern to get the intensity. Another useful formulation of this equation is to use the grating eq. 3-7 solved for λ and substituted into eq. 3-10. After applying a half angle formula we obtain

$$I(\beta) = \left(\frac{\sin(n\pi b/d)\{\cos\phi - [\sin\phi/\tan\tfrac{1}{2}(\alpha + \beta)]\}}{(n\pi b/d)\{\cos\phi - [\sin\phi/\tan\tfrac{1}{2}(\alpha + \beta)]\}} \right)^2. \tag{3-11}$$

For a given grating, b, d, and ϕ are fixed so that we can plot its "universal blaze curve" giving $I(\beta)$ as a function of $\alpha + \beta$ for any order n. These curves are shown in Fig. 3-11 for the case of $d = b$ and $\phi = 20°$. Notice how the higher orders have a blaze width inversely proportional to their order number. The maximum appears at $\alpha + \beta = 2\phi$, which of course is the configuration for specular reflection off the facet surfaces.

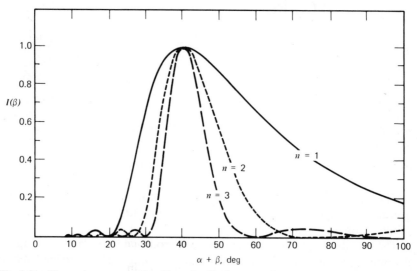

Fig. 3-11 The scalar wave grating blaze distribution according to eq. 3-11. The first three orders are shown for $\phi = 20°$.

The blaze distribution with wavelength is easily calculated once the configuration of the spectrograph is specified. The value of λ associated with each $\alpha + \beta$ of the curves in Fig. 3-11 is calculated from the grating equation. If we choose $\alpha = \phi = 20°$ we obtain the results in Fig. 3-12. A comparison of this scalar wave theory result, according to eq. 3-11, is made with the observed blaze distribution in unpolarized light in Fig. 3-13. The agreement is reasonable. However the efficiency does depend on the vector

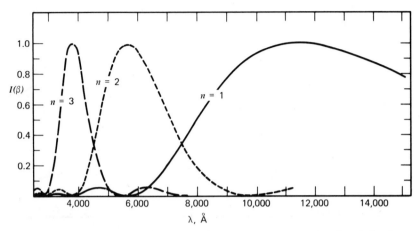

Fig. 3-12   The grating blaze distribution of Fig. 3-11 plotted on a wavelength scale for the case where the grating is used with $\alpha = \phi = 20°$.

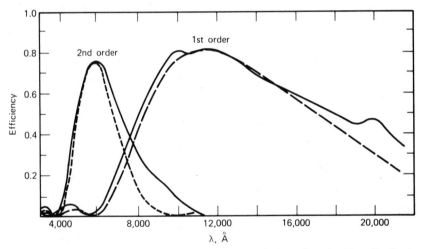

Fig. 3-13   The scalar wave blaze equation (-----) compared to a real grating blaze distribution (——). (Jarrell–Ash grating #1300 data courtesy of Jarrell–Ash.)

properties of the light, namely, the polarization. Radiation polarized with the electric vector perpendicular to the rulings generally shows an overall higher efficiency and a shift of the blaze to longer wavelengths compared to the opposite orientation (see Fig. 3-14). The polarizing or analyzing property as seen in Fig. 3-14 depends on wavelength and, at about 9600 Å for this particular grating, the sense of the polarization reverses. We conclude

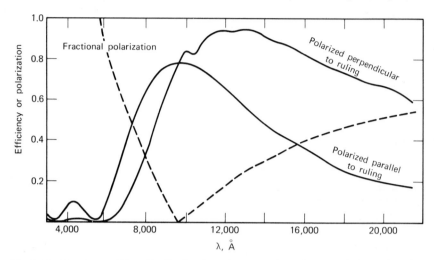

Fig. 3-14 The observed blaze distribution depends very much on the state of polarization of the incident light. The sum of the solid curves gives the solid curve used in Fig. 3-13. If the incident light is unpolarized, the grating will polarize it (-----). The sense of the polarization reverses as it goes through zero. (Adapted from data supplied by Jarrell–Ash.)

that near the peak of the blaze, linear polarization of a few per cent can be expected, but several tens of per cent can arise somewhat farther from the center of the blaze. The polarization properties are further discussed by Breckenridge (1971).

THE WAVELENGTH OF THE TRUE BLAZE

Manufacturers of gratings give the blaze wavelength for the $\alpha = \beta$ (called Lithrow) configuration used in first order. The true blaze will appear at a shorter wavelength if the grating is employed in some other orientation. We let β_b be the diffraction angle at which the blaze is directed, λ_0 the blaze wavelength for $\alpha = \beta$, and λ_b the wavelength of the true blaze. The ray configuration is shown in Fig. 3-15. Since the facet reflection gives the blaze, we write

$$\alpha - \phi = \phi - \beta_b$$

or

$$\alpha + \beta_b = 2\phi. \tag{3-12}$$

The grating equation for the general case is

$$\frac{n\lambda_b}{d} = \sin \alpha + \sin \beta_b$$

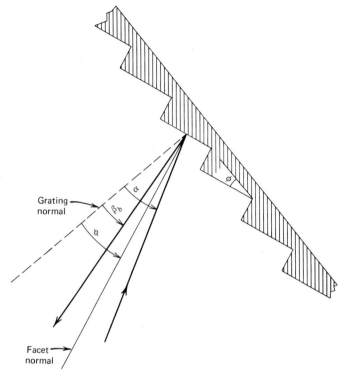

Fig. 3-15 The specular reflection at the facet determines the direction of the blaze, β_b. All angles shown in the figure are positive.

and for the $\alpha = \beta$ case it is

$$\frac{n\lambda_0}{d} = \sin \alpha + \sin \beta_b$$

$$= 2 \sin \phi$$

$$= 2 \sin \tfrac{1}{2}(\alpha + \beta_b)$$

where eq. 3-12 has been used. Dividing these two expressions gives

$$\frac{\lambda_b}{\lambda_0} = \frac{\sin \alpha + \sin \beta_b}{2 \sin \tfrac{1}{2}(\alpha + \beta_b)}.$$

By a trigonometric identity the right side of the equation is replaced giving

$$\frac{\lambda_b}{\lambda_0} = \cos \tfrac{1}{2}(\alpha - \beta_b). \tag{3-13}$$

For the case where the blaze is to be directed 40° from the incident light, cos 20° = 0.9397 and a grating that is said to be blazed at 8000 Å(λ_0) will have its blaze at 7517 Å(λ_b).

SHADOWING

Most gratings can be used in two orientations. The gratings can be tilted so α is on the same or the opposite side of the grating normal to the facet normal. These two possibilities are illustrated in Fig. 3-16. We see, upon

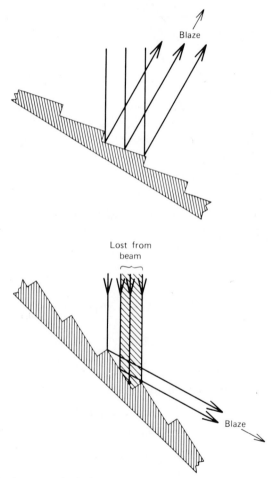

Fig. 3-16 Shadowing occurs in the bottom case but not in the top case. The angles are exaggerated to enhance the effect.

careful inspection of the diagram, that the second case may involve considerable light loss, especially when the diffraction angles are large. This is shadowing. The light may in fact cause more harm than just being lost from the beam; it may increase the level of scattered light in the spectrograph (see Chapter 12 for a discussion of scattered light). Shadowing is larger for those orientations of the grating where α is negative or at least smaller than ϕ. A diffracted monochromatic beam has a smaller width than the incident collimator beam and the angular dispersion is greater for this orientation. These results follow directly from the grating equation and their verification is left as an exercise for the reader. While shadowing can readily be demonstrated for real gratings, its behavior is more complex than the foregoing geometrical arguments indicate.

GRATING GHOSTS, SATELLITES, AND ANOMALIES

Spurious lines are generated by various imperfections in the gratings. One class, the ghosts, arises from periodic errors in the position of the rulings. These errors can usually be traced back to flaws in the ruling engine such as minute deviations in the pitch of the carriage advance screw used to space the rulings. The multiple line pattern of the type shown in Fig. 3-17 appears in place of each single emission line in the spectrum. Insight into the ghost phenomenon can be gained by viewing the grating in the light of a ghost.

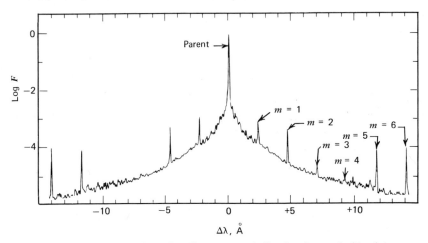

Fig. 3-17 The ghosts are weak spurious lines symmetrically placed on each side of the parent line as shown. (Adapted from Griffin 1968.)

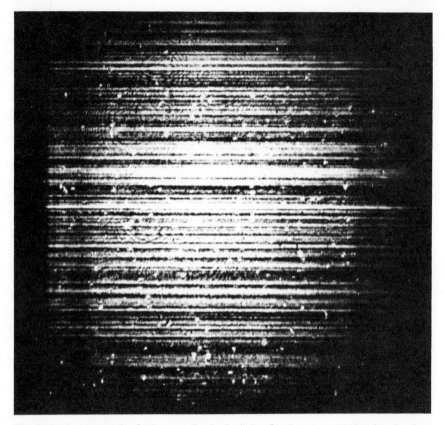

Fig. 3-18 A photograph of a large grating in the light of a ghost ($m = 1$) showing the ghost grating as a series of horizontal stripes superimposed across the face of the grating (courtesy of J. B. Breckinridge).

We find that the grating appears to be striped (Fig. 3-18), that is, the light forming the ghost comes from a series of bright strips. Each of these strips corresponds to one cycle of the periodic error in the rulings. The set of strips forms a ghost grating. The grating aperture distribution of eq. 3-4 should be replaced by

$$G'(x) = G(x)H(x)$$

where $H(x)$ is the periodic modulation by the ghost grating. If the spacing of the periodic error cycle is D, then

$$H(x) = B_3(x) * \mathrm{III}(x)B_2(x).$$

The function $B_3(x)$ describes the illumination across one strip of the ghost grating, $B_2(x)$ is the box corresponding to the full width of the grating [the same $B_2(x)$ as in eq. 3-4], and $III(x)$ is the Shah function $\sum \delta(x - mD)$, with δ functions spaced D apart with m being the order integer.

The amplitude spectrum formed is then

$$g'(\theta) = g(\theta) * h(\theta)$$

with $g(\theta)$ given by eq. 3-5 and $h(\theta)$ being the spectrum from the coarse ghost grating. The convolution brings the spectrum of this grating to the position of each line in $g(\theta)$ thus giving the ghosts. The dispersion of the ghost grating is low because of the large spacing D and in fact we can write

$$\Delta\theta_g = \frac{m\lambda_g}{D} \tag{3-14}$$

by analogy with eq. 3-7. Since $\Delta\theta_g = 0$ at the center of the parent line according to the convolution above, $\Delta\theta_g$ is a measure of the angular separation of the mth ghost from the parent line. We can write the deviation from the parent as

$$\Delta\theta_g = \Delta(\sin \alpha + \sin \beta_g)$$
$$= \cos \beta_g \, \Delta\beta_g$$

or, from eq. 3-14,

$$\Delta\beta_g = \frac{m\lambda_g}{D \cos \beta_g} \tag{3-15}$$

is the angular separation of the ghost from the parent. Since the dispersion of the main spectrum is $d\beta/d\lambda = n/(d \cos \beta)$, we can write

$$\Delta\lambda_g = \frac{\Delta\beta_g}{d\beta/d\lambda}$$

or

$$\Delta\lambda_g = \lambda_g \frac{m}{n} \frac{d}{D}. \tag{3-16}$$

This ghost spacing equation is important and useful in dealing with the ghost spectrum. The ghosts associated with each real line are found at the positions given by letting m run through the $\pm$ integers starting at $m = \pm 1$.

The spacing can be quite large. Take $d = 1/600$ mm, $n = 1$, $\lambda = 6000$ Å, and $D = 1$ mm (which is typical of a ruling engine screw pitch). We find $\Delta\lambda_g = \pm 10, \pm 20$ Å, and so forth.

The color of the light comprising the ghost images is the same as that of the parent line. This follows from the fact that the different order ghosts (m) are really different orders of the same λ formed by the ghost grating. Equation 3-16 gives the spacing of the ghosts but certainly does not imply the photons in the ghost spectrum have changed their wavelength from that of the parent line.

When more than one periodic error occurs in the rulings, the interference of the two ghost diffraction patterns can cause ghosts very distant from the parent line. (These are often called Lyman ghosts. Those near the parent and positioned according to eq. 3-16 are called Rowland ghosts.) The distant ghost still has the color of the parent. Visual inspection can be used to identify them on this basis.

The run of ghost intensity with m is usually irregular. Often the $m = \pm 1$ ghosts are the largest, and the general trend is for decreasing strength with increasing m, which must follow from general considerations of the diffraction envelope. The irregular behavior (as seen in Fig. 3-17) arises because there is more than one periodic error in the rulings and the interference between them reduces the intensity of some of the ghosts. Nonsymmetry in pairs of ghosts results if the ghost grating shows blaze characteristics.

Rowland (1893) and Meyer (1934) give a derivation of the ghost strength dependence with grating order, n. They find

$$I_g = I_p \left(\frac{\pi n e}{d} \right)^2 \tag{3-17}$$

where I_g and I_p are the intensity of the ghost and parent line, respectively, and e is the amplitude of that Fourier component of the ruling error being considered and is expressed as a fraction of the average ruling spacing d. Ghosts are likely to be a bother in high orders since they increase quadratically with order, but for a good grating they are less than 10^{-4} as strong as the parent line in first order. Ghost strength has also been considered by Calatroni and Garavaglia (1973 and 1974).

A second class of grating imperfections gives rise to other spurious lines called satellites. Again looking in the light of a satellite feature reveals that certain portions of the grating surface are distorted. In effect, one has a little grating superimposed on the main grating but not in the same plane. Satellites are usually much closer to the parent line than the ghost spacing. They are not measurable in the best gratings.

The third type of undesirable behavior we consider here is called Wood's anomaly (Wood 1935, Breckinridge 1971). It consists of a relatively large discontinuity introduced into an otherwise smooth continuous spectrum. An example of one as measured by Breckinridge (1971) is shown in Fig. 3-19. The position of these anomalies can be calculated for any given grating using the expression given by Stewart and Gallaway (1962), for example. Fortunately they generally lie outside of or on the edge of the blaze distribution.

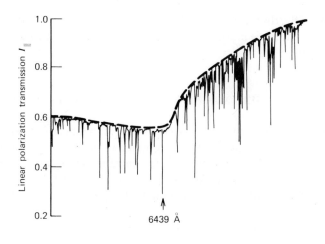

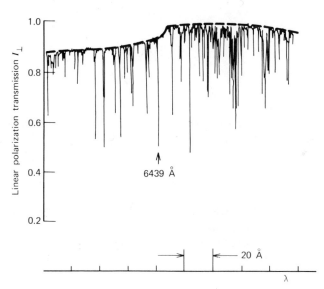

Fig. 3-19 A large Wood's anomaly recorded using the solar spectrum as a source. The heavy dashed line indicates the continuum which, if the anomaly were not present, would be a smooth horizontal line for both graphs. (Adapted from Breckinridge 1971.)

The amount of the discontinuity depends strongly on the polarization of the incident light and varies markedly from one Wood's anomaly to another for the same grating.

SPECTROGRAPH SLIT MAGNIFICATION AND SPECTRAL PURITY

The width of the image of the entrance slit as seen in the spectrum depends on the focal lengths of the collimator and camera and on the orientation of the grating. Let us define W' as the width of the entrance slit, f_{coll} the collimator focal length, and f_{cam} the camera focal length. The angular size of the entrance slit as seen from the grating (or the collimator) is

$$d\alpha = \frac{W'}{f_{coll}}.$$

From the grating equation (eq. 3-7) we know that

$$d\beta = -\frac{\cos \alpha}{\cos \beta} d\alpha$$

or

$$d\beta = -\frac{\cos \alpha}{\cos \beta} \frac{W'}{f_{coll}}$$

is the angular size of the monochromatic image of the entrance slit. If we call the slit image width w, then

$$w = -\frac{\cos \alpha \, f_{cam}}{\cos \beta \, f_{coll}} W'. \tag{3-18}$$

The magnification of the entrance slit width is w/W' and, as mentioned earlier in the chapter, the magnification along the slit is just f_{cam}/f_{coll}.

The image width w corresponds to a wavelength interval in the spectrum of

$$\Delta\lambda = w \frac{d\lambda}{dx} \tag{3-19}$$

and is called the spectral purity. When the slit width dominates the broadening of the instrumental profile, it is the spectral purity that determines the spectral resolution. By taking the expression for angular dispersion given in eq. 3-8 we can write the linear dispersion as in eq. 3-1.

$$\frac{d\lambda}{dx} = \frac{1}{f_{cam}} \frac{d \cos \beta}{n} \tag{3-20}$$

and the spectral purity is

$$\Delta\lambda = -\cos\alpha\,\frac{W'}{f_{\text{coll}}}\frac{d}{n}. \tag{3-21}$$

This is an important relation that ties together the grating parameters d and n with the spectrograph parameters W' and f_{coll}. We see that the camera focal length does not enter the expression. In Chapters 10 and 12 we apply these relations under more specialized conditions.

INTERFEROMETERS

Fourier spectroscopy, as this interferometer approach to spectroscopy is called, is a rapidly developing technique with certain advantages of importance to astronomical measurements. It is true that Fourier spectroscopy has not been employed extensively yet, although some excellent work has been done by Connes (1969, 1970), Marschall and Hobbs (1972), Beer et al. (1972), and others. What is presented here should be viewed as a primer on a young but potentially powerful field. A more extended treatment is given by Bell (1972).

Interferometers can be used to measure the Fourier transform of the spectrum. Several types of interferometers have been used, notably the Michelson and Fabry-Perot instruments. In the Michelson interferometer shown in Fig. 3-20, the path difference between the two arms is δ. The moveable mirror introduces a phase term $e^{2\pi i \delta \tilde{\nu}}$ for the monochromatic light of wave number $\tilde{\nu} = 1/\lambda$. The state of interference in the exit aperture varies sinusoidally with δ as the successive multiples of λ and $\lambda/2$ in path difference give maxima and minima. Take $B(\tilde{\nu})$ to be the spectrum of the light and $b(\delta)$ to be the signal recorded at the detector. These functions, for the monochromatic example, are shown in Fig. 3-21. We recognize that the two functions $B(\tilde{\nu})$ and $b(\delta)$ are Fourier transforms if we are willing to imagine that $B(-\tilde{\nu}) = B(\tilde{\nu})$ and $b(-\delta) = b(\delta)$. In this way $B(\tilde{\nu})$ and $b(\delta)$ are symmetric.

In polychromatic light we sum the monochromatic contributions and write

$$b(\delta) = \int_{-\infty}^{\infty} B(\tilde{\nu})e^{2\pi i \delta \tilde{\nu}}\,d\tilde{\nu}$$

$$= \int_{-\infty}^{\infty} B(\tilde{\nu})\cos 2\pi\delta\tilde{\nu}\,d\tilde{\nu}. \tag{3-22}$$

The cosine integral follows since $B(\tilde{\nu})$ is symmetric and real.

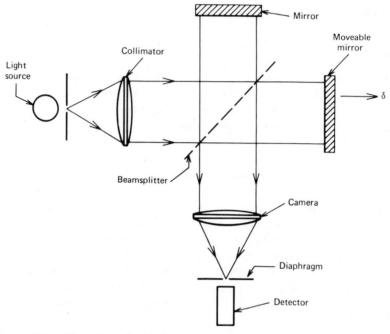

Fig. 3-20 A schematic diagram of a Michelson interferometer. The beamsplitter creates two beams. The path length and therefore the phase of the horizontal beam is changed by moving the mirror through the path difference δ. The interference ring pattern is focused in front of the detector by the camera lens.

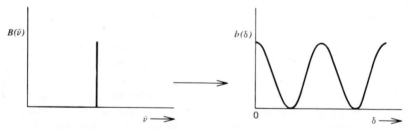

Fig. 3-21 The monochromatic spectrum on the left gives the cosinusoidal output on the right.

The interferogram, $b(\delta)$, is the detector output as a function of path difference. The spectrum of the light source is the Fourier transform of the interferogram. An interferogram and the resulting spectrum for Vega are shown in Figs. 3-22 and 3-23. At $\delta = 0$ all wavelengths are in phase and a large peak appears in $b(\delta)$. The phases rapidly become noncoherent as δ increases and the interference between different Fourier components reduces the amplitudes to lower values.

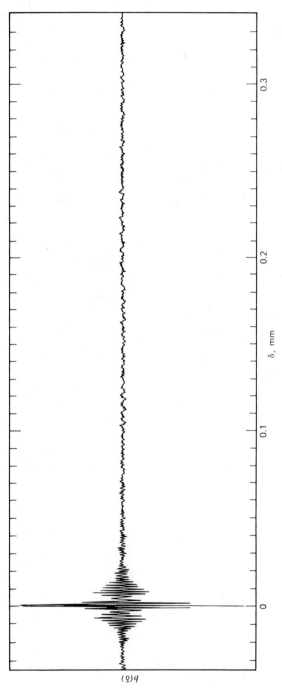

Fig. 3-22 An interferogram for the near IR spectrum of Vega (courtesy of R. P. Lowe).

$b(\delta)$

δ, mm

71

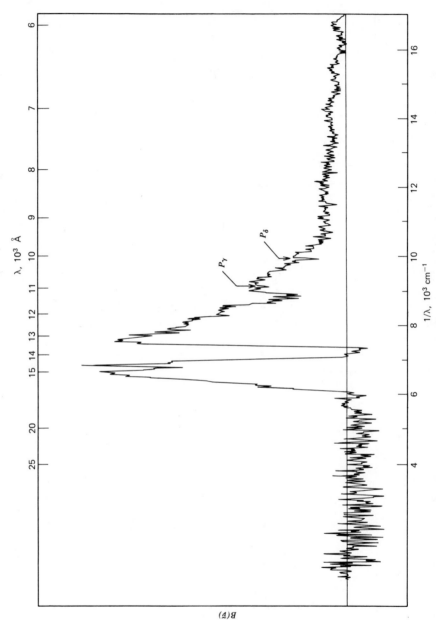

Fig. 3-23 The near IR spectrum of Vega obtained from the interferogram in Fig. 3-22. Two Paschen series lines are indicated. Most of the other features are telluric absorption bands.

The maximum path difference, δ_m, determines the resolution afforded by the interferometer in exact analogy with the diffraction grating width as described by eq. 3-6. That is, the full interferogram, which extends to $\delta = \infty$, is truncated at δ_m. This amounts to multiplying $b(\delta)$ in eq. 3-22 by a box of width δ_m. The spectrum computed from $b(\delta)$ is then convolved with sinc $\pi\delta_m\tilde{v}$, the width of which is $\Delta\tilde{v} = 1/\delta_m$, thereby setting the resolution at $\Delta\tilde{v}$. Because it is possible to construct interferometers having a path difference of 2 m, they allow nearly an order of magnitude higher resolution than the diffraction grating the largest of which is about 0.3 or 0.4 m in width.

In a grating spectrograph we saw how the spectral purity could only be kept high by using a small entrance slit (eq. 3-21). The slit allowed into the spectrograph only a small fraction of the light actually available in the seeing disc. An interferometer has no entrance slit. And given care in its design, the whole stellar seeing disc can be measured without loss in resolution. This potentially offers an efficiency gain of an order of magnitude over scanning monochrometers. On the other hand, image slicers (shown in Fig. 12-3) offset a significant portion of this advantage. Additional information can be found in the detailed work of Jaquinot (1960).

Detectors of infrared radiation are relatively noisy compared to detectors used for visible radiation. The signal to noise ratio can be enhanced in the IR case by increasing the photon flux. The interferometer, by lumping together all colors of light as in eq. 3-22, does just what is needed. It can be shown (Bell 1972, Mertz 1965) that if N points are to be measured in the IR spectrum, then a gain of $(N/2)^{1/2}$ is to be expected over a scanning monochrometer of the types discussed in Chapters 10 and 12. This gain is referred to as Fellgett's advantage or the multiplex advantage.

In the visible spectrum the detectors are much quieter and the noise is dominated by photon arrival fluctuations (see eq. 4-6). Increasing the light level increases the noise so there is no Fellgett's advantage. This explains why most Fourier spectroscopy has been done in the infrared region.

Photometric precision better than a few per cent has been difficult to attain with Fourier spectroscopy. Bell (1972) and Marschall and Hobbs (1972) discuss the situation more fully. Technical difficulties in measuring δ are gradually being overcome by the use of reference laser interferometers to measure the positions of the optical components.

REFERENCES

Beer, R., R. B. Hutchison, and R. H. Norton 1972. *Astrophys. J.* **172**, 89.

Bell, R. J. 1972. *Introductory Fourier Transform Spectroscopy*, Academic, New York.

Breckinridge, J. B. 1971. *Appl. Opt.* **10**, 286.

Calatroni, J. A. E., and M. Garavaglia 1973. *Appl. Opt.* **12**, 2298.

Calatroni, J. A. E., and M. Garavaglia 1974. *Appl. Opt.* **13**, 1009.

Connes, P. 1969. *Theory and Observation of Normal Stellar Atmospheres*, O. Gingerich, Ed., MIT Press, Cambridge, p. 323.

Connes, P. 1970. *Annu. Rev. Astron. Astrophys.* **8**, 209.

Harrison, G. R. 1973. *Appl. Opt.* **12**, 2039.

Jaquinot, P. 1960. *Rep. Prog. Phys.* **23**, 267.

Marschall, L. A., and L. M. Hobbs, 1972. *Astrophys. J.* **173**, 43.

Mertz, L. 1965. *Transformations in Optics*, Wiley, New York.

Meyer, C. F. 1934. *The Diffraction of Light, X-Rays, and Material Particles*, University of Chicago Press, Appendix F.

Piccirillo, J. 1973. *Publ. Astron. Soc. Pac.* **85**, 278.

Rowland, H. A. 1893. *Philos. Mag.* **35**, 397.

Stewart, J. E. and W. S. Gallaway 1962. *Appl. Opt.* **1**, 421.

Stroke, G. W. 1967. *Handb. Phys.* **29**, 426.

Wood, R. W. 1935. *Phys. Rev.* **48**, 928.

Young, A. T. 1974. *Astrophys. J.* **189**, 587.

CHAPTER FOUR

DETECTORS

A detector has the important job of converting the stellar photons into a recordable signal. It comes as no surprise that detectors take on a wide variety of forms ranging from thermopiles and bolometers through image tubes. Two detectors stand out as particularly significant in astronomical work. These are the photographic plate and the photomultiplier tube. Almost all the photometry of starlight done through the visible atmospheric window has involved one or the other of these detectors. It is for this reason and also because these two detectors illustrate most of the properties of detectors in general that this chapter is devoted mainly to them. In particular, the observer is interested in the quantum efficiency, spectral response, linearity, noise, and resolution of the detector. These properties can be defined for any detector and the reader will have little difficulty extending the concepts of this chapter to other detectors.

Astronomical light detectors are sensitive devices and should never be exposed to excessive illumination. In particular, the photographic plate and the photomultiplier can be destroyed by exposure to room lights. Even when a photomultiplier is unconnected it is best stored in darkness and viewed in subdued light.

WHAT IS A PHOTOGRAPHIC PLATE?

Crystals of silver halides (AgCl, AgI, but usually AgBr) can be processed in such a way that they are light sensitive. The crystals are suspended in a thin gelatin layer (about 15–30 μ thick) called the emulsion. The gelatin layer is

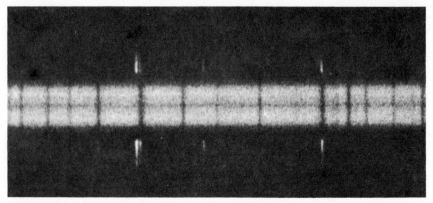

Fig. 4-1   A spectrogram is grainy when seen under magnification.

supported by a substrate of glass or celluloid. The absorption of several photons by agglomerations of silver halide crystals results in a photochemical change in the crystals. Details of the theory can be seen in the publications of Hanson (1962), Engel (1968), and James (1966). When the photograph is processed with a chemical solution called a developer, the photoactivated crystals are reduced to metallic silver much more rapidly than the unexposed crystals. The developing is done for a few minutes, in which time a large fraction of the light struck crystals are reduced but only a small fraction of the unexposed crystals are reduced. In addition to development, the plate is also placed successively in other solutions to stop the development at the proper stage and to remove the unused silver halide crystals.

The negative image is comprised of the precipitated silver crystals, which are black because of their light absorption properties.* All quantitative measurements are done directly on the negative. Inspection of the image shows it to be grainy in character as seen in Fig. 4-1. The number of grains per unit area and their size determine how dark the image is, and the number of grains in turn is related to the amount of light that has been absorbed in the plate. The coarseness of the grain depends on the emulsion type.

Quantitative reduction is made from the photograph by means of a microdensitometer, a discussion of which can be found in the book by Harrison et al. (1948) (see Latham 1971 for a summary of modern machines). This machine converts the image density into graphical or numerical form, which can then be analyzed in the usual way.

---

* Color photography has been useful only rarely in astronomy. The exposure times required are much longer than for black and white and the color balance is usually distorted by the low light levels.

PHOTOMULTIPLIER TUBES

The photomultiplier consists of the light sensitive surface, the cathode, followed by several electron multipliers, the dynodes, and ends in an electron collector, the anode. Figure 4-2 shows a photomultiplier and its schematic drawing.

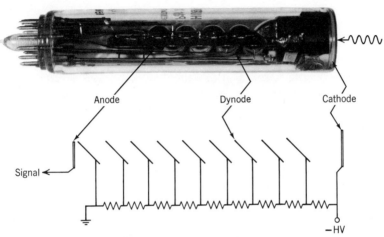

Fig. 4-2 The light strikes the photomultiplier from the right. Many electrons are generated in the dynode array for each photoelectron emitted by the cathode. The high voltage is distributed along the dynodes by the resistor string. The −HV terminal is connected to the power supply.

The detection of the light is based on the photoelectric effect. The photon of energy $h\nu$ strikes the cathode and if an electron is emitted, it has a kinetic energy $\frac{1}{2}mv^2 = h\nu - \phi$ where ϕ is a potential energy associated with the surface. When $h\nu = \phi$ we have the limiting case of the lowest energy photon that can release an electron from the cathode. Table 4-1 gives the limiting wavelength for several important elements. It is clear from these entries why Cs, K, and Rb are used in cathode construction. On the other hand, metals are generally found to show the photoelectric effect only weakly and it is necessary to use semiconductors composed of these metals and their oxides.

An electric field is impressed across the dynode array, usually with a voltage divider as indicated in Fig. 4-2. An electron emitted by the cathode is accelerated toward the dynode gaining $\sim 10^2$ eV from the field. The impact

Table 4-1   A Selection of Elements and Their Threshold Wavelengths*

Element	λ_{max}, Å	Element	λ_{max}, Å
Ag	2610	Na	5000
Cs	6600	Ni	2450
Fe	2700	Rb	5800
K	5500	Sb	3000
Li	5430	W	2650

* Data from Engstrom (1963) and Schonkeren (1970).

of the accelerated electron as it hits the dynode releases several secondary electrons. Each of these is accelerated to the next dynode and so on, so that after 10–15 stages the original electron has propogated between 10^6 and 10^9 electrons. This number is called the gain. The dynode array is a very good electronic amplifier. Comparison with other amplifiers has been made by Lallemand (1962) and Engstrom (1963).

Photomultipliers can be used in two distinct modes: pulse counting and current measurement. In the former the individual pulses of 10^6–10^9 electrons each are counted as one unit and correspond to one detected photon at the cathode. The individual pulses last $\sim 10^{-8}$ sec. If photon bombardment is not in excess of $\sim 10^7$/sec, the pulses remain distinguishable. In the second mode the anode current is measured without regard to the pulsed nature of the current. In both modes the output—number of pulses or electrical charge generated—is proportional to the amount of incident radiation.

Preference for one mode over the other depends upon the type of research (and perhaps the opinion of the individual). Pulse counting is better for weak sources while the current mode has advantages for bright sources (see Rolfe and Moore 1970, Baum 1962).

QUANTUM EFFICIENCY AND SPECTRAL RESPONSE

The quantum efficiency is

$$q(\lambda) \equiv \frac{\text{number of photons detected}}{\text{number of photons incident}} \tag{4-1}$$

and is to be thought of as the probability of a photon causing a recordable event. It is very important to have q as large as possible, especially when dealing with low light levels. The rate of darkening on the photograph or the size of the output from a photomultiplier is directly proportional to q.

The spectral response is a statement of the change in sensitivity of a detector with wavelength. Specifying $q(\lambda)$ is one way of defining the spectral response. Other quantities are often easier for the manufacturer to measure and are used instead of $q(\lambda)$. For the photographic plate, the "exposure" is defined as the product of the level of illumination (watts per square meter for example) times the exposure time. It is proportional to the total number of photons impressed upon a unit area of the plate. Then the spectral response of the photograph is the reciprocal of the exposure needed to give a specified level of darkening on the plate. Figure 4-3 gives the spectral response for selected emulsion types. Different spectral response is induced in the basic blue sensitive plate by introducing dyes into the emulsion (details may be seen in Keyes and Van Lare 1962 or Brooker 1966). Various dyes have been developed that extend the spectral response of photographic plates over the full visible window. For the observer it is merely a matter of ordering plates with the chosen spectral response. The maximum quantum efficiency of photographic plates is usually between 1.0 and 0.1%, depending on the emulsion type and treatment of the plate (Argyle 1955, Fellgett 1958, Jones 1958, and Perrin 1966). Values as high as 4% have been reported (Millikan 1974). The quantum efficiency is found to depend on the level of density, being larger for relatively weak exposures (Jones 1958, Ables et al. 1971, Fellgett 1958). In some cases it is possible to increase the quantum efficiency by baking the plate or treating it with hypersensitizing agents prior to exposure (see Morrison 1969 and Babcock et al. 1974, for example). One reason the quantum efficiency is so low is that a substantial fraction of the incident radiation is reflected and scattered by the emulsion rather than being absorbed. Furthermore, according to Webb (1948), as many as 10 photons may be needed per grain to make it darken during development. Photons absorbed in grains where the total photon input is too low to render the grain developable are also lost to the record.

Photomultiplier manufacturers list $q(\lambda)$ or, more commonly, photocathode current per watt of incident illumination. Suppose the radiation incident on the photomultiplier has a strength of P watts and consists of N photons/sec. Then

$$P = \frac{Nhc}{\lambda}$$

and the current generated by the cathode due to these photons is

$$i = q(\lambda)Ne$$

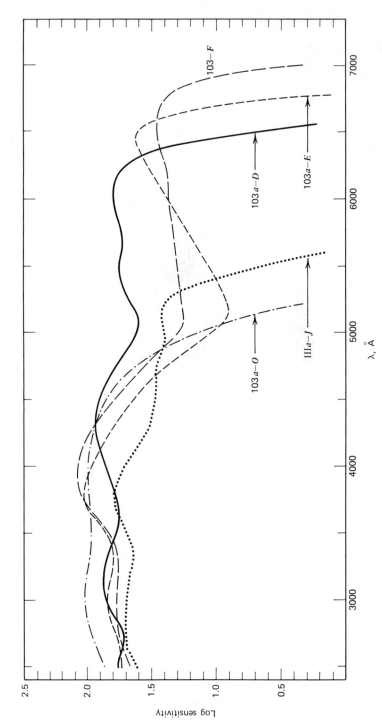

Fig. 4-3 The spectral responses of several useful emulsion types are shown. The O-type emulsion is the basic undyed emulsion and is most sensitive to blue light. The E, J, and D curves are adapted from Kodak (197), a copyrighted Eastman Kodak publication, with persmission. The O and F curves are courtesy of the Eastman Kodak Company.

where e is the electronic charge. Combining these last two equations we find the ratio of cathode current to radiant power is related to the quantum efficiency by

$$\frac{i}{P} = \frac{\lambda e}{hc} q(\lambda)$$

or

$$\frac{i}{P} = 8.07 \times 10^{-5} \lambda q(\lambda) \qquad (4\text{-}2)$$

for λ in angstroms. Conversion from one form of specifying the spectral response to the other is easily accomplished with eq. 4-2. Figure 4-4 shows some of the important spectral response types. Maximum quantum efficiencies occur in the blue and are in the range of 10–30%. It is difficult to obtain high quantum efficiency beyond 9000 Å making near IR observations difficult. New cathodes, termed types III and V because they are composed of elements from those columns of the periodic table, are likely to give improved performance at the longer wavelengths.

The spectral response is in general temperature dependent. Spectral sensitizing dyes used in photographic emulsions show a loss of effect at

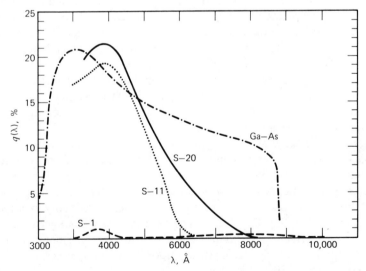

Fig. 4-4 The quantum efficiency is shown as a function of wavelength for some common cathode types. The curves are qualitative with substantial difference occurring from one tube to the next.

lower temperatures leading to a diminished long wavelength response. But enhancement of the quantum efficiency at other wavelengths has been accomplished in some instances by refrigeration of the plate during exposure (Hoag 1961, Hollars 1971). Photomultipliers show changes in the blue that are usually less than $0.5\%/°C$. Much larger changes may occur in the red. Refrigeration of the photomultiplier usually reduces its red response and increases its blue response (Lontie–Bailliez and Meessen 1959, Young 1963).

Photomultiplier quantum efficiency can be increased by multiple reflection devices that return to the cathode light reflected from the face of the tube. Oke and Schild (1968) have used reflecting hemispheres to increase the interaction of the light with the cathode while Gunter et al. (1970) have employed prisms. Gains of up to a factor of 4 have been obtained.

DETECTOR DARK OUTPUT

Common to all detectors is the output generated in the absence of illumination. The observer needs to measure the dark output for two reasons. The first is that it is only the detector output above the average dark level that corresponds to the light being measured. The second reason is that the dark component contributes noise. We defer the question of dark noise to later in the chapter and consider now the behavior of the *average* dark output.

Photographic plates show background "fog." Some of the fog is the result of cosmic rays and radioactive trace elements in the vicinity of the plate, but most of it results from thermal excitation of the silver halides in the emulsion. For this reason it is universal practice to refrigerate spectroscopic plates until they are used. The amount of fog that comes up during development depends on the emulsion type, the past history of the plate, the developing procedure, and the type of developer. At times the fog is not uniform over the plate, leading to enhanced errors.

Photoelectric detectors show a dark signal arising from leakage currents, discharge glow, and thermal emission. Thermal emission from the cathode dominates. Refrigeration of photomultipliers has consequently proved very effective in reducing the dark output. Most photometers are equipped with a "cold box" to house the photomultiplier. There are many methods of refrigeration including Dry Ice, thermoelectric heat exchangers, and compressor run refrigerators. Figure 4-5 illustrates the thermal response of photomultiplier dark signals for three important cathode types. For most of them cooling to $-30°C$ results in a reduction of the dark signal by a large fraction. Cooling to still lower temperatures results in only a small additional reduction. Exceptions occur for the red and IR sensitive detectors such as

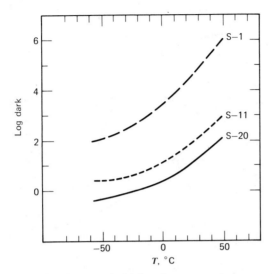

Fig. 4-5 The dark output of a photomultiplier shows a nearly exponential change with temperature but with a leveling off at cooler temperatures. Again, individual photomultipliers show unique behavior and the curves shown are qualitative.

S–1 and Ga–As cathodes, and for solid state detectors like silicone diodes for which much lower temperatures (-60 to $-100°C$ typically) are preferable.

A possible side effect of photomultiplier refrigeration is the formation of condensation or even frost on the photomultiplier, lenses, or windows of the cold box. In some cases electric leakage is greatly increased by the condensation, making the system very noisy and unusable. In other cases the light is simply prevented from reaching the photocathode. Resistive heaters may be placed around lenses and windows if they are not too close to the cathode. Dissipation of a watt or less is often sufficient. The inside of the cold box can be purged with dry nitrogen* either once before being sealed or continuously during operation.

The phototube dark signal is proportional to the cathode area. If the light level is extremely low, such that the stellar signal is comparable to the dark output, then the light should be imaged into a very small cathode (~ 2 mm $\times$ 2 mm, for example) to increase the signal to dark ratio. Tubes with small effective cathodes cannot be obtained from all photomultiplier producers however. A dark output for a good S-20 cathode tube properly refrigerated is less than 1 electron per second per square millimeter of cathode area.

---

* Nitrogen from a standard compressed "air" cylinder can be expected to have a relative humidity of $\lesssim 1\%$.

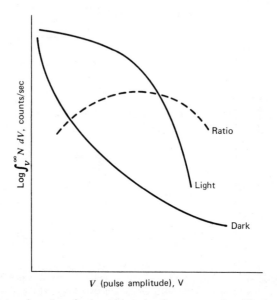

Fig. 4-6 The integral number of pulses with peak voltage greater than V is plotted as a function of V. The light curve will be vertically displaced depending on the level of illumination. The dark curve will be similarly displaced depending on the temperature of the photomultiplier. The ratio of light to dark count reaches a maximum at some voltage.

The pulse height distribution for dark output shows an enhanced number of small pulses compared to the distribution for signal pulses (Fig. 4-6). The ratio of light to dark pulses goes through a maximum. It is necessary to place immediately after the photomultiplier a pulse height discriminator that removes the many small noise pulses. The discriminator is set to pass pulses greater than a specified size and is adjusted to minimize the error (eq. 4-6 below). The proper setting of the discriminator level is more fully discussed by Morton (1968).

The component of the dark *current* arising from thermal emission from the cathode shows a dependence on high voltage that parallels the dependence of the signal current. Over a wide range of voltage, the ratio of dark current to signal current deviates only slightly from the optimum value.

It should not always be supposed that the dark signal is independent of the light level at which the tube is being used. High light levels can induce (even if the voltage was off during the exposure) an increase in dark output, and this disturbance can persist for several minutes or even hours. Equilibrium is eventually reached after the tube is operated in darkness for some time. The strength of the disturbance depends on the wavelength of the light and the exposure time, as well as the level of illumination (Schonkeren 1970).

LINEARITY AND THE PHOTOGRAPHIC PLATE

When the output of a detector is proportional to the amount of incident radiation, the detector is said to show linearity, which is a highly desirable property because a linear detector can be used to find the flux ratio between two sources without the need of calibration. The photographic process shows only an approximately linear behavior over a limited density range. For underexposed or overexposed conditions, the deviations from linearity can be pronounced. It is, as a consequence, necessary to calibrate each and every spectrogram to be used for quantitative measurement. One type of calibration is shown in Fig. 4-7. The light strips below the stellar spectrum differ

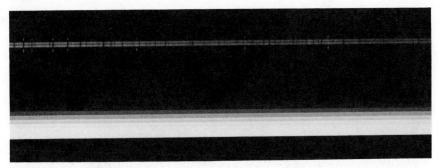

Fig. 4-7 The calibration step wedge is photographed on the same plate with the stellar spectrum and the comparison spectrum. (Dominion Astrophysical Observatory spectrogram.)

from each other by known ratios of exposure. Strips of this sort can be created by replacement of the entrance slit of the spectrograph with a series of slits of different width. (Alternatively the normal entrance slit can be covered with a step filter or rotating sector to accomplish the same end.) When these slits are illuminated with an extended light source so even the widest slit is fully and uniformly illuminated, the amount of light entering the spectrograph is proportional to the widths of the slits. By using a source having a continuous spectrum, the brightness of the images in the spectral plane and, therefore, the exposures are also proportional to the slit widths. The relation between the known slit widths and the density of the calibration strips on the plate is used to form the needed calibration. The calibration strips and the stellar spectrum are measured on the same densitometer with the same unaltered adjustments used for both. A plot of densitometer deflection for the calibration strips versus exposure ratios forms the calibration curve as in Fig. 4-8. Each deflection in the stellar spectrum record is then

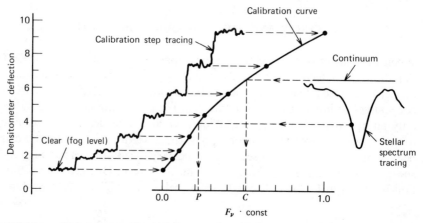

Fig. 4-8 The calibration device was constructed to make the levels of illumination differ by a factor of 1.585 (0.2 dex) from one step to the next. The position of the points defining the calibration curve are spaced in the relative flux coordinate, $F_\nu \cdot$ const, by this factor. The highest point is arbitrarily chosen as unity. Points on the spectrum tracing are then read into the calibration curve to obtain relative flux values. The continuum is usually chosen as the normalizing value. The flux ratio F_ν/F_{cont} for the point indicated on the stellar tracing is then equal to the ratio P/C.

read into the calibration curve* and translated into a relative light flux ratio, that is, flux times a constant. The relative values are usually normalized to a continuum of unity. A new calibration curve should be constructed every 200 or 300 Å along the spectrum according to Wright (1966).

The slope of the quasilinear portion of the calibration (called the constrast, and often denoted by γ) depends on the exposure, the treatment during development, the emulsion type, and the wavelength. Sometimes it is found advantageous to have a relatively small density difference between the stellar continuum and the deepest part of an absorption line; this can be accomplished by exposing the spectrogram for a longer time and developing for a shorter time. The slope of the calibration curve is lowered.

It is possible to reduce the data through the calibration curve automatically as the plate is measured in the densitometer. Analog devices are described by Oke (1957) and Phillips (1959), for example. In the newer densitometers the data is digitized and run through the calibration curve by the small computer

---

* When the ordinate in Fig. 4-8 is expressed as the logarithm of the ratio of incident to transmitted light and the abscissa is expressed in logarithmic physical units such as watt-seconds, then the calibration curve is called the Hurter–Driffield (1890) curve or HD curve. Mathematical transformations that linearize the calibration are discussed by de Vaucouleurs (1968) and Anderson (1956).

that is an integral part of the machine. A wealth of details on spectrogram calibration including technique and uncertainties can be found in the collection of papers edited by Hoag (1969) and in the work of Dunham (1956) and Wright et al. (1964).

Photography shows another type of nonlinear behavior in regions of strong density gradient. Figure 4-9 is an example of the so-called adjacency effects (other examples can be found in Kodak 1973 and Miller 1971). In part the effect results from under- and overdevelopment of low and high density sides, respectively, of a sharp density gradient and as such can be reduced by careful agitation of the developer during processing of the plate. The actual amount of distortion that occurs in a spectral line profile is difficult to evaluate. A quantitative discussion is presented by Edmonds (1965).

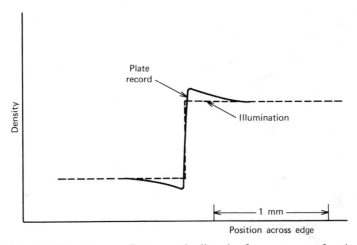

Fig. 4-9 The adjacency effect causes the distortion from a pure step function.

The photographic process has other interesting characteristics. The memory of the emulsion fades during long exposures to faint light, making the developed image weaker than if the same number of photons had been impressed upon the plate in a shorter time (called the reciprocity failure). Interrupting the light during exposure or shining on the emulsion lights of unequal brightness does not produce the same density image as a uniform constant exposure of the same total number of photons (called the intermittency and Clayden effects). These characteristics are probably important in weakening the photometric precision of the spectrogram. Seeing and scintillation cause a strong modulation in the amount of light reaching the emulsion, and the stellar seeing disc may be trailed along the spectrograph slit several times to widen the spectrogram uniformly. These are just the conditions needed to enhance the Clayden and intermittency effects.

LINEARITY AND OTHER PHOTOMULTIPLIER CHARACTERISTICS

It is generally assumed that photomultipliers do exhibit linear behavior. If the tube is operated according to the manufacturer's specifications and reasonable care is exercised the linearity assumption is probably justified. But there are a number of ways in which the linear relation between output and input can be destroyed. Let us look briefly at the more important effects.

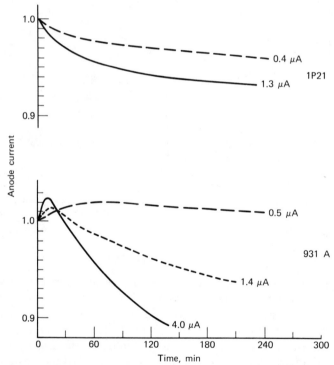

Fig. 4-10 Photomultiplier fatigue can take on several forms. The top panel illustrates the simple fatigue (positive fatigue); photomultiplier 1P21. The output decreases with time under constant illumination. The bottom panel illustrates more complicated behavior where the output first increases (negative fatigue) before it drops; photomultiplier 931A.

Fatigue may occur for higher output currents. Fatigue appears either as an increase or a decrease in tube sensitivity induced by the incident radiation. Figure 4-10 shows the two types of fatigue. The output should remain constant since the illumination is not changed. After the tube is in darkness for some time it recovers its original sensitivity. The cause is thought to lie with a change in the cesium equilibrium on the dynode surface according to how heavily the dynodes are bombarded with electrons. A simple model is given by Cathey (1958) that explains why there should be positive as well as negative fatigue. His model is illustrated in Fig. 4-11, which shows in a qualitative way how the dynode's efficiency in generating secondary electrons depends on the cesium content in the surface layers. The electron bombardment decreases the cesium level. If the dynode is overcesiated during manufacture, it will show increased sensitivity. Undercesiated dynodes show the reverse. The overcesiated dynode may make a transition to the undercesiated case if the tube current is large enough or of long duration. It is likely that other phenomena such as the buildup of charge on insulators inside the tube also contribute to fatigue. Dynodes of different chemical composition in general show different behavior. The less common silver–magnesium and the copper–beryllium dynodes are more stable against fatigue.

Fatigue is observed to decrease rapidly with decreasing levels of illumination, as is evident from Fig. 4-11. The level beyond which fatigue is negligible

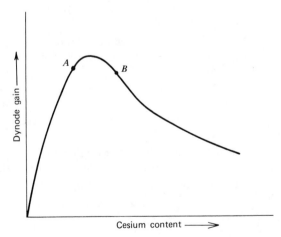

Fig. 4-11 The dynode gain reaches an optimum value as a function of the cesium level in the dynode surface. Electron bombardment reduces the cesium level. If the dynode starts at A, it will show normal positive fatigue. If it starts at B, it will show antifatigue followed by fatigue. (After Cathey 1958.)

is best found by experiment. Fatigue is temperature sensitive and varies greatly from tube to tube. Do not be fooled into believing that high photon fluxes are never encountered in stellar work. Consider Vega, the primary photometric standard (see Chapter 10), for example. It gives about 10^3 photons/sec cm² Å at 5556 Å outside the atmosphere. A medium sized telescope has a collecting area of 10^4 cm². If we adopt a 10^2 Å band pass, a 20% quantum efficiency for the photomultiplier, and include a 90% light loss from reflections and atmospheric extinction, we have 2×10^7 counts/sec. A photomultiplier might have a gain of 10^7 and using 1.6×10^{-19} C/electron, this translates into 32 μA. Fatigue can easily occur under such conditions. Usually the anode current should be held below 10^{-7} A.* Various aspects of fatigue have been considered by Youngbluth (1970), Engstrom (1947), Cathey (1958), and Marshall et al. (1947). General aging effects have been measured by Budde (1973).

Gain stability is essential for both pulse counting and current modes of operation. The gain depends on the voltage between the dynodes and, consequently, on the high voltage power supply used to run the photomultiplier. In the current mode the output is given by the product of the gain and the cathode current. The gain is strongly dependent on the voltage supplied to the dynode voltage divider (Fig. 4-12). One can approximate the relation between the gain, G, and the voltage, V, as

$$G = a_0 V^n \qquad (4\text{-}3)$$

where a_0 is a constant and n is the number of dynodes. From this relation we deduce that if the gain is to be stable to 0.1%, then the voltage must be stable to $0.1\%/n \sim 0.01\%$.

In the pulse counting mode, it is the size of the pulse that is proportional to the gain. Instability in the gain means instability in the fraction of pulses (as in Fig. 4-6) passed by the pulse height discriminator. Consequently photometric operations done with different gains are not comparable with each other regardless of whether the current or pulse mode is being used. Fortunately high quality power supplies have no difficulty meeting the stability requirements.

Deviations from linearity occur when the rate of pulses is greater than the counting speed of the electronic circuitry. If we let L represent the number of light counts per second and ΔL the error, then

$$\frac{\Delta L}{L} = \frac{Lt_0}{1 - Lt_0} \qquad (4\text{-}4)$$

---

* This may be accomplished by reducing the photon rate with a neutral density filter, diaphragming or otherwise reducing the collecting aperture, or by lowering the photomultiplier voltage. The last alternative may affect the spectral response.

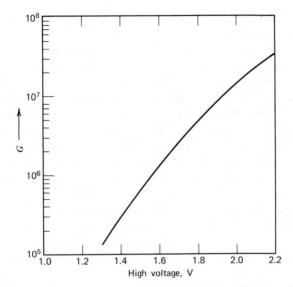

Fig. 4-12 The photomultiplier gain is nearly exponentially dependent on the voltage impressed across the voltage divider. Equation 4-3 is used to approximate the curve.

where t_0 is the cycle or dead time of the system associated with each recorded photon (derivations are done by Schonkeren 1970 and Enge 1966, for example). If the counters can reach 10^6 counts/sec and we observe a star giving 10^5 counts/sec, then $\Delta L/L = 11\%$ must be applied as a correction.

Other considerations include the effect of the earth's magnetic field. Systematic errors may be introduced into observations done at the Cassegrain focus due to the changing orientation of the field relative to the photomultiplier as stars in different parts of the sky are measured (see Smyth 1952). And there are technical considerations such as the need to have the ratio of anode current to voltage divider current be at least as small as the fractional error one expects to achieve in the photometry. Many design details can be found in Schonkeren (1970), Sharpe (1961), Engstrom (1963), and Schlueter (1967).

SPACIAL RESOLUTION

The ability of a detector to distinguish between closely spaced images diminishes as the spacing decreases. In order to speak of spacial resolution at all, we presuppose the detector is a multielement detector such as a photographic plate, a bank of photomultipliers, the phosphor on a television screen or image tube, or an array of photodiodes. Each detector has a graininess that filters the signal impressed upon it.

We can address the question of spacial resolution in a general way by thinking of an image composed of many Fourier frequencies. The slow contrast variations cover a large area and are described by low spacial frequencies. The sharp image features need the high spacial frequencies to describe them. For those Fourier components with wavelengths smaller than the graininess of the detector, the information is lost. But the transition from the lost information to the recorded information is a continuous one when considered as a function of Fourier frequency. A specification of the fraction of the Fourier amplitude that is actually recorded, as a function of spacial frequency, tells us exactly how this transition occurs.

The function can be obtained from measurements in either Fourier domain. The classical method is to impress upon the detector a very narrow line image (a δ function) and record the signal strength as a function of position across the line image. The measured function is called the instrumental profile or δ function response.* Often a characteristic width, say the half width, is given as a measure of the detector's resolution. A numerical Fourier transform of the instrumental profile gives the spacial frequency response of the detector. (Recall that the transform of the δ function is a constant as a function of frequency. It gives a "white" test spectrum.) The Fourier transform of the instrumental profile is often called the modulation transfer function (MTF).

Measurements can be made directly in the spacial frequency domain instead. Suppose we place in front of the detector a mask having a sinusoidal variation in density along one coordinate. The amplitude of the modulation as recorded by the detector will be diminished according to how much that spacial frequency is attenuated by the lack of resolution in the detector. Measurements on many masks covering a wide range of spacial frequency allows the filtering properties of the detector to be mapped out. Many common test patterns are made of black and white strips or lines rather than a sinusoidal variation in density. The fundamental periodicity of the strip pattern is the last Fourier component to fade however, so that both types of patterns give the same result. A typical mask is shown in Fig. 4-13. In Fig. 4-14 the modulation transfer function of a three stage electrostatic image tube is given as an example. Notice the frequency coordinate has units of lines per millimeter. It is equivalent to cycles per millimeter.

Suppose we wish to measure a feature in a spectral line that is characterized by frequencies near 10 cycles/mm. The detector having the filtering properties in Fig. 4-14 would reduce the amplitude of the feature to 48% of its original value. If the noise is white, the signal to noise ratio for the 10

---

* In Chapter 12 the instrumental profile of the whole spectrometer is considered in detail.

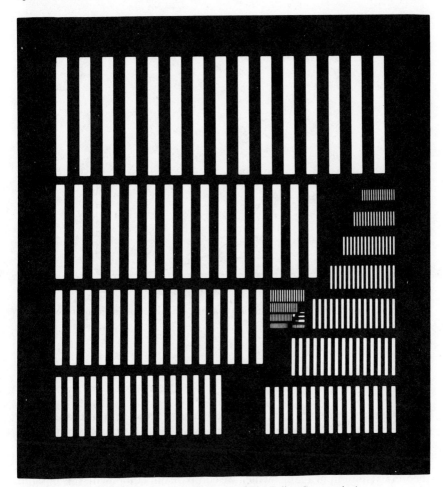

Fig. 4-13   A bar resolution test pattern (courtesy of The Ealing Corporation).

cycles/mm spacial frequency is worse by a factor of 2 compared to a much higher resolution detector.

Modulation transfer functions usually show a monotonically decreasing amplitude with increasing spacial frequency (as seen in Fig. 4-14). In lieu of a full description, some manufacturers specify a single point on the curve, like 50 lines/mm at 30% contrast. But this is truly minimal information since the shape of the transfer functions can vary widely. Indeed photographic emulsions may not even show a maximum at zero when the adjacency effects enhance certain frequencies above their true value (see Perrin 1966).

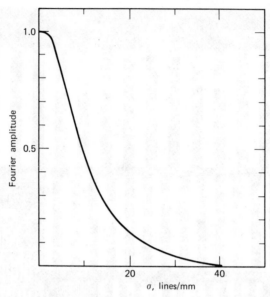

Fig. 4-14 The modulation transfer function of a three stage electrostatic image tube.

NOISE

The origins of noise are many. We can usually class them into one of the three categories: (1) photon noise in the starlight, (2) photon noise in the sky background, and (3) equipment noise. The combined fluctuation in output due to the inevitable sources of noise in categories 1 and 2 represents a theoretical minimum against which we compare our equipment. It is possible to come within a few per cent of this minimum under favorable conditions. A system that shows 2–3 times the minimum needs some attention.

Photon noise arises from the irregular times of arrival of the photons. The fluctuations in photon rate are described by Bose–Einstein statistics. For a thermal source of characteristic temperature T, the mean square fluctuation, Δn photons/sec, is

$$\Delta n = n^{1/2}\left(1 + \frac{1}{e^{h\nu/kT} - 1}\right)^{1/2}$$

and for nonthermal sources a slightly different relation may hold (Smith et al. 1957, Fellgett 1955). Since $h\nu$ is appreciably greater than kT in most cases, we choose to neglect these deviations from classical Poisson statistics* and write

$$\Delta n = n^{1/2}. \qquad (4\text{-}5)$$

The fluctuations in the output due to equipment noise can usually be taken to be random events and therefore describable with Poisson statistics (eq. 4-5 again). In this case we think of n as the equivalent number of detected photons that would cause the same noise.

We can combine the main sources of noise as follows. We define the rate of photon counts for the light source to be L per second and the rate of the counts corresponding to the equipment noise and other background (sky, cosmic rays, ...) to be B. The number of recorded counts in an integration time or time constant interval t is the sum

$$n = (L + B)t.$$

Our signal is comprised of the number of light counts, which is $N = Lt$, and in practice is found by subtracting the background signal from the total,

$$N = n - Bt$$
$$= n - b$$

where b is defined to be Bt. The statistical uncertainty in N is denoted by the standard deviation, ΔN, and is comprised of the standard deviations in n and b, again defined as Δn and Δb, by

$$\Delta N = [\Delta n^2 + \Delta b^2]^{1/2}$$

where we assume that the deviations are sufficiently well represented by a Gaussian distribution so that eq. 2-19 can be applied.

Using eq. 4-5 we can substitute $\Delta n = n^{1/2}$, and $\Delta b = b^{1/2}$ and can write the fractional error as

$$\varepsilon \equiv \frac{\Delta N}{N} = \frac{(n + b)^{1/2}}{N}.$$

---

* While the deviation may not be large for normal stars observed through the visible window, it can amount to several per cent. This may be important in the assessment of the performance of an instrument on the basis of the observed data error versus the theoretical minimum of pure photon noise. Further, when n is not small, the skewness and kertosis of the Poisson distribution are small, so that for practical purposes it can be represented by a Gaussian.

Or, if we substitute the equivalent count *rates* defined above,

$$\varepsilon = \frac{[(L + B)t + Bt]^{1/2}}{Lt}$$

$$= \left(\frac{1 + 2B/L}{Lt}\right)^{1/2}. \tag{4-6}$$

The equation is fundamental. Consider the regime where $B/L \ll 1$ (the desirable case). Then $\varepsilon \doteq (Lt)^{-1/2} = N^{-1/2}$, which is pure signal photon noise. This error is reducible according to the square root of the integration time, t, for a fixed light level L. In pulse counting a total count of $N = 10^4$ gives a 1 % error as the theoretical minimum error. On the other hand, if we think of working at a fixed level of precision, then L and t are inversely related. This is a useful fact for estimating observation time needed for some new star based on earlier experience.

In the opposite regime where the background count exceeds the light count, $B/L \gg 1$, we can write $\varepsilon \doteq [(2B/L)/Lt]^{1/2}$. Since B/L is a large number, ε remains large unless t is made very long. Observations carried out under a heavy background are very consuming of telescope time and it follows that only a few such measurements can be made. In the situation where the observations are made with a single channel spectral line scanner (discussed in Chapter 12)—and dozens of measured points are needed to define a single line profile—it is completely impossible to work in this regime. By the same reasoning multielement detectors increase the efficiency over a single detector in proportion to the number of elements, but the faint star limit shows only modest improvement. There are only two ways to get back to the $B/L \ll 1$ regime. One is to increase L by using a larger telescope, a machine with lower photon loss, or a detector with higher quantum efficiency. The other is to decrease B by using a cooled detector, a detector that has a smaller cathode or a cathode that is intrinsically less noisy, or an emulsion with less graininess, and by decreasing the nonstellar photons from the sky and elsewhere. The sky may be important in energy distribution measurements but is negligible for coudé slit spectroscopy.

When using photoelectric detectors in the pulse counting mode, the rates B and L in eq. 4-6 can be measured directly. In the current mode of operation the same principles hold, but the anode currents for the stellar and the background measurements must first be converted to detected photons by dividing by the electron charge and the gain. Additional material dealing with noise in photomultipliers can be found in the works of Rolfe and Moore (1970), Young (1966), Eberhardt (1967), Whitford (1962), and Baum (1962).

Noise in photographic measurements is more difficult to analyze. While relation 4-6 still holds, it is difficult to deduce the rates B and L from the spectrogram (see Fellgett 1958). More commonly the performance in terms of the signal to noise ratio is specified by the detective quantum efficiency (DQE). The detective quantum efficiency is obtained by taking repeated measurements, computing the internal error, deducing the number of photons needed according to eq. 4-6 to give that error, and then dividing by the real number of incident photons. Obviously the DQE parameter can be computed for any detector but it is more commonly used with those detectors with which individual photon counting is difficult. The background noise is often dominated by the plate grain and its clumpiness. Tracing a portion of uniformly exposed plate with a densitometer gives one measure of the grain noise. Plate averaging is reasonably effective in reducing the grain noise as shown, for example, by Bonsack (1971) and Stoeckley and Morris (1974). The relation between photographic noise and various reduction procedures is considered by Brandt and Nesis (1973). Additional noise arises from the variation in thickness of the emulsion causing a density variation in the image. Fourier filtering to remove grain noise and other high frequency noise can be very useful (e.g., Rusconi and Sedmak 1971; see also Chapter 12).

OTHER DETECTORS

Image tubes, electronic cameras, television cameras, and diode arrays are all being used with some success as detectors in spectral photometry. Advances have been made on several fronts: simultaneous photoelectric recording of hundreds or thousands of spectral elements, increased photometric precision, and the ability to record fainter sources. Beaver et al. (1972) have employed a Digicon for coude spectro-photometry. Walker et al. (1972) and Kinman and Green (1974) describe successful television systems. Electronic cameras are described and used by Kron (1971) and Ables et al. (1974). An image tube spectrum scanner is described by Robinson and Wampler (1972), McNall et al. (1972), and Anderson (1972). A review of some of these detectors is given by Livingston (1973). The amazing developments in detector technology are having a wonderful impact on the collection of stellar data.

REFERENCES

Ables, H. D., A. V. Hewitt, and K. A. Janes 1971. *Am. Astron. Soc. Bull.* 1971, No. 1, 18.

Ables, H. D., E. B. Newell, and E. J. O'Neil Jr. 1974. *Publ. Astron. Soc. Pac.* **86**, 311.

Anderson, C. M. 1972. *Astrophys. J. Lett.* **177**, L121.

Anderson, J. W. 1956. *Appl. Spectrosc.* **10**, 195.

Argyle, P. E. 1955. *J. Astron. Soc. Can.* **49**, 19.

Babcock, T. A., P. M. Ferguson, and T. H. James 1974. *Astron. J.* **79**, 92.

Baum, W. A. 1962. *Astronomical Techniques*, W. A. Hiltner, Ed., University of Chicago Press, Chicago, p. 1.

Beaver, E. A., E. M. Burbidge, C. M. McIlvan, H. W. Epps, P. A. Strittmatter 1972. *Astrophys. J.* **178**, 95.

Bonsack, W. K. 1971. *Astron. Astrophys.* **15**, 374.

Brandt, P. N. and A. Nesis 1973, *Sol. Phys.* **31**, 75.

Brooker, L. G. S. 1966. *The Theory of the Photographic Process*, 3rd ed. T. H. James, Ed., Macmillan, New York, p. 198.

Budde, W. 1973. *Appl. Opt.* **12**, 2109.

Cathey, L. 1958. *IRE Trans. Nucl. Sci.* December, p. 109.

de Vaucouleurs, G. 1968. *Appl. Opt.* **7**, 1513.

Dunham, T. Jr. 1956. *Vistas in Astronomy*, Vol. 2, A. Beer, Ed., Pergamon Press, London, p. 1223.

Eberhardt, E. H. 1967. *Appl. Opt.* **6**, 251.

Edmonds, F. N. Jr., 1965. *Astrophys. J.* **142**, 278.

Enge, H. A. 1966. *Introduction to Nuclear Physics*, Addison-Wesley, Reading, Mass.

Engel, C. E. 1968. *Photography for the Scientist*, Academic, New York.

Engstrom, R. W. 1947. *J. Opt. Soc. Am.* **37**, 420.

Engstrom, R. W. 1963. *Phototubes and Photocells*, RCA Technical Manual PT-60, pp. 6, 28.

Fellgett, P. B. 1955. *Vistas in Astronomy*, Vol. 1, A. Beer, Ed., Pergamon, London, p. 475.

Fellgett, P. 1958. *Mon. Not. Roy. Astron. Soc.* **118**, 224.

Gunter, W. D. Jr., G. R. Grant, and S. A. Shaw 1970, *Appl. Opt.* **9**, 251.

Hanson, W. T. 1962. *Photography: Its Materials and Processes*, 6th ed., C. B. Neblette, Ed., Van Nostrand, Princeton, p. 180.

Harrison, G. R., R. C. Lord, and J. R. Loofbourow 1948. *Practical Spectroscopy*, Prentice-Hall, Englewood Cliffs, N.J., p. 350.

Hoag, A. A. 1961. *Publ. Astron. Soc. Pac.* **73**, 301.

Hoag, A. A. 1969. *Bull. Am. Astron. Soc.* **1**, 141.

Hollars, D. R. 1971. *Am. Astron. Soc. Photo-Bull. 1971*, No. 2, 18.

Hurter, F. and V. C. Driffield 1890. *J. Soc. Chem. Ind. (London)* **9**, 455.

James, T. H. 1966. *The Theory of the Photographic Process*, 3rd ed., Macmillan, New York.

Jones, R. C. 1958. *Photog. Sci. Eng.* **2**, 57.

Keyes, G. and E. Van Lare 1962. *Photography: Its Materials and Processes*, 6th ed., C. B. Neblette, Ed., Van Nostrand, Princeton, p. 196.

Kinman, T. D. and M. Green 1974. *Publ. Astron. Soc. Pac.* **86**, 334.

Kodak 1973. *Plates and Films for Scientific Photography*, Kodak Publication No. P-315, Eastman Kodak Co., Rochester, N.Y.

Kron, G. E. 1971. *Astronomical Use of Television-Type Image Sensors*, V. Bascarino, Ed., NASA SP-256, p. 207.

Lallemand, A. 1962. *Astronomical Techniques*, W. A. Hiltner, Ed., University of Chicago Press, p. 126

Latham, D. W. 1971. *Am. Astron. Soc. Photo-Bull. 1971*, No. 1, 28.

Livingston, W. D. 1973. *Annu. Rev. Astron. Astrophys.* **11**, 95.

Lontie-Bailliez, M. and A. Meessen, 1959. *Ann. Soc. Sci.* Brux. **73**, 390.

Marshall, F. H., J. W. Coltman, and L. P. Hunter 1947. *Rev. Sci. Inst.* **18**, 504.

McNall, J. F., D. E. Michalski, and T. L. Miedaner 1972. *Publ. Astron. Soc. Pac.* **84**, 176.

Miller, W. C. 1971. *Am. Astron. Soc. Photo-Bull. 1971*, No. 2, 3.

Millikan, A. G. 1974. *Am. Sci.* **62**, 324.

Morrison, D. 1969. *Am. Astron. Soc. Photo-Bull. 1969*, No. 1, 12.

Morton, G. A. 1968. *Appl. Opt.* **7**, 1.

Oke, J. B. 1957. *J. Roy. Astron. Soc. Can.* **51**, 133.

Oke, J. B. and R. Schild 1968. *Appl. Opt.* **7**, 617.

Perrin, F. H. 1966. *The Theory of the Photographic Process*, J. H. James, Ed., Macmillan, New York, Chapter 23.

Phillips, J. G. 1959. *J. Opt. Soc. Am.* **49**, 972.

Robinson, L. B. and E. J. Wampler 1972. *Publ. Astron. Soc. Pac.* **84**, 161.

Rolfe, J. and S. E. Moore 1970. *Appl. Opt.* **9**, 63.

Rusconi, L. and G. Sedmak 1971. *Astron. Astrophys.* **10**, 469.

Schlueter, P. 1967. *Appl. Opt.* **6**, 239.

Schonkeren, J. M. 1970. *Photomultipliers*, Phillips Application Book, Eindhoven, Netherlands, pp. 6, 28, 116.

Sharpe, J. 1961. *Photoelectric Cells and Photomultipliers*, EMI Electronics Ltd. Whitaker Corporation, Plainview, N.Y.

Smith, R. A., F. E. Jones, and R. P. Chasmar 1957. *The Detection and Measurement of Infrared Radiation*, Oxford 1968, p. 209.

Smyth, M. J. 1952. *Mon. Not. Roy. Astron. Soc.* **112**, 88.

Stoeckley, T. R. and C. S. Morris 1974, *Astrophys. J.* **188**, 579.

Walker, G. A. H., J. R. Auman, V. L. Buchholz, B. A. Goldberg, A. C. Gower, B. C. Isherwood, R. Knight, and D. Wright 1972. *Adv. Electron. Electron Phys.* **33B**, 819.

Webb, J. H. 1948. *J. Opt. Sci. Am.* **38**, 312.

Whitford, A. E. 1962. *Handb. Astrophys.* **54**, 240.

Wright, K. O. 1966. *Abundance Determinations in Stellar Spectra*, International Astronomical Union Symposium 26, H. Hubenet, Ed., Academic, New York, p. 15.

Wright, K. O., E. K. Lee, T. V. Jacobson, and J. L. Greenstein 1964. *Publ. Dom. Astron. Observ. Victoria* **12**, No. 7.

Young, A. T. 1963. *Appl. Opt.* **2**, 51.

Young, A. T. 1966. *Rev. Sci. Inst.*, **37**, 1472.

Youngbluth, O. Jr. 1970. *Appl. Opt.* **9**, 321.

CHAPTER FIVE

RADIATION TERMS AND DEFINITIONS

We must have a carefully defined terminology to properly describe light and its interaction with the photospheric material. Therefore this short chapter introduces these basic quantities.

We shall consider the distribution of photons with wavelength by specifying measurable macroscopic quantities in the limiting case where these quantities become infinitesimal. We take the meaning of this limit in a practical sense such that infinitesimal means smaller than our resolving power in measuring the variable under consideration. In this way we can ignore physical complications due to the uncertainty principle or other quantum aspects of the radiation.

SPECIFIC INTENSITY

Let us consider the radiating surface depicted in Fig. 5-1. This surface might be a physical boundary or it could be an imaginary surface inside a radiating gas. We denote a small portion of this surface by ΔA. The specific intensity, I_v, is defined by

$$I_v \equiv \lim \frac{\Delta E_v}{\cos \theta \, \Delta A \, \Delta \omega \, \Delta t \, \Delta v}$$

$$\equiv \frac{dE_v}{\cos \theta \, dA \, d\omega \, dt \, dv}. \tag{5-1}$$

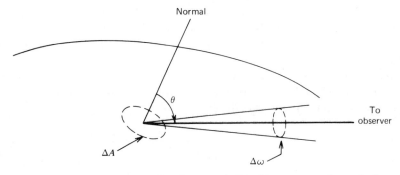

Fig. 5-1 The specific intensity is defined geometrically with respect to the angle from the normal (θ), the solid angle ($\Delta\omega$), and the projected area ($\Delta A \cos \theta$).

The limit is taken as all the small quantities ΔA, $\Delta\omega$ = increment of solid angle, Δt = increment in time, and Δv = spectral band all diminish toward zero. Of course ΔE_v diminishes to zero as well. In this way we define the specific intensity at a "point" on the surface, in a direction θ, at a frequency v in the spectrum. The adjective "specific" is usually omitted unless needed to avoid ambiguity.

One has a choice of units for specific intensity. Equation 5-1 uses frequency units with Δv in the denominator. We could equally well define I_λ by replacing Δv by the wavelength increment $\Delta\lambda$. The relation between these two spectral distributions is

$$I_v \, dv = I_\lambda \, d\lambda, \tag{5-2}$$

which follows directly from eq. (5-1). The cgs units of I_v are ergs/sec cm^2 rad^2 c/sec = ergs/rad^2 cm^2. The units of I_λ are ergs/sec cm^2 rad^2 Å.

It is easy enough to see that the spectral distributions I_v and I_λ have different shapes for the same spectrum. Characteristic features such as intensity maxima appear at different wavelengths in the two expressions. The solar spectrum, for example, has a maximum in the green (~ 4500 Å) for I_λ but the maximum is in the near infrared (~ 8000 Å) for I_v. This curious state of affairs can be understood by looking at the relation between spectral bands for the two cases. Since $c = \lambda v$, $dv = -(c/\lambda^2) \, d\lambda$. The minus sign refers to the direction of the coordinates, as λ increases, v decreases. We see that equal intervals in λ correspond to very different intervals in v across the spectrum. With increasing λ, a constant $d\lambda$ (the I_λ case) corresponds to a smaller and smaller dv. These smaller dv intervals obviously contain diminishing energy compared to the constant dv intervals associated with I_v.

The prism and grating spectrographs portray (approximately) I_v and I_λ, respectively. This results directly from the fact that the dispersion varies approximately as λ^{-2} for the prism (Hartmann 1898) and is nearly constant with λ for the grating (eq. 3-8). The total energy allowed through a spectral window is, as can be shown by integration of eq. 5-1, the same regardless of the units.

The mean intensity is defined as the directional average of the specific intensity. We shall follow Kourganoff (1963) and use $\bar{I}_\nu$ to denote this quantity (J_ν is often used instead). Then

$$\bar{I}_\nu \equiv \frac{1}{4\pi} \oint I_\nu \, d\omega \qquad (5\text{-}3)$$

where the circle on the integral indicates that the integration is done over the whole unit sphere centered on the point of interest.

FLUX

The flux is a measure of the net energy flow across an area ΔA, in a time Δt, in a spectral range $\Delta \nu$ as these quantities are taken to the infinitesimal limit. The only directional significance is whether the energy crosses ΔA from the top or from the bottom. This total net energy is the sum of all the ΔE_ν's we would need to describe the same radiation using the specific intensity terminology and we therefore denote it by $\sum \Delta E_\nu$. We write the definition of flux*

$$\mathcal{F}_\nu \equiv \lim \frac{\sum \Delta E_\nu}{\Delta A \, \Delta t \, \Delta \nu}$$

$$= \frac{\oint dE_\nu}{dA \, dt \, d\nu} \qquad (5\text{-}4)$$

where again the circle on the integral indicates a complete integration over all directions. We can relate flux and intensity using eq. 5-1 to substitute for $dE_\nu/(dA \, dt \, d\nu)$,

$$\mathcal{F}_\nu = \oint I_\nu \cos \theta \, d\omega \qquad (5\text{-}5)$$

where θ is shown in Fig. 5-1. When the radiation is isotropic, $\mathcal{F}_\nu = 0$. In this sense the flux is a measure of the flow of radiation and the anisotropy of the radiation field. If we look at a point on the physical boundary of a

---

* In theoretical stellar atmospheres, a quantity called "astrophysical flux" is often defined. It is equal to the real flux, $\mathcal{F}_\nu$, divided by π.

radiating sphere, we can write, using eq. 5-5

$$\mathcal{F}_v = \int_0^{2\pi} d\phi \int_0^{\pi} I_v \sin \theta \cos \theta \, d\theta$$

$$= \int_0^{2\pi} d\phi \int_0^{\pi/2} I_v \sin \theta \cos \theta \, d\theta + \int_0^{2\pi} d\phi \int_{\pi/2}^{\pi} I_v \sin \theta \cos \, d\theta,$$

that is, the flux leaving the surface plus the flux entering the surface. In the case at hand the second term is zero. If, in addition, there is no azimuthal dependence for I_v, we have

$$\mathcal{F}_v = 2\pi \int_0^{\pi/2} I_v \sin \theta \cos \theta \, d\theta. \tag{5-6}$$

This equation is fundamental in computing the spectrum expected from a normal star. It is used later in several instances of application. In the computation of model photospheres (Chapter 9), I_v is obtained from the transfer equation. The study of this computation comprises the heart of the theory of stellar atmospheres. Some of the remainder of this chapter and portions of several others are preparation for obtaining I_v for given boundary conditions and specified geometry.

The simplest configuration occurs when I_v in eq. 5-6 is independent of direction (as long as $0 \le \theta \le \pi/2$). Then

$$\mathcal{F}_v = \pi I_v \tag{5-7}$$

since the integral of $\sin \theta \cos \theta$ gives a value of $\frac{1}{2}$.

The student who is meeting these definitions for the first time will do well to reflect on the difference between I_v and $\mathcal{F}_v$. For example, I_v is independent of distance from the source whereas $\mathcal{F}_v$ obeys the standard inverse square law. I_v can only be measured directly if we can resolve the radiating surface; otherwise we necessarily measure $\mathcal{F}_v$. It is $\mathcal{F}_v$ we measure for most stars for this reason. These differences arise because of the solid angle $\Delta\omega$ appearing in eq. 5-1 but not in eq. 5-4.

Consider an angularly large and uniformly bright source at a distance r being measured with the telescope and detector shown in Fig. 5-2. The small circle marked A_D represents a resolution element of the detector. For convenience assume it is a photomultiplier cathode of area A_D. The lens (or telescope) images the source on the detector and the source area A_S matches A_D. The solid angle in which radiation is coming to A_D is A_S/r^2. The surface in Fig. 5-1 corresponds to the lens. We clearly are measuring I_v because it is the radiation per solid angle the detector records. As r increases, the source diminishes in angular size, but the area A_D is still fully illuminated and the

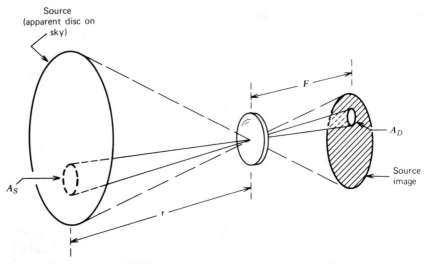

Fig. 5-2 A portion of the uniformly bright source, A_s, corresponds to the detector resolution element, A_D. The focal length of the lens is F. The distance to the source is $r \gg F$. As r increases, the image of the source becomes smaller until it is completely contained within A_D. When the image is smaller than the detector resolution, only the flux of the source can be measured.

output signal from this detector element remains constant since the source is uniformly bright over its surface. Eventually, as r increases, the whole source will subtend an area equal to and then less than A_S. At this distance the source is no longer resolved and there is a transition from intensity measurement to flux measurement.

THE K INTEGRAL AND RADIATION PRESSURE

It is found useful to define, in addition to the mean intensity

$$\bar{I}_v = \frac{1}{4\pi} \oint I_v \, d\omega$$

and the flux

$$\mathscr{F}_v = 2\pi \oint I_v \cos \theta \, d\omega,$$

a third equation using $\cos \theta$, that is,

$$K_v = \frac{1}{4\pi} \oint I_v \cos^2 \theta \, d\omega. \tag{5-8}$$

Physically this integral is related to the radiation pressure.

Electromagnetic radiation has momentum equal to energy divided by c. Consider photons transferring momentum to a solid wall in a manner analogous to the usual discussion of particles in kinetic gas theory. The component of momentum normal to the wall taken per unit time and area is the pressure,

$$dP_\nu = \frac{1}{c} \frac{dE_\nu \cos \theta}{dt \, dA}$$

where θ is the angle the photon trajectory forms with the normal to the wall. Casting this in terms of specific intensity (eq. 5-1) gives

$$dP_\nu = P_\nu \, d\nu \, d\omega = \frac{I_\nu}{c} \cos^2 \theta \, d\nu \, d\omega.$$

This in turn is integrated over solid angle leaving

$$P_\nu \, d\nu = \oint \frac{I_\nu}{c} \cos^2 \theta \, d\nu \, d\omega. \tag{5-9}$$

This "monochromatic" pressure refers to the pressure exerted by photons with frequency in the range $d\nu$. We see immediately the similarity between eqs. 5-8 and 5-9 from which we can write

$$P_\nu = \frac{4\pi K_\nu}{c}.$$

A special case that is usually used in stellar computations involves the assumption that I_ν is independent of direction so that it can be factored out of eq. 5-9, and upon integration of $\cos^2 \theta$ one finds

$$P_\nu = \frac{4\pi}{3c} I_\nu$$

and the total radiation pressure is

$$P_R = \frac{4\pi}{3c} \int_0^\infty I_\nu \, d\nu. \tag{5-10}$$

In certain applications a temperature is defined using eqs. 5-7 and 6-3 to give

$$\pi \int_0^\infty I_\nu \, d\nu = \sigma T^4$$

so that the radiation pressure is

$$P_R = \frac{4\sigma}{3c} T^4. \tag{5-11}$$

In practice, if eq. 5-11 is used in place of eq. 5-10, T is taken to be the kinetic gas temperature. Radiation pressure becomes quite important in early-type stars. In solar-type and cooler stars it can be safely neglected (see Table 9-1 for numerical values).

THE ABSORPTION COEFFICIENT AND OPTICAL DEPTH

We consider radiation in a specified direction shining on a small thickness of material which itself is so cool that it does not radiate measurably. The intensity of light is found experimentally to be diminished upon passage through the layer by an amount dI_v where

$$dI_v = -\kappa_v \rho I_v \, dx. \tag{5-12}$$

Here ρ is the density in grams per cubic centimeter and dx is in centimeters. The absorption coefficient, κ_v, has units of square centimeters per gram and is therefore a mass absorption coefficient.

There are two physical processes contributing to κ_v. The first is true absorption where the photon is destroyed and the energy thermalized. The second is scattering where the photon is deviated in direction and removed from the solid angle being considered. As such, the absorption coefficient, κ_v, is really an extinction coefficient.

The radiation sees neither $\kappa_v \rho$ nor dx alone, but it is the combination of the two over some path length L given by

$$\tau_v = \int_0^L \kappa_v \rho \, dx \tag{5-13}$$

that is important. Here τ_v is the *optical depth* and x is the geometrical depth. With this equation, eq. 5-12 can be written

$$dI_v = -I_v \, d\tau_v.$$

The solution is $I_v = I_v^0 \, e^{-\tau_v}$, the usual simple extinction law. This is the simplest example of the radiative transfer equation and its solution.

THE EMISSION COEFFICIENT AND THE SOURCE FUNCTION

We treat the emission much the same as we did the absorption. The increment of radiation emitted in a specified direction is defined as

$$dI_v = j_v \rho \, dx. \tag{5-14}$$

Hence ρ and dx are the same as for absorption, and the emission coefficient j_ν has units of ergs per second per square rad per c per second per gram.

Again the physical processes contributing to j_ν are (1) real emission, the creation of a photon, and (2) scattering of a photon into the direction being considered. Scattering in this context rarely refers to diffraction and the like but is an absorption followed immediately by a reemission of a photon from the same atomic transition.

The ratio of emission to absorption has the same units as I_ν. As such, it is convenient to use this ratio when discussing the transfer of radiation. It is given a special name, "the source function," and is defined by

$$S_\nu \equiv \frac{j_\nu}{\kappa_\nu}. \tag{5-15}$$

The physics of calculating S_ν is complicated in all but the simplest of cases. We delay treating most of these complications until Chapter 13 where spectral lines are considered, for it is there that the complications really become important. For now let us consider the two cases of scattering and absorption as examples of the source function.

PURE ISOTROPIC SCATTERING

In this case all the "emitted" energy is due to photons being scattered into the direction under consideration. We imagine the state of affairs to be as depicted in Fig. 5-3. The contribution, dj_ν, to the emission from the solid angle $d\omega$ is proportional to $d\omega$ and to the absorbed energy $\kappa_\nu I_\nu$. It is isotropically reradiated so the fraction per unit solid angle is $\frac{1}{4\pi}$. In this way

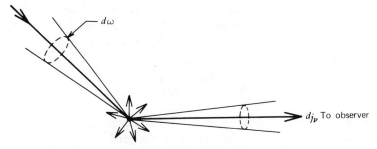

Fig. 5-3 The amount of radiation scattered toward the observer is the sum of the contributions from all increments of solid angle like the $d\omega$ shown. The arrows around the scattering center indicate that the radiation from $d\omega$ is scattered in all directions. Only the radiation scattered into the infinitesimal solid angle around the direction toward the observer contribute to j_ν.

we can write

$$dj_v = \frac{1}{4\pi} \kappa_v I_v \, d\omega.$$

Since j_v and κ_v are both per unit density, the factor of ρ on both sides has been omitted. We obtain the full value of j_v by integrating over the solid angle

$$j_v = \frac{1}{4\pi} \oint \kappa_v I_v \, d\omega.$$

But κ_v is usually independent of ω so

$$S_v = \frac{j_v}{\kappa_v} = \oint \frac{I_v \, d\omega}{4\pi}$$

or

$$S_v = \bar{I}_v. \tag{5-16}$$

In a word, the source function for pure isotropic scattering is the mean intensity. For this case the source function depends completely on the radiation field.

PURE ABSORPTION

When all the absorbed photons are destroyed and all the emitted photons are newly created with a distribution governed by the physical state of the material, we have the case of pure absorption. Common usage of the term has narrowed its meaning to that physical state of the material called thermodynamic equilibrium. The emission from a gas in thermodynamic equilibrium is described by the laws of the black body radiator, which is the topic of the next chapter. The source function for this case is given by Planck's radiation law

$$S_v = \frac{2h v^3}{c^2} \frac{1}{e^{hv/kT} - 1}. \tag{5-17}$$

According to this equation, the source function depends only on the frequency of the radiation and the temperature of the material.

It is instructive to combine scattering with pure absorption. We add the photons contributed from the two mechanisms, so the emission coefficient is

$$j_v = \kappa_v^S \bar{I}_v + \kappa_v^A B_v(T)$$

with the superscripts S and A meaning scattering and absorption. Similarly we add the absorption coefficients to obtain the total absorption coefficient. The source function for this combination is then the ratio of the total emission coefficient to the total absorption coefficient,

$$S_v = \frac{\kappa_v^S}{\kappa_v^S + \kappa_v^A} \bar{I}_v + \frac{\kappa_v^A}{\kappa_v^S + \kappa_v^A} B_v(T). \tag{5-18}$$

The two source functions $\bar{I}_v$ and $B_v(T)$ are added in a weighted average according to the relative strength of the absorption at the point being considered.

THE EINSTEIN COEFFICIENTS

When we deal with spectral lines or bound–bound transitions, we can describe the spontaneous probabilities for emission in terms of the atomic constants. Consider the spontaneous transition between an upper level, u, and a lower level, l, separated by an energy hv. We assume the emission is isotropic. Then the probability that the atom will emit its quantum of energy in a time dt and in a solid angle $d\omega$ is $A_{ul} dt d\omega$. The proportionality constant A_{ul} is the Einstein probability coefficient for spontaneous emission. If there are N_u excited atoms per unit volume, the contribution of spontaneous emission to the emission coefficient is

$$j_v \rho = N_u A_{ul} hv. \tag{5-19}$$

Additional emission is induced if a radiation field is present that has photons corresponding to the energy difference between the levels u and l. The induced emission is proportional to the amount of this radiation and each newly emitted photon shows phase coherence and a direction of propagation that is the same as that of the inducing photon. The process of stimulated emission is often referred to as a process of negative absorption for this reason. We adopt this terminology here. The probability for stimulated emission giving a quantum in the time interval dt in the solid angle $d\omega$ is $B_{ul} I_v dt d\omega$ where the directional dependence of the specific intensity becomes very important since it fixes the directional dependence of these new photons. B_{ul} is the Einstein probability coefficient for stimulated emission.

The true absorption probability is defined in the same way and the proportionality constant is denoted by B_{lu}.

The mass absorption coefficient for this bound–bound transition can be expressed in terms of these B's by considering the radiation absorbed per unit path length from the intensity beam I_v, namely,

$$\kappa_v \rho I_v = N_l(B_{lu} I_v)hv - N_u(B_{ul} I_v)hv$$

where N_l is the population of the lower level per unit volume. The I_v can be canceled in the above equation giving

$$\kappa_v \rho = N_l B_{lu} h\nu - N_u B_{ul} h\nu. \tag{5-20}$$

The amount of reduction in absorption from the second term turns out to be only a few per cent in the visible spectrum (refer to Table 8-1). The numerical calculation of κ_v is the topic of Chapters 8 (continuous absorption) and 11 (line absorption).

REFERENCES

Hartmann, J. 1898. *Astrophys. J.* **8**, 218.

Kourganoff, V. 1963. *Basic Methods in Transfer Problems*, Dover Pub., New York.

CHAPTER SIX

THE BLACK BODY AND ITS RADIATION

Let us begin by supposing we have a container that is completely closed except for a very small hole in one wall. Any light entering the hole has a very small probability of finding its way out again and eventually will be absorbed by the walls of the container or the gas inside the container. The photon's predicament is illustrated in Fig. 6-1. Embodied in such a container we have the concept of a perfect absorber—all the light that enters the small hole is absorbed inside. Of course there is a very small probability that the photon will find the hole and get out. This probability is related to the size of the hole relative to the wall area of the container and the condition of the wall surface, namely, its roughness and reflection coefficient.

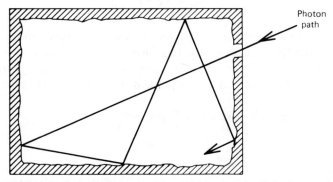

Fig. 6-1 The black body cavity acts as a perfect absorber because light that enters the small opening in the cavity wall is almost certain to be absorbed inside.

So the perfect absorber is not truly perfect but clearly can be made sufficiently so that we could not *measure* the difference. This type of container, or more specifically the hole, is called a black body.

If the above container is heated, the walls emit photons filling the inside with radiation. Each of these photons is reabsorbed inside the container—the small hole being negligible. Lengthy arguments can be put forward to support what seems obvious to the physically minded, namely that an equilibrium condition exists inside the chamber provided the temperature of the walls remains constant and uniform. Each physical process is balanced by its inverse. The container is said to be in thermodynamic equilibrium.

A small fraction of the internal radiation leaks out the hole and we can measure its spectrum, but this leakage produces no measurable difference on the thermodynamics of the container. It was with a container of this type that the laws of black body radiation were discovered and studied.

OBSERVED RELATIONS

The observations show that the black body spectrum is continuous, isotropic, and unpolarized. The intensity of the continuum is found to depend only on frequency and the temperature of the black body. Two laws are found through observation.

The first of these laws is a scaling relation and can be written

$$I_v = v^3 F\left(\frac{v}{T}\right) \tag{6-1a}$$

or

$$I_\lambda = \frac{c^4}{\lambda^5} F\left(\frac{c}{\lambda T}\right) \tag{6-1b}$$

where c denotes the velocity of light and F is a unique function that can be tabulated from the measurements. Equation 6-1 says that when we observe at a given frequency, the temperature dependence of I_v is always the same, namely, F, provided we scale the argument of F by v and the absolute intensity by v^3.

Alternately, we could take $u = v/T$ and write

$$I_v = T^3 u^3 F(u).$$

Then we see that when the temperature is held constant, the frequency dependence of I_v is always the same, namely, $u^3 F(u)$, but scaled by the cube of the temperature of the black body. The maximum $u^3 F(u)$ always occurs at the same u since $F(u)$ is a unique function. The observations show* that the maximum I_v is at $u = u_m = 5.8789 \times 10^{10}$ c/sec °K, or

$$\lambda'_m T = 0.50995 \text{ cm °K} \tag{6-2a}$$

where λ'_m is the wavelength at which the frequency distribution I_v has its maximum. The corresponding relation for I_λ is

$$\lambda_m T = 0.28978 \text{ cm °K}. \tag{6-2b}$$

The relation 6-1 is called "Wien's law," although the special case 6-2 often goes by the same name. The fact that the maximum appears at a different place in the spectrum for frequency and wavelength units has been discussed in Chapter 5.

The second law, the Stefan–Boltzmann law, says that the total power output is specified purely by the temperature according to

$$\int_0^\infty \mathscr{F}_v \, dv = \sigma T^4 \tag{6-3}$$

where $\mathscr{F}_v$ is the flux and σ is the Stefan–Boltzmann constant of value

$$\sigma = 5.6703 \times 10^{-5} \text{ erg/sec cm}^2 \text{ deg}^4.$$

The Stefan–Boltzmann relation follows directly from Wien's law. Again we let $u = v/T$. Then $I_v = u^3 T^3 F(u)$ and

$$\int_0^\infty I_v \, dv = \int_0^\infty u^3 T^3 F(u) T \, du = T^4 \int_0^\infty u^3 F(u) \, du = \text{const } T^4.$$

As is shown below, the constant has a value of σ/π. Thermodynamic derivations of these two laws can be seen in Richtmyer, Kennard, and Lauritsen (1955).

The shapes of the spectra are illustrated in Fig. 6-2. At low frequencies the curves are found to have the form

$$I_v = \frac{2kTv^2}{c^2} \tag{6-4}$$

called the Rayleigh–Jean's approximation. At high frequencies

$$I_v = \text{const } v^3 e^{-\text{const } v/T}.$$

The name "Wien approximation" is associated with this equation.

* The numbers from Appendix A have been adopted in calculating or specifying the constants of this chapter. Because of the comprehensive adjustment approach used to obtain the numbers in Appendix A, some of the constants may be specified with a higher number of significant figures than could be obtained from one direct experiment. The expected errors are given after each number in Appendix A.

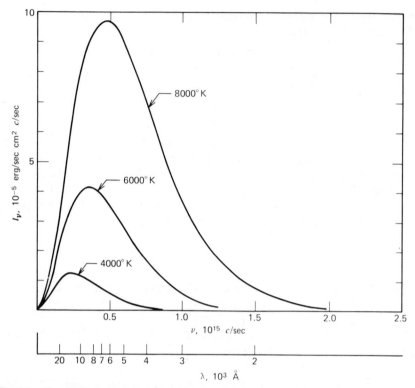

Fig. 6-2 The specific intensity I_v (emitted by a black body) is shown for the three temperatures indicated. The scaling of Wien's law (eq. 6-1) can be seen in the shape of the curves. The area under the curves obeys the Stefan–Boltzmann law (eq. 6-3).

PLANCK'S RADIATION LAW

The history of the black body and its fundamental part in starting quantum theory is well known. We shall omit the historic details here except to point out that any derivation of the relation describing the spectra as in Fig. 6-2, relies on the quantum approach.

We derive Planck's law using a two-level atom.* The upper level has a population N_u, the lower N_l, and since these atoms are in a container in a state of thermodynamic equilibrium, their populations are related by eq. 1-14,

$$\frac{N_u}{N_l} = \frac{g_u}{g_l} e^{-hv/kT}$$

* M. Born (1957) gives a derivation based on atomic oscillators. Cox and Giuli (1968) give a thermodynamic derivation.

where the energy difference between the levels is $\chi_u - \chi_l = hv$. The equilibrium condition implies that all the ways for the electron to go from u to l must be balanced by the return paths.

The number of spontaneous emissions per second per unit solid angle per unit volume is $N_u A_{ul}$ where A_{ul} is the Einstein coefficient defined in Chapter 5. Similarly, the rate of stimulated emission per unit solid angle and volume is $N_u B_{ul} I_v$, and I_v must be taken at v appropriate to the transition. Finally, absorption pumps electrons upward, so for this rate we write $N_l B_{lu} I_v$. In addition to radiative transitions we could have collision induced transitions, but in equilibrium there are as many up as there are down and they cancel out of the equilibrium equation. And so we have

$$N_u A_{ul} + N_u B_{ul} I_v = N_l B_{lu} I_v.$$

If we now solve this for I_v, we find

$$I_v = \frac{A_{ul}}{B_{lu} \dfrac{N_l}{N_u} - B_{ul}} \tag{6-5}$$

which is just the ratio of emission to absorption as in eq. 5-15, that is, eq. 6-5 has the form of a source function. If we also substitute for the population ratio according to the excitation equation above, eq. 6-5 becomes

$$I_v = \frac{A_{ul}}{(g_l/g_u)B_{lu}e^{hv/kT} - B_{ul}}.$$

But we know that this expression must revert to the Rayleigh–Jean approximation (eq. 6-4) in the limit of small v. Therefore we expand the exponential keeping only the first order term and

$$I_v = \frac{A_{ul}}{(g_l/g_u)B_{lu} - B_{ul} + (g_l/g_u)B_{lu}(hv/kT)} \qquad \text{for} \quad hv/kT \ll 1 \tag{6-6}$$

which must be made equal to $2kTv^2/c^2$. This can be done only* if

$$B_{ul} = \frac{B_{lu}g_l}{g_u} \tag{6-7a}$$

and

$$A_{ul} = \frac{2hv^3}{c^2} B_{ul}. \tag{6-7b}$$

* It was through this type of argument that the stimulated emission was first discovered.

The result of substituting these relations back into eq. 6-5 is

$$I_\nu = \frac{2h\nu^3}{c^2} \frac{1}{e^{h\nu/kT} - 1}. \qquad (6\text{-}8)$$

Equation 6-8 is Planck's radiation law. It is also common to denote this expression as $B_\nu(T)$, meaning the black body source function. It gives the proper fit to the observed curves in Fig. 6-1 and has the correct limiting forms of eqs. 6-4 and the Wien approximation. It has the functional form giving the observed relations expressed in eq. 6-1. The maximum is found in the usual way be setting the derivative equal to zero. We use $x = h\nu/kT$ and obtain

$$x = \frac{3(e^x - 1)}{e^x}.$$

Solution by iteration gives $x = 2.8214$ or $\lambda'_m T = 0.50995$ cm deg in accord with eq. 6-2. Integration of Planck's law leads to the Stefan–Boltzmann law in the following way. From eq. 5-5 the flux is given by

$$\mathcal{F}_\nu = \oint I_\nu \cos\theta \, d\omega$$

and in this case I_ν is Planck's law. Since no significant radiation is entering the hole in the black body chamber, and the escaping radiation is isotropic, we have $\mathcal{F}_\nu = \pi I_\nu$ according to eq. 5-7. It then follows that

$$\int_0^\infty \mathcal{F}_\nu \, d\nu = \pi \int_0^\infty \frac{2h\nu^3}{c^2} \frac{1}{e^{h\nu/kT} - 1} \, d\nu$$

$$= \frac{2\pi h}{c^2}\left(\frac{kT}{h}\right)^4 \int_0^\infty \frac{x^3}{e^x - 1} \, dx$$

where $x \equiv h\nu/kT$. Evaluating the integral gives $\pi^4/15$ and

$$\int_0^\infty \mathcal{F}_\nu \, d\nu = \frac{2\pi^5 k^4}{15h^3c^2} T^4 \equiv \sigma T^4.$$

Finally, notice that the relations 6-7 are between fundamental atomic constants. These relations must hold independent of the physical situation the atoms are encountering and therefore are perfectly general relations. The transitions probabilities are proportional to the oscillator strengths of the levels, the details of which are given in Chapter 11.

CALCULATION OF BLACK BODY RADIATION

In many applications it is quicker and more convenient to compute from eq. 6-8 than it is to use tabulations such as those of McDonald (1955) or Pivovonsky and Nagel (1961). We write, using the B_v notation

$$B_v(T) = \frac{2hv^3}{c^2} \frac{1}{e^{hv/kT} - 1}$$

$$= \frac{c_3 v^3}{e^{c_4 v/T} - 1} \tag{6-9}$$

and

$$B_\lambda(T) = \frac{2hc^2}{\lambda^5} \frac{1}{e^{hc/\lambda kT} - 1}$$

$$= \frac{c_1/\lambda^5}{e^{c_2/\lambda T} - 1} \tag{6-10}$$

where $c_1 = 2hc^2 = 1.19106 \times 10^{-5}$ erg cm^2/sec rad^2 for λ in centimeters
$\qquad\qquad = 1.19106 \times 10^{27}$ erg Å^4/sec cm^2 rad^2 for λ in angstroms,
$c_2 = hc/k = 1.43879$ cm deg
$\qquad\qquad = 1.43879 \times 10^8$ Å deg,
$c_3 = 2h/c^2 = c_1/c^4 = 1.4745 \times 10^{-47}$ erg sec^3/cm^2 rad^2,
$c_4 = h/k = c_2/c = 4.7993 \times 10^{-11}$ deg sec.

As usual c is the velocity of light.

Approximate values of B_v can be obtained from Fig. 6-3. This graph is generated by using the Wien scaling relation to reduce Planck's law to its unique form as in eq. 6-1. This is accomplished with eq. 6-9 by letting $u = v/T$ and writing

$$\frac{B_v}{B_{max}} = \frac{e^{c_4 u_m} - 1}{u_m^3} \frac{u^3}{e^{c_4 u} - 1}$$

$$= 7.78 \times 10^{-32} \frac{u^3}{e^{c_4 u} - 1} \tag{6-11}$$

where B_{max} is the largest value of B_v and, as we saw earlier, occurs at a value of $u = u_m = 5.8789 \times 10^{10}$ c/sec °K. The value of B_{max} can be calculated from

$$B_{max} = \frac{c_3 u_m^3 T^3}{e^{c_4 u_m} - 1}$$

$$= 1.90 \times 10^{-16} T^3. \tag{6-12}$$

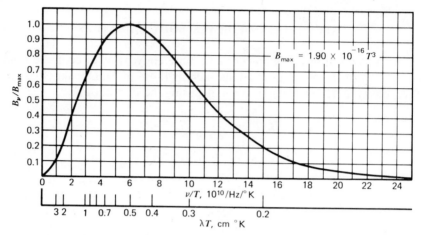

Fig. 6-3 The relative spectral distribution of B_ν is shown as a function of ν/T in accordance with eqs. 6-11 and 6-12.

To use Fig. 6-3, read B_ν/B_{max} from the graph at the ν/T of interest, then multiply by B_{max} given in eq. 6-12. A similar graph can be constructed for B_λ expressed as a function of λT.

THE BLACK BODY AS A RADIATION STANDARD

The black body radiator is used as a primary standard of radiation against which many types of secondary standards are calibrated. The usefulness of the black body in this context stems from the simple unique dependence of the radiation on temperature, (i.e., eq. 6-8).

A black body suitable for this type of work is illustrated in Fig. 6-4. The cavity in the center is surrounded by gold of high purity (< 10 ppm impurity) and is contained in a graphite cell. Outside the gold chamber are electrical heating coils. The small hole from which the black body radiation is emitted can be seen in the left wall of the central cavity. The apparatus is used by first slowly heating the gold until it melts. (The liquid metal makes a particularly useful mechanism for uniformly heating the walls of the central cavity.) The system is then allowed to cool to the gold freezing point. A stable plateau lasts several minutes, as shown in Fig. 6-5, and it is during this interval that the calibrations are done. Operation at a metallic freezing point results in good reproducibility and stability. The temperature of freezing gold is 1337.6°K. The black body can be operated at other temperatures by using metals other than gold. Platinum, for example, has a freezing point at 2045°K; copper at 1358°K. The more inert the metal, the less it

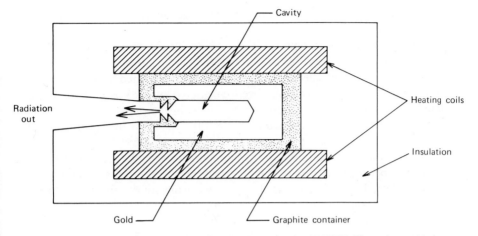

Fig. 6-4 A black body for operation at the gold freezing point (1338°K). The molten gold gives uniformly heated cavity walls with a well defined temperature. Radiation leaks out from the small hole in the cavity and can be measured by a spectrophotometer placed at the left of the diagram. (Adapted from Kostkowski and Lee 1962.)

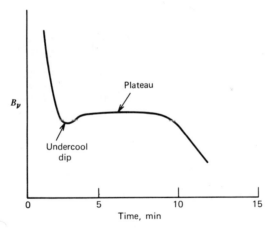

Fig. 6-5 The cooling curve for a gold "freeze" shows the stable plateau in time. The measurements are taken during these few minutes of stable operation. (Adapted from Kostkowski and Lee 1962.)

becomes contaminated with impurities. The freezing temperature is a function of the purity.

The precision of photometric calibrations can be limited by improper construction or operation of the black body. Uniform heating of the cavity walls is necessary. The effect of the finite hole size compared to the size of the cavity and the type of wall has been calculated by DeVos (1954). Useful details of black body operation can be found in the paper of Kostkowski and Lee (1962). The most fundamental limit though is the uncertainty in the physical temperature scales. Bless et al. (1968) and Griem (1964) find the UV calibration uncertain (relative to the standard wavelength $\lambda 5556$ radiation) by several per cent because of the temperature uncertainty. This limitation we see reflected again in Chapter 10 where we deal with the absolute calibration of stellar continua.

REFERENCES

Bless, R. C., A. D. Code, and D. J. Schroeder 1968. *Astrophys. J.* **153**, 545.

Born, M. 1957. *Atomic Physics*, 6th ed., Hafner, New York.

Cox, J. P., and R. T. Giuli 1968. *Stellar Structure*, Gordon & Breach, New York, p. 108.

DeVos, J. C. 1954. *Physica* **20**, 669.

Griem, H. R. 1964. *Plasma Spectroscopy*, McGraw-Hill, New York, p. 257.

Kostkowski, H. J., and R. D. Lee 1962. *Natl. Bur. Stand. Monogr. (U.S.)* **41**.

McDonald, J. K. 1955. *Publ. Dom. Astrophys. Observ. Victoria* **10**, 127.

Pivovonsky, M., and M. R. Nagel 1961. *Tables of Black Body Radiation Functions*, Macmillan, New York.

Richtmyer, F. K., E. H. Kennard, and T. Lauritsen 1955. *Introduction to Modern Physics*, 5th ed., McGraw-Hill, New York.

CHAPTER SEVEN

RADIATIVE AND CONVECTIVE ENERGY TRANSPORT

The major mode of energy transport through the surface layer of the star is by radiation. Convective transport rarely, if ever, carries a large fraction of the outward flux in a stellar photosphere.

The solution of the radiative transfer problem is the point at which the physical parameters for the material of which the star is composed and the spectrum that we ultimately see are coupled. We start by setting up the differential equation describing the flow of radiation through an infinitesimal volume. The integration of the equation can then be accomplished for the geometry of the situation. Unfortunately the step from a differential to an integral equation is not a physical solution to the problem because the integrand depends on the atomic excitation of the material, which itself depends on the temperature of the material and the radiation field in the material. Both the thermal (collisional) excitation and the nonthermal (radiative) excitation are dependent on depth in the photosphere. In most applications of the theory to real stars, a model of the star's photosphere is postulated from which the integrand and thence the outcoming spectrum of the star can be calculated.

THE TRANSFER EQUATION AND ITS FORMAL SOLUTION

We consider radiation traveling in a direction s. The change in specific intensity, dI_ν, over an increment of path length, ds, is the sum of the losses and gains,

$$dI_\nu = -\kappa_\nu \rho I_\nu \, ds + j_\nu \rho \, ds.$$

121

Now we divide the equation by $\kappa_\nu \rho\, ds$, which we call $d\tau_\nu$ in accord with eq. 5-13, and

$$\frac{dI_\nu}{d\tau_\nu} = -I_\nu + \frac{j_\nu}{\kappa_\nu}$$

or using eq. 5-15,

$$\frac{dI_\nu}{d\tau_\nu} = -I_\nu + S_\nu. \tag{7-1}$$

This is the equation of radiative transfer.

The integration follows from a standard integrating factor scheme. Noticing that the variable τ_ν appears alone in eq. 7-1, we are led to try a solution of the form

$$I_\nu(\tau_\nu) = f e^{b\tau_\nu} \tag{7-2}$$

where f is a function to be determined. Substitution into eq. 7-1 gives

$$\frac{dI_\nu}{d\tau_\nu} = fb e^{b\tau_\nu} + e^{b\tau_\nu} \frac{df}{d\tau_\nu}$$

or

$$bI_\nu + e^{b\tau_\nu} \frac{df}{d\tau_\nu} = -I_\nu + S_\nu.$$

We therefore make the first terms on each side of the equation equal by setting $b = -1$ and for the second terms we require

$$e^{-\tau_\nu} \frac{df}{d\tau_\nu} = S_\nu$$

or

$$f = \int_0^{\tau_\nu} S_\nu(t_\nu) e^{t_\nu}\, dt_\nu + c_0$$

with t_ν serving as a dummy variable. Then eq. 7-2 becomes

$$I_\nu(\tau_\nu) = e^{-\tau_\nu} \int_0^{\tau_\nu} S_\nu(t_\nu) e^{t_\nu}\, dt_\nu + c_0 e^{-\tau_\nu}. \tag{7-3}$$

The integration constant can be written as $c_0 = I_\nu(0)$, as seen by setting $\tau_\nu = 0$ in eq. 7-3.

It is instructive to rewrite eq. 7-3 bringing the $e^{-\tau_\nu}$ inside the integral,

$$I_\nu(\tau_\nu) = \int_0^{\tau_\nu} S_\nu(t_\nu) e^{-(\tau_\nu - t_\nu)}\, dt_\nu + I_\nu(0) e^{-\tau_\nu}. \tag{7-4}$$

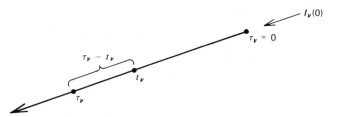

Fig. 7-1   Radiation along the line at the point τ_ν is composed of the sum of intensities, S_ν, originating at the points t_ν along the line but suffering extinction dependent upon the optical separation $\tau_\nu - t_\nu$. Any radiation incident at $\tau_\nu = 0$, $I_\nu(0)$, suffers the extinction $e^{-\tau_\nu}$.

Figure 7-1 illustrates the line along which radiation is being considered and the positions of the variables for this equation. In particular, at a point τ_ν, the "original" intensity, $I_\nu(0)$, has suffered an exponential extinction $e^{-\tau_\nu}$ and in a similar way, the intensity generated at any point t_ν, which we denote as $S(t_\nu)$, undergoes an exponential extinction $e^{-(\tau_\nu - t_\nu)}$ before being summed at the point τ_ν.

Equation 7-4 is the basic integral form of the transfer equation. If we can solve this equation we have the answer we expect from the theory. However, in order to actually do the integration, $S_\nu(t_\nu)$ must be a known function. In some situations this is a complicated function. In others it may be simple, as shown in the examples of Chapter 5.

THE TRANSFER EQUATION FOR DIFFERENT GEOMETRIES

Equation 7-4 expresses I_ν as a function of optical depth along a line. In many applications, particularly in studies of interstellar material, this equation can be applied directly since the observation of a point on the celestial sphere corresponds to observation along such a line. On the other hand, for stellar atmospheres it is conventional to define the optical depth relative to the star along a stellar radius, and not along the line of sight. The appropriate projection factor must then be incorporated as discussed below.

We assume *spherical* stars, thereby making spherical coordinates appropriate as depicted in Fig. 7-2. The z axis is chosen toward the observer. Because we are concerned with the geometry, we write the transfer equation in the form

$$\frac{dI_\nu}{\kappa_\nu \rho \, dz} = -I_\nu + S_\nu.$$

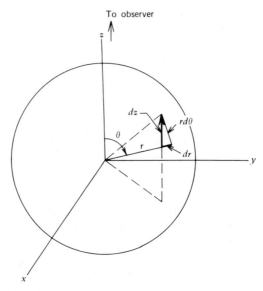

Fig. 7-2   The conversion from line of sight to polar coordinates. The arrows are toward the observer.

In general

$$\frac{dI_\nu}{dz} = \frac{\partial I_\nu}{\partial r}\frac{dr}{dz} + \frac{\partial I_\nu}{\partial \theta}\frac{d\theta}{dz}. \qquad (7\text{-}5)$$

From the figure we see that

$$dr = \cos\theta\, dz$$

$$r\, d\theta = -\sin\theta\, dz.$$

By substitution of these expressions, the transfer equation becomes

$$\frac{\partial I_\nu}{\partial r}\frac{\cos\theta}{\kappa_\nu\rho} - \frac{\partial I_\nu}{\partial\theta}\frac{\sin\theta}{\kappa_\nu\rho r} = -I_\nu + S_\nu. \qquad (7\text{-}6)$$

This form of the equation is used in stellar interiors and in the calculation of very thick stellar atmospheres such as those found in supergiants and possibly giants. Fortunately in most stars the geometrical thickness of the photosphere is very small compared to the stellar radius. The solar photosphere is ~700 km thick or ~0.1 % of the solar radius (refer to Fig. 1-1). Under these conditions a plane parallel approximation can be made, namely, θ does not depend upon z so there is no second term in eq. 7-5. Then

$$\cos\theta\,\frac{dI_\nu}{\kappa_\nu\rho\, dr} = -I_\nu + S_\nu.$$

Custom has been to adopt a new geometrical depth variable, x, for the plane geometry case defined by $dx = -dr$. Adopting this convention and writing $d\tau_\nu$ for $\kappa_\nu \rho \, dx$ gives the basic form of the transfer equation used in the central arena of stellar atmospheres,

$$\cos\theta \frac{dI_\nu}{d\tau_\nu} = I_\nu - S_\nu. \tag{7-7}$$

The optical depth defined in this way is measured along x and not along the line of sight which is at some angle, θ, as shown in Fig. 7-3. This amounts

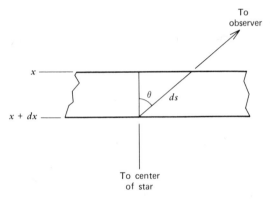

Fig. 7-3 A segment of the plane parallel photosphere of thickness dx. The increment of path length along the line of sight is $ds = dx \sec \theta$.

to replacing τ_ν in eq. 7-1 with $-\tau_\nu \sec \theta$. The negative sign arises from choosing $dx = -dr$. The integrated form corresponding to eq. 7-4 is then

$$I_\nu(\tau_\nu) = -\int_c^{\tau_\nu} S_\nu(t_\nu) e^{-(t_\nu - \tau_\nu)\sec\theta} \sec\theta \, dt_\nu. \tag{7-8}$$

The integration limit, c, replaces the $I_\nu(0)$ integration constant of eq. 7-4. This is done because the boundary conditions are clearly different for radiation having $\theta \geq 90°$ (radiation inward) than for that having $\theta \leq 90°$ (radiation outward). In the first case $I_\nu(0) = 0$ for a normal isolated star, where $\tau_\nu = 0$ is taken to be the outer boundary of the atmosphere. The radiation from other stars, galaxies, and so forth is completely negligible compared to the star's own radiation. Hence for inward directed radiation

$$I_\nu^{\text{in}}(\tau_\nu) = -\int_0^{\tau_\nu} S_\nu e^{-(t_\nu - \tau_\nu)\sec\theta} \sec\theta \, dt_\nu. \tag{7-9}$$

In the second case we consider radiation at the depth τ_v and deeper until no more radiation can be seen at our station. In other words, the integration limit is $\tau_v = \infty$. Then we have for the outward directed radiation

$$I_v^{out}(\tau_v) = \int_{\tau_v}^{\infty} S_v e^{-(t_v - \tau_v)\sec\theta} \sec\theta \, dt_v. \qquad (7\text{-}10)$$

The full intensity at the position τ_v is then

$$I_v(\tau_v) = I_v^{out}(\tau_v) + I_v^{in}(\tau_v)$$

$$= \int_{\tau_v}^{\infty} S_v e^{-(t_v - \tau_v)\sec\theta} \sec\theta \, dt_v - \int_{0}^{\tau_v} S_v e^{-(t_v - \tau_v)\sec\theta} \sec\theta \, dt_v. \qquad (7\text{-}11)$$

The mathematically minded will usually add that one must require $\lim_{\tau_v \to \infty} S_v e^{-\tau_v} = 0$ to ensure that the integral exists. Real stars apparently know how to meet this condition without effort.

An important special case of eq. 7-9 and 7-10 occurs at the stellar surface. In this case

$$I_v^{in}(0) = 0$$

$$I_v^{out}(0) = \int_{0}^{\infty} S_v e^{-t_v \sec\theta} \sec\theta \, dt_v. \qquad (7\text{-}12)$$

This is especially relevant to solar work where intensity measurements can be made as a function of θ. For most stars, however, we must deal with the flux.

THE FLUX INTEGRAL

We expand the defining eq. 5-5 in spherical polar coordinates and assume no azimuthal dependence in I_v. Then

$$\mathcal{F}_v = 2\pi \int_{0}^{\pi} I_v \cos\theta \sin\theta \, d\theta$$

$$= 2\pi \int_{0}^{\pi/2} I_v^{out} \cos\theta \sin\theta \, d\theta + 2\pi \int_{\pi/2}^{\pi} I_v^{in} \cos\theta \sin\theta \, d\theta.$$

Using the definitions of I_v^{out} and I_v^{in} from eqs. 7-9 and 7-10 gives

$$\mathcal{F}_v = 2\pi \int_{0}^{\pi/2} \int_{\tau_v}^{\infty} S_v e^{-(t_v - \tau_v)\sec\theta} \sin\theta \, dt_v \, d\theta$$

$$- 2\pi \int_{\pi/2}^{\pi} \int_{0}^{\tau_v} S_v e^{-(t_v - \tau_v)\sec\theta} \sin\theta \, dt_v \, d\theta$$

or

$$\mathcal{F}_v = 2\pi \int_{\tau_v}^{\infty} S_v \int_0^{\pi/2} e^{-(t_v - \tau_v)\sec\theta} \sin\theta \, d\theta \, dt_v$$

$$- 2\pi \int_0^{\tau_v} S_v \int_{\pi/2}^{\pi} e^{-(t_v - \tau_v)\sec\theta} \sin\theta \, d\theta \, dt_v \qquad (7\text{-}13)$$

where we have assumed S_v is isotropic. If we now let $w = \sec\theta$ and $x = t_v - \tau_v$, the first inner integral can be written

$$\int_0^{\pi/2} e^{-(t_v - \tau_v)\sec\theta} \sin\theta \, d\theta = \int_1^{\infty} \frac{e^{-xw}}{w^2} \, dw. \qquad (7\text{-}14)$$

Integrals of this form are well known. They are called exponential integrals. Their properties are discussed later in this chapter. For now we simply use

$$E_n(x) \equiv \int_1^{\infty} \frac{e^{-xw}}{w^n} \, dw$$

realizing that $E_n(x)$ is a monotomically diminishing function of x (see Fig. 7-4). And for eq. 7-13 we can write

$$\mathcal{F}_v(\tau_v) = 2\pi \int_{\tau_v}^{\infty} S_v E_2(t_v - \tau_v) \, dt_v - 2\pi \int_0^{\tau_v} S_v E_2(\tau_v - t_v) \, dt_v. \qquad (7\text{-}15)$$

To obtain the $E_2(\tau_v - t_v)$ in the second integral it is necessary to make the substitution $w = -\sec\theta$ and $x = \tau_v - t_v$ in the inner integral of the second term of eq. 7-13.

Equation 7-15 is the basic relation we have sought. It is analogous to eq. 7-11 for intensity. The theoretical stellar spectrum to be compared to observation is $\mathcal{F}_v(0)$, that is,

$$\mathcal{F}_v(0) = 2\pi \int_0^{\infty} S_v(t_v) E_2(t_v) \, dt_v. \qquad (7\text{-}16)$$

It says that the surface flux is composed of the sum of the source function at each depth multiplied by an extinction factor, $E_2(t_v)$, appropriate to that depth below the surface and the sum is taken over all depths contributing a significant amount of radiation at the surface. Notice that $\mathcal{F}_v$ is the flux per unit area. The total radiation at the frequency v is $4\pi R^2 \mathcal{F}_v$ where R is the star's radius. This integral is considered again in several places, especially in Chapter 9.

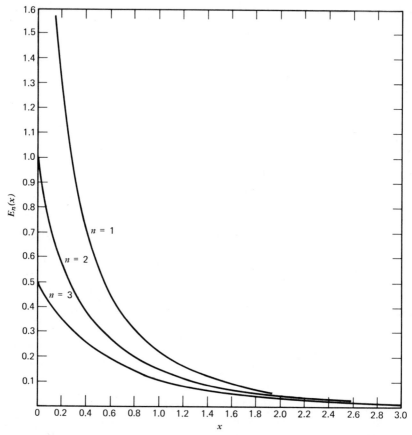

Fig. 7-4 The exponential integrals $E_1(x)$, $E_2(x)$, and $E_3(x)$ are shown. At $x = 0$, $E_1(x)$ is unbounded, $E_2(x)$ is 1.0, and $E_3(x)$ is 0.5. At large x, all the functions behave the same.

THE MEAN INTENSITY AND K INTEGRALS

We can easily derive expressions for $\bar{I}_v$ and K_v as functions of τ_v and obtain expressions similar to eq. 7-15 for flux. The result is

$$\bar{I}_v(\tau_v) = \frac{1}{2} \int_{\tau_v}^{\infty} S_v E_1(t_v - \tau_v) \, dt_v + \frac{1}{2} \int_0^{\tau_v} S_v E_1(\tau_v - t_v) \, dt_v \qquad (7\text{-}17)$$

and

$$K_v(\tau_v) = \frac{1}{2} \int_{\tau_v}^{\infty} S_v E_3(t_v - \tau_v) \, dt_v + \frac{1}{2} \int_0^{\tau_v} S_v E_3(\tau_v - t_v) \, dt_v. \qquad (7\text{-}18)$$

The details of the derivation are left to the reader.

EXPONENTIAL INTEGRALS

Let us repeat the following equation for the exponential integrals, $E_n(x)$, of nth order

$$E_n(x) = \int_1^\infty \frac{e^{-xt}}{t^n} \, dt. \tag{7-19}$$

The value at the origin is found directly by setting $x = 0$, leaving

$$E_n(0) = \int_1^\infty \frac{dt}{t^n} = \frac{1}{n-1}.$$

The first three functions are shown in Fig. 7-4 where it can be seen that $E_1(0)$ is infinite, $E_2(0)$ is unity, and $E_3(0) = \frac{1}{2}$.

Differentiation leads to a useful relation.

$$\frac{dE_n}{dx} = \int_1^\infty \frac{1}{t^n} \frac{d}{dx} e^{-tx} \, dt = - \int_1^\infty \frac{e^{-tx}}{t^{n-1}} \, dt$$

or

$$\frac{dE_n}{dx} = -E_{n-1}. \tag{7-20}$$

One can also show the recurrence formula

$$nE_{n+1}(x) = e^{-x} - xE_n(x) \tag{7-21}$$

holds. This expression is particularly useful for generating numerical values of $E_{n+1}(x)$ from $E_n(x)$. Abramowitz and Stegun (1964) give the following approximation which is suited for computer calculation

$$E_1(x) = -\ln x - 0.57721566 + 0.99999193x - 0.24991055x^2$$
$$+ 0.05519968x^3 - 0.00976004x^4 + 0.00107857x^5 \qquad \text{for} \quad x \le 1$$

and

$$E_1(x) = \frac{x^4 + a_1x^3 + a_2x^2 + a_3x + a_4}{x^4 + b_1x^3 + b_2x^2 + b_3x + b_4} \frac{1}{xe^x} \qquad \text{for} \quad x > 1$$

where $a_1 = 8.5733287401,$ $\qquad b_1 = 9.5733223454,$
$\qquad a_2 = 18.0590169730,$ $\qquad b_2 = 25.6329561486,$
$\qquad a_3 = 8.6347608925,$ $\qquad b_3 = 21.0996530827,$
$\qquad a_4 = 0.2677737343,$ $\qquad b_4 = 3.9584969228.$

These polynomials fit E_1 with an error of less than 2×10^{-7}. Then from eq. 7-21 the higher order exponential integrals can be computed.

The asymptotic behavior is given by the expansion

$$E_n(x) = \frac{1}{xe^x}\left[1 - \frac{n}{x} + \frac{n(n+1)}{x} - \cdots\right]$$

$$\doteq \frac{1}{xe^x}.$$

This is a very sharp drop with x and lends confidence to the convergence of the integral in eq. 7-10. Notice that the asymptotic form is ultimately independent of n.

RADIATIVE EQUILIBRIUM

Conservation of energy must apply to the flow of radiation upward through the stellar photosphere. It is generally assumed that there are no sources or sinks of energy in the photosphere and the energy generated in the core of the star is simply flowing outward to the outer boundary. The general condition expressing the lack of sources and sinks is that the divergence of the flux be zero everywhere in the photosphere. However virtually all the calculations we consider in later chapters are for the plane parallel geometry. In this geometry the divergence condition is

$$\frac{d}{dx}\mathcal{F}(x) = 0 \quad \text{or} \quad \mathcal{F}(x) = F_0, \quad \text{a constant} \tag{7-22}$$

where $\mathcal{F}$ is the total energy flux in ergs/per square centimeter. In that case where all the energy is carried via radiation we have

$$\mathcal{F} = \int_0^\infty \mathcal{F}_v \, dv$$

with $\mathcal{F}_v$ given by eq. 7-15. This special case is called radiative equilibrium and the first condition for radiative equilibrium is, explicitly

$$\int_0^\infty \mathcal{F}_v \, dv = F_0. \tag{7-23}$$

This condition must be satisfied at every depth in the photosphere. If convection plays a significant role as a mode of energy transport, then

$$\Phi(x) + \int_0^\infty \mathcal{F}_v \, dv = F_0 \tag{7-24}$$

must hold where $\Phi(x)$ is the convective flux. Equation 7-24 is a more general expression of flux constancy and the term radiative equilibrium is used only to denote eq. 7-23.

Suppose now we take eq. 7-15 and substitute it into eq. 7-23. This leads immediately to the relation

$$\int_0^\infty \left(\int_{\tau_v}^\infty S_v E_2(t_v - \tau_v)\, dt_v - \int_0^{\tau_v} S_v E_2(\tau_v - t_v)\, dt_v \right) dv = \frac{F_0}{2\pi}. \qquad (7\text{-}25)$$

This is referred to as Milne's second equation; the other two Milne equations appear shortly. In essence it says that, in the case of radiative equilibrium, the solution of the transfer equation is found when an S_v is known that satisfies this equation.

The other radiative equilibrium conditions that have been used are derived from the transfer equation, which we write as

$$\cos\theta\, \frac{dI_v}{dx} = \kappa_v \rho I_v - \kappa_v \rho S_v. \qquad (7\text{-}26\text{a})$$

First, an integration over solid angle gives us

$$\frac{d}{dx} \oint I_v \cos\theta\, d\omega = \kappa_v \rho \oint I_v\, d\omega - \kappa_v \rho \oint S_v\, d\omega$$

assuming $\kappa_v \rho$ is independent of direction. If S_v is also independent of direction, and we substitute for the first and second integrals the definitions of flux and mean intensity, the equation becomes

$$\frac{d\mathcal{F}_v}{dx} = 4\pi\kappa_v \rho \bar{I}_v - 4\pi\kappa_v \rho S_v. \qquad (7\text{-}26\text{b})$$

Second, an integration over frequency gives

$$\frac{d}{dx} \int_0^\infty \mathcal{F}_v\, dv = 4\pi\rho \int_0^\infty \kappa_v \bar{I}_v\, dv - 4\pi\rho \int_0^\infty \kappa_v S_v\, dv. \qquad (7\text{-}27)$$

In radiative equilibrium, the left hand side is zero, leaving the relation

$$\int_0^\infty \kappa_v \bar{I}_v\, dv = \int_0^\infty \kappa_v S_v\, dv. \qquad (7\text{-}28)$$

This particular form of the radiative equilibrium condition is interesting because the value of the flux constant does not appear.

The final radiative equilibrium condition is obtained in a similar way. We multiply eq. 7-26a by $\cos\theta$ before integrating over solid angle and frequency with the result

$$\int_0^\infty \frac{dK_\nu}{d\tau_\nu} \, d\nu = \frac{F_0}{4\pi}. \tag{7-29}$$

The Milne equations corresponding to eqs. 7-28 and 7-29 follow from substituting eqs. 7-17 and 7-18 into them and are

$$\int_0^\infty \kappa_\nu \left(\frac{1}{2} \int_{\tau_\nu}^\infty S_\nu E_1(t_\nu - \tau_\nu) \, dt_\nu + \frac{1}{2} \int_0^{\tau_\nu} S_\nu E_1(\tau_\nu - t_\nu) \, dt_\nu - S_\nu \right) d\nu = 0 \tag{7-30}$$

and

$$\int_0^\infty \frac{d}{d\tau_\nu} \left(\frac{1}{2} \int_{\tau_\nu}^\infty S_\nu E_3(t_\nu - \tau_\nu) \, dt_\nu + \frac{1}{2} \int_0^{\tau_\nu} S_\nu E_3(\tau_\nu - t_\nu) \, dt_\nu \right) d\nu = \frac{F_0}{4\pi}. \tag{7-31}$$

The three Milne equations 7-30, 7-25, and 7-31 are not independent. The S_ν that is a solution of one will be the solution of all three.

The flux constant, F_0, is often expressed in terms of an effective temperature (ref. eq. 1-2),

$$F_0 \equiv \sigma T_{\text{eff}}^4.$$

When model photospheres are constructed using flux constancy as a condition to be fulfilled by the model, the effective temperature becomes one of the fundamental parameters characterizing the model.

Equation 7-26b leads to another interesting result. In very deep layers where the optical depth at all frequencies is large, we can be assured that, compared to the other terms in the equation, $d\mathscr{F}_\nu/dx$ is negligible. We conclude that $\bar{I}_\nu = S_\nu$ in these deep layers.

THE GREY CASE

There is one simplification that deserves special comment because of the historic spot it occupies and because it is useful in some cases as a starting point in iterative calculations. The simplification is that κ_ν is frequency independent hence the name grey. The only known opacity source relevant to stellar atmospheres that is grey is electron scattering (see Chapter 8). The grey case is, as a consequence, not very realistic.

Let us integrate the transfer equation over frequency,

$$\cos \theta \frac{d}{dx} \int_0^\infty I_\nu \, d\nu = \rho \int_0^\infty \kappa_\nu I_\nu \, d\nu - \rho \int_0^\infty \kappa_\nu S_\nu \, d\nu.$$

In this grey case κ_ν can be factored out of the integrals, and we denote it by κ without a subscript. If, in addition, we define $I = \int_0^\infty I_\nu \, d\nu$ and $S = \int_0^\infty S_\nu \, d\nu$ the above equation becomes

$$\cos\theta \, \frac{dI}{dx} = \kappa\rho I - \kappa\rho S$$

or

$$\cos\theta \, \frac{dI}{d\tau} = I - S. \tag{7-32}$$

This is a remarkable result because the total radiation is described by a single transfer equation, whereas in the general case we had an infinite number— one for each frequency.

This type of simplification runs through the radiative equilibrium equations and Milne's equations. Thus the equivalents of eqs. 7-23, 7-28, and 7-29 are

$$\left.\begin{array}{c} \mathcal{F} = F_0 \\[6pt] \bar{I} = S \\[6pt] \dfrac{dK}{d\tau} = \dfrac{F_0}{4\pi}. \end{array}\right\} \tag{7-33}$$

Milne's equations simplify to

$$\left.\begin{array}{c} \dfrac{1}{2}\displaystyle\int_\tau^\infty SE_1(t - \tau) \, dt + \dfrac{1}{2}\displaystyle\int_0^\tau SE_1(\tau - t) \, dt - S = 0 \\[12pt] \displaystyle\int_\tau^\infty SE_2(t - \tau) \, dt - \displaystyle\int_0^\tau SE_2(\tau - t) \, dt = \dfrac{F_0}{2\pi} \\[12pt] \dfrac{d}{d\tau}\displaystyle\int_\tau^\infty SE_3(t - \tau) \, dt + \dfrac{d}{d\tau}\displaystyle\int_0^\tau SE_3(\tau - t) \, dt = \dfrac{F_0}{2\pi}. \end{array}\right\} \tag{7-34}$$

The solution of the grey case has been obtained in several ways. We notice that according to eq. 7-33, S and $\bar{I}$ are the same. The solution of the nongrey case amounts to finding the function S. In the grey case this is the same as obtaining $\bar{I}$. One of the first solutions was the approximate scheme of Eddington (1926). From the third radiative equilibrium condition (eq. 7-33) we know, upon integration, that

$$K(\tau) = \frac{F_0\tau}{4\pi} + \text{const.} \tag{7-35}$$

The solution is effected by casting both K and $\mathcal{F}$ in terms of the intensity, which Eddington assumed, as a first approximation, could be represented by a constant inward term and a constant outward term given by

$$I(\tau) = I_{in}(\tau) \qquad \text{for all} \quad \theta > \frac{\pi}{2}$$

$$I(\tau) = I_{out}(\tau) \qquad \text{for all} \quad 0 \le \frac{\pi}{2}.$$

That is, at any τ, I is constant over each hemisphere. As a result of this hemispherical isotropy restriction on I, we can write

$$\bar{I}(\tau) = \oint I \frac{d\omega}{4\pi} = \frac{1}{2} \int_0^{\pi/2} I_{out} \sin\theta \, d\theta + \frac{1}{2} \int_{\pi/2}^{\pi} I_{in} \sin\theta \, d\theta$$

$$= \frac{1}{2} I_{out} \int_0^{\pi/2} \sin\theta \, d\theta + \frac{1}{2} I_{in} \int_{\pi/2}^{\pi} \sin\theta \, d\theta$$

or

$$\bar{I}(\tau) = \tfrac{1}{2}[I_{out}(\tau) + I_{in}(\tau)]. \tag{7-36}$$

Further,

$$\mathcal{F} = \oint I \cos\theta \, d\omega = 2\pi \int_0^{\pi/2} I_{out} \cos\theta \sin\theta \, d\theta + 2\pi \int_{\pi/2}^{\pi} I_{in} \cos\theta \sin\theta \, d\theta$$

or

$$\mathcal{F}(\tau) = \pi[I_{out}(\tau) - I_{in}(\tau)]. \tag{7-37}$$

Finally,

$$K = \oint I \cos\theta \frac{d\omega}{4\pi} = \frac{1}{2} \int_0^{\pi/2} I_{out} \cos^2\theta \sin\theta \, d\theta + \frac{1}{2} \int_{\pi/2}^{\pi} I_{in} \cos^2\theta \sin\theta \, d\theta$$

or

$$K(\tau) = \tfrac{1}{6}[I_{out}(\tau) + I_{in}(\tau)]. \tag{7-38}$$

Comparing eqs. 7-36 and 7-38 gives

$$K(\tau) = \tfrac{1}{3}\bar{I}(\tau). \tag{7-39}$$

A similar comparison of eqs. 7-36 and 7-37 when evaluated at the outer boundary gives

$$2\pi\bar{I}(0) = \mathcal{F}(0) = F_0. \tag{7-40}$$

Thus we have in eqs. 7-39 and 7-40 the relations we seek and wish to apply to the general grey case solution expressed in eq. 7-35. Doing this gives

$$\bar{I}(\tau) = \frac{3}{4\pi}\left(\tau + \frac{2}{3}\right)F_0$$

or

$$S(\tau) = \frac{3}{4\pi}\left(\tau + \frac{2}{3}\right)\mathcal{F}(0). \tag{7-41}$$

This then is the source function for the grey case as solved using the Eddington approximation. The source function varies linearly with optical depth.

It is instructive to follow the discussion one more step by writing the frequency integrated source function $S = \bar{I}$ for the case of pure absorption. We use the frequency integrated form of Planck's law instead of eq. 5-17 and along with eq. 5-7

$$\bar{I}(\tau) = \frac{\sigma}{\pi}\,T^4.$$

We have already written $F_0 = \sigma T_{\text{eff}}^4$ so when these are put into eq. 7-41 it becomes

$$T^4(\tau) = \frac{3}{4}\left(\tau + \frac{2}{3}\right)T^4$$

or

$$T(\tau) = \left[\frac{3}{4}\left(\tau + \frac{2}{3}\right)\right]^{1/4} T_{\text{eff}}. \tag{7-42}$$

Notice that at $\tau = \frac{2}{3}$ the temperature is equal to the effective temperature and that $T(\tau)$ scales in proportion to the effective temperature. At this same optical depth, eq. 7-41 shows that $\mathcal{F}(0) = \pi S(\frac{2}{3})$.

Complete and rigorous solutions of the grey case (Chandrasekhar 1957) lead to a solution differing only slightly from eq. 7-41. It is usually written in the form

$$\bar{I}(\tau) = \frac{3}{4\pi}[\tau + q(\tau)]F_0$$

or

$$T(\tau) = \{\tfrac{3}{4}[\tau + q(\tau)]\}^{1/4}T_{\text{eff}} \tag{7-43}$$

where $q(\tau)$ is a slowly varying function ranging from 0.577 at $\tau = 0$ to 0.710 at $\tau = \infty$.

The usefulness of the grey scale is small when it comes to interpreting real stellar spectra. It is worthwhile recalling Eddington's (1926) comment concerning the grey case, "This, however, is a lazy way of handling the problem and it is not surprising that the result fails to accord with observation. The proper course is to find the spectral distribution of the emergent radiation by treating each wave-length separately using its own proper value of j and κ."

CONVECTIVE TRANSPORT

Convection plays a minor role in the energy transport through the photosphere. Convection functions because some volume of gas that is rising through the atmosphere possesses an excess of heat compared to its surroundings. Such a cell or "energy bucket" must have sufficient optical depth across its dimensions to prevent complete radiative loss of the excess energy to the surroundings. It is immediately clear that convection cannot convey much flux in the optically thin upper photosphere.

The flux carried by convection can be expressed in terms of the average cell characteristics: the temperature excess, ΔT, of the cell over its terminal surroundings; the specific heat at constant pressure, C_p; the cell density, ρ; and the upward velocity, v, of the cell,

$$\Phi = C_p \rho \, \Delta T \, v. \tag{7-44}$$

The convective flux Φ is incorporated in the flux constancy condition eq. 7-24 during the calculation of a model photosphere.

In the phenomenon of the solar granulation we see the top of the convective cell pattern. This region is at the bottom of the photosphere where $\rho \sim 10^{-7}$ g/cm^3. The observed ΔT and v are of the order of $10^2 \, ^\circ$K and 10^5 cm/sec. Using $C_p \simeq \frac{5}{2}R = 8 \times 10^7$ erg/g $^\circ$K for hydrogen, we find $\Phi \sim 2 \times 10^8$ erg/sec cm^2 compared to $\sim 10^{11}$ erg/sec cm^2 of radiative flux. Convection is a small contributor to the energy flow.

On the other hand, the physical motions of the convective cells are especially significant for spectral lines where even small Doppler shifts alter the line profiles. These aerodynamic behaviors are often called turbulence. The momentum of the convective cells is not dissipated when the optical depth of the cells becomes small. The turbulence may extend well into the upper photosphere even though the original cells are in complete radiative exchange with the adjacent gases. Furthermore, a variety of waves are generated by convection (see Schwartzschild's 1961 review). The acoustical waves generated by convection appear to be the source of energy needed to cause the temperature rise above the photosphere. Convection is also an efficient mixing mechanism producing chemical homogeneity in the photosphere.

We are only beginning to obtain observational information on convection generated motions in stars (see Chapter 18). Theoretical calculations of convection are approximate.

CONDITION FOR CONVECTIVE FLOW

The convective cell must be buoyed upward at each depth if it is to continue to rise. Since the density of the photosphere is less in higher layers, and the cell can be expected to stay in pressure equilibrium with its surroundings, it expands as it rises. We calculate the density change to see if the expanded cell will continue to rise or if it will sink, stopping the convection.

The classical Schwarzschild (1906) criterion is obtained by assuming the convective cell behaves adiabatically (no leakage whatever from the energy bucket). Then, if we use γ to denote the ratio of specific heats and P for the total pressure,

$$P\rho^{-\gamma} = \text{const}$$

or

$$\frac{d \log \rho}{d \log P} = \frac{1}{\gamma}.$$

We compare this to $d \log \rho/d \log P$ in the surrounding gas. In order to have convection the density of the cell must decrease at least as rapidly as the photospheric density. Thus for convection,

$$\frac{1}{\gamma} = \left.\frac{d \log \rho}{d \log P}\right|_{cell} > \left.\frac{d \log \rho}{d \log P}\right|_{phot} \tag{7-45}$$

must hold. In many calculations we prefer to use temperature rather than density in the photospheric gradient. Using the gas law in the form $P = (\rho/\mu)kT$ where μ is the mean molecular weight in grams,

$$\log P = \log \rho - \log \mu + \log k + \log T$$

and

$$\left.\frac{d \log \rho}{d \log P}\right|_{phot} = 1 + \frac{d \log \mu}{d \log P} - \frac{d \log T}{d \log P}.$$

Placing this into inequality 7-45 gives

$$\left.\frac{d \log T}{d \log P}\right|_{phot} > 1 - \frac{1}{\gamma} + \frac{d \log \mu}{d \log P} \tag{7-46}$$

as the condition needed for convection. In situations where hydrogen and helium do not change their ionization stage and the gases are chemically homogeneous, $d \log \mu / d \log P = 0$. The value of γ for a monatomic gas is $\frac{5}{3}$. The Schwarzschild criterian would then be $d \log T / d \log P > 0.4$ for convection. Polyatomic gases have γ approaching unity as the degrees of freedom increase with the complexity of the molecules. Therefore convection is encouraged in cool stars where molecules are numerous.

Should the transition from ionized to neutral hydrogen occur in the photosphere, μ will increase by about a factor of 2. The gradient $d \log \mu / d \log P$ is negative, which, according to inequality 7-46 encourages convection. Recombination of the electron and proton releases 13.6 eV of energy per atom as well. The ionization energy heats the gas in the adiabatic cell causing an increase in volume, again encouraging convection. The ionization is a new degree of freedom and the effect can be incorporated by redefining γ as done by Unsöld (1955) and Swamy (1961).

The value of γ may also be decreased from the monatomic value of 1.67 by the presence of radiation. From the radiation pressure equation 5-11 and the first law of thermodynamics, one can show (Chandrasekhar 1957, for example) that the radiation trapped in the convective cell obeys an adiabatic gas law with $\gamma = \frac{4}{3}$.

In practice the stability condition 7-46 is introduced into the construction of model photospheres by first computing the model in radiative equilibrium and then at each depth testing the model's $d \log T / d \log P$ against $1 - 1/\gamma + d \log \mu / d \log P$ (or the equivalent in a more comprehensive treatment). When the radiative temperature gradient is larger, the convective flux is then included in the flux constancy constraint, eq. 7-24. Some evidence has been presented by Defouw (1970a and 1970b) that indicates that convection may occur even when the radiative gradient is smaller than the adiabatic gradient. But in the classical approach the actual temperature gradient in convective regions is between the radiative gradient* and the adiabatic gradient. Radiation loss from the convective cells as they travel upward, that is, failure of the adiabatic assumption, reduces the efficiency of convection and causes the actual temperature gradient to be closer to the radiative gradient. Radiative leakage has been studied by Vitense (1953) and Öpik (1950).

Calculation of the convective flux, eq. 7-44, has not yet been done rigorously, but approximate values of v and ΔT can be estimated. Examples can be found in Böhm-Vitense (1958), Schwarzschild (1958), Spiegel (1963), Henyey et al. (1965), Mihalas (1970), and Travis and Matsushima (1973). A compari-

---

* In the dense regions below the photosphere, a convection zone has the adiabatic gradient, but convection in the photosphere does not carry enough of the total flux to allow us to use this limiting case.

son of computational methods has been made by Nordlund (1974). In each case there remains a free parameter commonly referred to as the mixing length. Typical values corresponding to $\Delta \ln P \sim 1$ are chosen for it, but no actual value is defined by the theory. Such a small portion of the surface flux originates in the convective layers of most stars that models of photospheres are only weakly dependent on the chosen mixing length.

REFERENCES

Abramowitz, M. and I. A. Stegun 1964. *Handbook of Mathematical Functions*, National Bureau of Standards Applied Mathematics Series 55, p. 227.

Böhn-Vitense, E. 1958. *Z. Astrophys.* **46**, 108.

Chandrasekhar, S. 1957. *An Introduction to the Study of Stellar Structure*, Dover, New York, p. 55.

Defouw, R. J. 1970a. *Sol. Phys.* **14**, 42.

Defouw, R. J. 1970b. *Astrophys. J.* **160**, 659.

Eddington, A. S. 1926. *The Internal Constitution of the Stars*, 1959 printing, Dover, New York, pp. 322, 325.

Henyey, L., M. S. Vardya and P. Bodenheimer 1965. *Astrophys. J.* **142**, 841.

Mihalas, D. 1970. *Stellar Atmospheres*, W. H. Freeman, San Francisco, p. 201.

Nordlund, A. 1974. *Astron. Astrophys.* **32**, 407.

Öpik, E. J. 1950. *Mon. Not. Roy. Astron. Soc.* **110**, 559.

Schwarzschild, K. 1906. *Akad. Wiss. Goett. Nachr.* **1**, 41.

Schwarzschild, M. 1958. *Structure and Evolution of Stars*, Princeton University Press, Princeton, p. 47.

Schwarzschild, M. 1961. *Astrophys. J.* **134**, 1.

Spiegel, E. A. 1963. *Astrophys. J.* **138**, 216.

Swamy, K. S. 1961. *Astrophys. J.* **134**, 1017.

Travis, L. D. and S. Matsushima 1973. *Astrophys. J.* **180**, 975.

Unsöld, A. 1955. *Physik der Sternatmosphären*, Springer-Verlag, Berlin, p. 227.

Vitense, E. 1953. *Z. Astrophys.* **32**, 135.

CHAPTER EIGHT

THE CONTINUOUS ABSORPTION COEFFICIENT

As a precursor to calculating the transfer of radiation through a model stellar photosphere, we now look at the continuous absorption coefficient. The wavelength dependence of the continuous absorption coefficient shapes the continuous spectrum emitted by the star. It follows that we must know the mass absorption coefficient, κ_ν, as used in the previous chapters, with good accuracy if we expect to do a precise job of fitting model computed spectra to real spectra.

ORIGINS OF CONTINUOUS ABSORPTION

The total continuous absorption coefficient is the sum of the absorption resulting from many physical processes. These are considered here one by one. All of them can be classed into one of two categories. The first is an ionization process where a bound–free transition occurs, and the second is a so-called free–free transition, which is the acceleration of one charge as it passes close to another charge. The remaining possibility is a bound–bound transition which, of course, gives a spectral line. Spectral lines are not normally included in κ_ν, but when there are many overlapped and crowded lines, they act much like the continuous absorption processes.

Most of the continuous opacity is due to hydrogen in some form or other. This is a direct result of hydrogen's overwhelming dominance of the chemical composition. Pecker (1965) has reviewed the continuous absorbers and his article may be consulted for additional references.

The atomic absorption coefficient with units of square centimeters, per absorber we call α. It is possible to express α as a function of λ or of ν, but since it is not a distribution in the sense that I_λ and I_ν are, we shall not use a λ or ν subscript on α. The wavelength versus frequency units question does not arise for α and one could write $\alpha_\nu = \alpha_\lambda$. On the other side, it is frequently useful, particularly when dealing with spectral lines (Chapter 11), to describe the amount of power abstracted in the frequency interval $d\nu$ from a unit intensity (I_ν) beam. Such a power is denoted by $\alpha\, d\nu$, which is a distribution and has units of ergs/(per second per square rad per square centimeter per cycle/per second per absorber). The corresponding quantity in wavelength units is $\alpha\, d\lambda$, which is abstracted from a unit I_λ beam. The relation between them is the usual c/λ^2, thus $\alpha\, d\nu = (c/\lambda^2)\alpha\, d\lambda$.

The continuous absorption coefficient per neutral hydrogen atom is denoted by κ. Like α, κ is not a distribution. At the end of the chapter κ is converted to units of square centimeters per gram and is denoted by κ_ν. The subscript ν simply denotes the important frequency dependence of the continuous absorption coefficient.

Let us not dwell on the atomic theory but instead move directly to the question of how to calculate κ_ν. In each case we start with α and then multiply by the number of absorbers per hydrogen atom.

THE STIMULATED EMISSION FACTOR

The absorption processes we are about to consider, usually do not include the stimulated emission (exceptions are noted explicitly). A simple allowance for stimulated emissions follows from eqs. 5-20, 6-7a, and 1-14,

$$
\begin{aligned}
\kappa_\nu \rho &= N_l B_{lu} h\nu - N_u B_{ul} h\nu \\
&= N_l B_{lu} h\nu \left(1 - \frac{N_u}{N_l}\frac{B_{ul}}{B_{lu}}\right) \\
&= N_l B_{lu} h\nu (1 - e^{-h\nu/kT}).
\end{aligned}
\tag{8-1}
$$

The absorption actually produced is lower by the factor $(1 - e^{-h\nu/kT})$ because of stimulated emissions. This factor need only be applied to the stated absorption coefficients and we do so in eq. 8-16, for example.* Table 8-1 shows the type of numbers that are involved.

When the validity of eq. 1-14 is in doubt, a calculation more detailed than eq. 8-1 may need to be done. Such added complications are sometimes necessary when dealing with spectral lines.

---

* It is true that we have derived eq. 8-1 only for the case of bound–bound absorption, but it has been shown by Milne (1924) that the same factor also holds with continuous absorption.

Table 8-1   Stimulated Emission Factors

λT, cm °K	$1 - e^{-h\nu/kT}$	Decrease, %	λ, Å	
			$T = 5000$°K	$T = 10,000$°K
0.2	0.999	0.08	4,000	2,000
0.4	0.973	2.82	8,000	4,000
0.6	0.909	10.0	12,000	6,000
0.8	0.834	19.8	16,000	8,000
1.0	0.763	31.1	20,000	10,000

NEUTRAL HYDROGEN

Both the bound–free and the free–free absorption of hydrogen are important. Bound–free transitions can occur from any level to the continuum. Ionization from levels of principal quantum numbers, n, correspond to the following positions in the spectrum,

n	λ, Å
1	912
2	3,647
3	8,206
4	14,588

Pressure lowering of the effective continuum may increase these wavelengths a few angstroms. Some numerical examples are given by Mihalas (1967). Photons of wavelength shorter than any given ionization threshold impart a kinetic energy to the freed electron,

$$\tfrac{1}{2}mv^2 = h\nu - \frac{hRc}{n^2} \tag{8-2}$$

where h is Planck's constant and R is the Rydberg constant for hydrogen (1.0968×10^5 cm^{-1}), making hRc/n^2 the energy of the bound level below the continuum.* This is shown graphically in Fig. 8-1. The excitation poten-

---

* You will recall that the wavelength, λ, and the Rydberg constant are related for bound–bound transitions by $1/\lambda = R(n^{-2} - u^{-2})$ where u is the principal quantum number of the upper level.

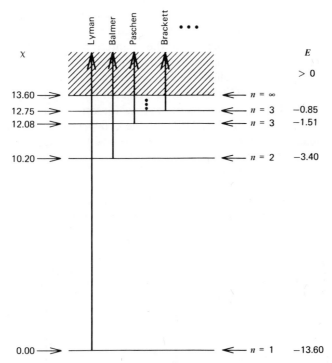

Fig. 8-1 The simplified energy level diagram for hydrogen showing binding energy, E, and excitation potential, χ, in electron volts for the first four levels and the continuum.

tial, χ, is defined to be the energy above the ground level and one can write

$$\chi = I - \frac{hRc}{n^2} = hRc - \frac{hRc}{n^2}$$

$$= 13.60\left(1 - \frac{1}{n^2}\right). \tag{8-3}$$

The original derivation done by Kramers (1923) and modified by Gaunt (1930) leads to an atomic absorption coefficient given by

$$\alpha_n = \frac{32}{3\sqrt{3}} \frac{\pi^2 e^6}{h^3} \frac{R}{n^5 \nu^3} g_n'$$

$$= \frac{32}{3\sqrt{3}} \frac{\pi^2 e^6}{h^3 c^3} R \frac{\lambda^3}{n^5} g_n'$$

$$= \frac{\alpha_0 g_n' \lambda^3}{n^5} \quad \text{with} \quad \alpha_0 = 1.044 \times 10^{-26} \tag{8-4}$$

for λ in angstroms. In this expression the electron charge has a value $e = 4.803 \times 10^{-10}$ esu and g'_n is the Gaunt factor,* which brings Kramer's original equation into agreement with the quantum mechanical results. It is of the order of unity and is discussed at the end of the section.

At the ionization limits, where the kinetic energy of the released electron is zero, eq. 8-2 gives $\nu = hRc/n^2$ so that α_n at these limits is proportional to n. The cross section for absorption increases as λ^3 for ionization from any given level. This is depicted in Fig. 8-2. Although the λ^3 dependence is very rapid, it is not so rapid that α_n becomes negligible when considering α_{n-1}. The overlap is shown in Fig. 8-2.

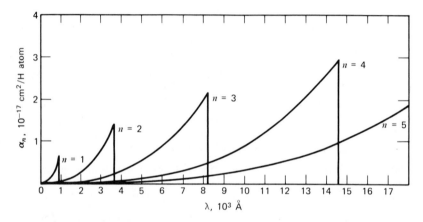

Fig. 8-2   The bound–free atomic absorption coefficient of neutral hydrogen.

The sum of the numbers of absorbers in each level times α_n is what we need. We write

$$\frac{N_n}{N} = \frac{g_n}{u_0(T)} e^{-\chi/kT}$$

where N_n/N is the number of hydrogen atoms excited to the level n per neutral hydrogen atom, $g_n = 2n^2$ is the statistical weight, $u_0(T) = 2$ is the partition function, and χ is the excitation potential given by eq. 8-3. The absorption

---

* The Gaunt factor, g'_n, is primed to distinguish it from g_n, which we have already used for the statistical weight of the nth atomic level.

coefficient in square centimeters per neutral H atom for all continua starting at n_0 is

$$\kappa(H_{bf}) = \sum_{n_0}^{\infty} \frac{\alpha_n N_n}{N} = \alpha_0 \sum_{n_0}^{\infty} \frac{\lambda^3}{n^3} g'_n e^{-\chi/kT}$$

$$= \alpha_0 \sum_{n_0}^{\infty} \frac{\lambda^3}{n^3} g'_n 10^{-\theta\chi}. \tag{8-5}$$

Unsöld (1955) has shown that the small contribution of the higher terms can be replaced by an integral,

$$\sum_{n_0+3}^{\infty} e^{-\chi/kT} \frac{1}{n^3} = -\frac{1}{2} \int_{n_0+3}^{\infty} e^{-\chi/kT} d\left(\frac{1}{n^2}\right).$$

And according to eq. 8-3, we have $d\chi = -I\, d(1/n^2)$ so that

$$\sum_{n_0+3}^{\infty} e^{-\chi/kT}\left(\frac{1}{n^3}\right) = \frac{1}{2} \int_{\chi_3}^{I} e^{-\chi/kT} \frac{d\chi}{I}$$

$$= \frac{kT}{2I} (e^{-\chi_3/kT} - e^{-I/kT})$$

where

$$\chi_3 = I\left[1 - \frac{1}{(n_0+3)^2}\right].$$

We are justified in neglecting the n dependence of g'_n because the integral is small compared to the first three terms we retain explicitly in the summation. Finally, eq. 8-5 becomes

$$\kappa(H_{bf}) = \alpha_0 \lambda^3 \left[\sum_{n_0}^{n_0+2} \left(\frac{g_n}{n^3} 10^{-\theta\chi}\right) + \frac{\log e}{2\theta I} (10^{-\chi_3\theta} - 10^{-I\theta})\right]. \tag{8-6}$$

Here $\log e = 0.43429$ and α_0 is given in eq. 8-4.

The free–free absorption is much smaller. When the free electron has a collision with a proton its orbit (unbound of course) is altered. A photon may be absorbed during such a collision with the orbital energy of the electron increased by the photon energy. The strength of the absorption depends on the electron velocity, v, because a slow encounter increases the probability of a photon passing by during the collision. According to Kramers (1923) the atomic absorption coefficient is

$$d\alpha_f = \frac{2}{3\sqrt{3}} \frac{h^2 e^2 R}{\pi m^3} \frac{1}{v^3 v} dv,$$

which is the cross section in square centimeters per H atom for the fraction of electrons in the velocity interval v to $v + dv$. The complete free–free absorption per electron is given by integrating this expression over velocity. In almost every case it is reasonable to use a Maxwell-Boltzmann speed distribution (eq. 1-10), which gives

$$\alpha_f = \frac{2}{3\sqrt{3}} \frac{h^2 e^2 R}{\pi m^3} \frac{1}{v^3} \int_0^\infty \left(\frac{2}{\pi}\right)^{1/2} \left(\frac{m}{kT}\right)^{3/2} v e^{-mv^2/2kT} \, dv$$

$$= \frac{2}{3\sqrt{3}} \frac{h^2 e^2 R}{\pi m^3} \frac{1}{v^3} \left(\frac{2m}{\pi kT}\right)^{1/2}. \tag{8-7}$$

The quantum mechanical derivation by Gaunt (1930) leads to nearly the same expression but modified by the correction factor g_f, which we call the free–free Gaunt factor.

The absorption coefficient in square centimeters per neutral hydrogen atom is proportional to the number density of electrons, N_e, and protons, N_i, giving

$$\kappa(\text{H}_{\text{ff}}) = \frac{\alpha_f g_f N_i N_e}{N_0}.$$

The number density of neutral hydrogen is denoted by N_0. By a slight modification of eq. 1-17 we can write

$$\kappa(\text{H}_{\text{ff}}) = \alpha_f g_f \frac{(2\pi m kT)^{3/2}}{h^3} e^{-I/kT}$$

and

$$\kappa(\text{H}_{\text{ff}}) = \alpha_0 \lambda^3 g_f \frac{\log e}{2\theta I} 10^{-\theta I}. \tag{8-8}$$

The constant α_0 is defined in eq. 8-4. We have used the relations $I = hcR$ and $R = 2\pi^2 me^4/h^3 c$ in the algebra leading to eq. 8-8.

The Gaunt factors needed to properly evaluate eqs. 8-6 and 8-8 have been tabulated in detail by Karzas and Latter (1961). Gingerich (1964) has given polynomial approximations for g_1', g_2', g_3', and g_f. The general expressions based on the work of Menzel and Pekeris (1935) are adequate for most purposes. They are

$$g_n' = 1 - 0.3456(\lambda R)^{-1/3}\left(\frac{\lambda R}{n^2} - \frac{1}{2}\right) \tag{8-9}$$

and

$$g_f = 1 + 0.3456(\lambda R)^{-1/3}\left(\frac{\lambda kT}{hc} + \frac{1}{2}\right)$$

$$= 1 + 0.3456(\lambda R)^{-1/3}\left(\frac{\log e}{\theta\chi_\lambda} + \frac{1}{2}\right) \qquad (8\text{-}10)$$

where in this last expression $\chi_\lambda = h\nu = 1.2398 \times 10^4/\lambda$ is the photon energy when λ is in angstroms.

In stars of B, A, and F spectral types, neutral hydrogen is a dominant continuous absorber. The energy distributions of such stars show strong modulation at the edges of an increased absorption (3647 and 8206 Å in the visible window). Examples can be seen in Figs. 10-6, 10-9, and 10-17. But in cooler stars the absorption from neutral hydrogen diminishes and the absorption from the negative hydrogen ion increases.

THE NEGATIVE HYDROGEN ION

The hydrogen atom is capable of holding a second electron in a bound state because the simple electron–proton combination is highly polarized. The ionization of the extra electron from the bound level requires 0.754 eV. All photons with $\lambda < 16{,}450$ Å have sufficient energy to ionize the H^- ion back to the neutral hydrogen atom and a free electron. The extra electrons needed to form H^- come from the ionized metals. A free–free absorption can also occur when there is a collision between a neutral hydrogen atom and an electron. Following the suggestion by Wildt (1939), calculations as well as laboratory measurements confirm the negative hydrogen ion as a major source of absorption in the sun. A short review of the historical development is given by Massey (1969). At temperatures somewhat higher than those in the solar photosphere, H^- is ionized to such an extent that it ceases to be a strong absorber. For cooler stars it is the dominant absorber but drops off rapidly in the very coolest stars because of the lack of free electrons.

The bound–free atomic absorption coefficients have been calculated by Krogdahl and Miller (1967), Doughty et al. (1966), and Geltman (1962). A comparison of results is shown in Fig. 8-3. There is general agreement but differences in detail. On a logarithmic scale, the absorption coefficient is nearly parabolic, with λ having a maximum near 8500 Å. The following polynomials fit Geltman's data to better than $\frac{1}{3}\%$ (probably an order of

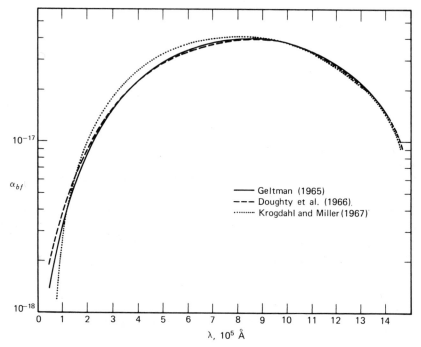

Fig. 8-3 The bound–free absorption coefficient of the H^- ion in square centimeters per ion per unit electron pressure.

magnitude better than the uncertainty in α_{bf} itself).

$1500 \text{ Å} < \lambda < 5250 \text{ Å}$

$$\log \alpha_{bf} = -16.20450 + 0.17280 \times 10^{-3}(\lambda - 8500)$$
$$+ 0.39422 \times 10^{-7}(\lambda - 8500)^2 + 0.51345 \times 10^{-11}(\lambda - 8500)^3$$

$5250 \text{ Å} < \lambda < 11{,}250 \text{ Å}$

$$\log \alpha_{bf} = -16.40383 + 0.61356 \times 10^{-6}(\lambda - 8500)$$
$$- 0.11095 \times 10^{-7}(\lambda - 8500)^2 + 0.44965 \times 10^{-13}(\lambda - 8500)^3$$

$11{,}250 \text{ Å} < \lambda < 15{,}000 \text{ Å}$

$$\log \alpha_{bf} = -15.95015 - 0.36067 \times 10^{-3}(\lambda - 8500)$$
$$+ 0.86108 \times 10^{-7}(\lambda - 8500)^2 - 0.90741 \times 10^{-11}(\lambda - 8500)^3.$$

The units are square centimeters per H^- ion per unit electron pressure and the stimulated emission factor is not included. The H^- ionization is given by eq. 1-17 with $u_0(T) = 1$ and $u_1(T) = 2$, namely,

$$\log \frac{N(\text{H})}{N(\text{H}^-)} = -\log P_e - \frac{5040}{T} I + 2.5 \log T + 0.1248.$$

Then

$$\kappa(\text{H}_{\text{bf}}^-) = 4.158 \times 10^{-10} \alpha_{bf} P_e \theta^{5/2} 10^{0.754\theta} \qquad (8\text{-}12)$$

with units of square centimeters per neutral hydrogen atom.

The free–free component of the H$^-$ absorption becomes dominant in the infrared. Calculations have been made by John (1966), Geltman (1965), Doughty and Frazer (1966), and Stilley and Callaway (1970). The behavior of the absorption coefficient with wavelength and temperature can be seen in Fig. 8-4.

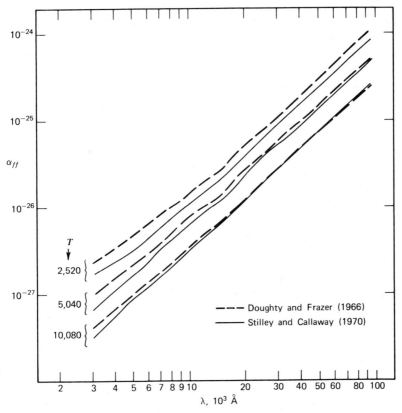

Fig. 8-4 The free–free absorption coefficient of the H$^-$ ion in square centimeters per ion.

The following polynomial fits the data of Stilley and Callaway to typically $\pm 1\%$ for θ in the interval 0.5–2.0 and λ in the interval 3038–91130 Å.

$$\kappa(H_{ff}^-) = P_e 10^{f_0 + f_1 \log \theta + f_2 \log^2 \theta} \tag{8-13}$$

where

$$f_0 = -31.3602 + 0.48735 \log \lambda + 0.296586 \log^2 \lambda - 0.0193562 \log^3 \lambda,$$
$$f_1 = 15.3126 - 9.33651 \log \lambda + 2.000242 \log^2 \lambda - 0.1422568 \log^3 \lambda,$$
$$f_2 = -2.6117 + 3.22259 \log \lambda - 1.082785 \log^2 \lambda + 0.1072635 \log^3 \lambda.$$

The stimulated emission factor is included.

There are, in addition to the bound–free and free–free absorption, some resonances of the H^- ion. These amount to photo transitions from the bound state to unstable levels at energies in the continuum. Their absorption extends over a few angstroms but is probably unimportant as a stellar opacity.

The forbidden H^- continuum was found by Myerscough (1968) to be negligible with regard to its effect on the spectrum.

OTHER HYDROGEN CONTINUOUS ABSORBERS

Hydrogen molecules appear in large numbers in cool stars. The H_2 molecules are more numerous than atomic hydrogen for stars cooler than mid-M spectral type, and while the molecule itself does not absorb in the visible spectrum, various ions do. The H_2^+ absorption has been studied by Bates (1951, 1952), Buckingham et al. (1952), Bates et al. (1953), and Matsushima (1964) among others. It is a significant absorber in the UV, but Matsushima has shown the H_2^+ absorption to be $<10\%$ of the H^- absorption for $\lambda > 4000$ Å.

Doyle (1968) has investigated the absorption due to the so-called quasi-H_2 and finds it significant for $\lambda < 2500$ Å in solar type stars.

In the infrared the free–free absorption of H_2^- becomes important, equaling a substantial fraction of the H^- absorption. Absorption coefficients can be obtained from Somerville (1964) and Dalgarno and Lane (1966).

THE NEGATIVE HELIUM ION

The bound–free absorption is generally considered to be negligible for He^- because there is only one bound level with an excitation potential of 19 eV. The population of such a level is, under normal conditions, too small to warrant consideration.

The free–free absorption can be important, particularly at long wavelengths in cool stars. Figure 8-5 shows the atomic absorption coefficient according to McDowell et al. (1966). It includes the stimulated emission. A useful polynomial fit is

$$\kappa(He_{ff}^{-}) = P_e 10^{a_0 + a_1 \lambda^{1/3} + a_2 \lambda^{1/2} + a_3 \lambda} A(He)\left[1 + \frac{\Phi(He)}{P_e}\right]^{-1}. \qquad (8\text{-}14)$$

The abundance of helium relative to hydrogen is $A(He)$, and the factor $[1 + \Phi(He)/P_e]^{-1}$ takes into account the ionization of helium making κ per hydrogen atom [$\Phi(He)$ is given by eq. 1-18]. The coefficients in the exponent are

$a_0 = -0.3183 \times 10^2 + 0.1358 \times 10^1 \theta - 0.1047 \theta^2 - 0.2819 \times 10^{-1} \theta^3,$

$a_1 = +0.4332 - 0.1268 \theta - 0.3535 \times 10^{-2} \theta^2 + 0.6869 \times 10^{-2} \theta^3,$

$a_2 = -0.4808 \times 10^{-1} + 0.1960 \times 10^{-1} \theta + 0.4305 \times 10^{-4} \theta^2 - 0.9558 \times 10^{-3} \theta^3,$

$a_3 = 0.1788 \times 10^{-4} - 0.1192 \times 10^{-4} \theta + 0.7870 \times 10^{-6} \theta^2 + 0.3936 \times 10^{-6} \theta^3$

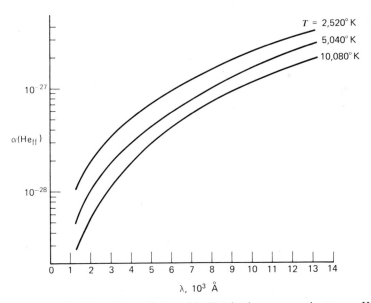

Fig. 8-5 The free–free absorption coefficient of the He⁻ ion in square centimeters per H atom per unit electron pressure. (Data from McDowell et al. 1966.)

giving $\kappa(He_{ff}^{-})$ to a precision of better than, or of the order of, 2% for $\lambda > 3000$ Å. Slightly more recent calculations have been published by John (1968).

THE METALS

Elements such as carbon, silicon, aluminum, magnesium, and iron produce bound–free opacity, which is important in the ultraviolet. The best way to incorporate these absorptions is to use well determined experimental cross sections multiplied by the number of absorbers. This can actually be done in a few cases where the cross sections have been measured. Rich (1967), for example, measured silicon (37 × 10^{-18} cm² per absorber for the ground state) and his results were used by Gingerich and Rich (1966) to show that silicon dominates the metal opacity in the 1300–2000 Å region with absorption edges at 1526 Å (ground level ionization) and 1682 Å (first excited level ionization). Some observational confirmation has been obtained, for example, by Porter et al. (1967).

But in many cases less dependable theoretical estimates must be used. For those cases in which a single outer electron is involved in the transitions,

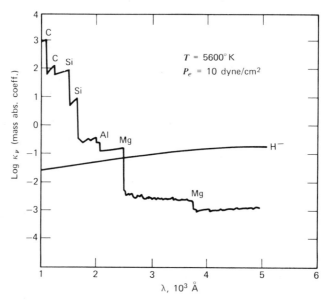

Fig. 8-6 A sample of the cumulative absorption of metals through the visible and the near UV. Various absorption edges are labeled. (Adapted from Travis and Matsushima 1968 with permission from the Astrophysical Journal. Copyright 1968 by the University of Chicago.)

one may expect the ionization behavior to resemble that of hydrogen (see Vitense 1951 or Unsöld 1955). An empirically adjusted nuclear charge is usually used in the computations in place of the actual charge. This method leads to a silicon absorption, for instance, which is one-tenth the observed value mentioned above.

A better theoretical approach was given by Burgess and Seaton (1960) called the quantum defect method. It can only be applied to situations of LS coupling. Application has been made by Peach (1967) and Travis and Matsushima (1968) among others. Extensive tables of absorption coefficients have been published by these last authors. The accuracy of the quantum defect method has been considered by Matsushima and Travis (1973).

The nature of the metal absorption coefficient is shown in Fig. 8-6 taken from the work of Travis and Matsushima (1968). The strength of the individual components obviously depends on the mix of the elements. The results of Fig. 8-6 are calculated using the abundances of Goldberg et al. (1060); however in 1969 the iron abundance measurement was revised upward by one order of magnitude (see Chapter 14). The metal absorption of Fig. 8-6 is actually greater as a result of this, especially in the 3000–3500 Å region. Changes of this nature may account for the unknown opacity source discussed by Matsushima (1968).

SCATTERING

Free electrons scatter radiation with the same efficiency at all wavelengths. The absorption coefficient is

$$\alpha_e = 0.6655 \times 10^{-24} \text{ cm}^2/\text{electron.}$$

It has generally been assumed that the intrinsic nonisotropy in the scattering process (phase function, of the form $1 + \cos^2 \theta$) averages to zero over the symmetrical geometry of the star and many scatterings through the atmosphere. We write

$$\kappa(e) = \frac{\alpha_e N_e}{N_H} = \frac{\alpha_e P_e}{P_H}$$

where P_H is the partial pressure of hydrogen. P_H is related to the gas and electron pressure in the following way. If N is taken to represent the total number of particles per cubic centimeter, then

$$N = \sum N_j + N_e = N_H \sum A_j + N_e$$

with N_j particles of the jth element per cubic centimeter and $A_j \equiv N_j/N_H$. We solve for N_H and write

$$N_H = \frac{N - N_e}{\sum A_j} \quad \text{or} \quad P_H = \frac{P_g - P_e}{\sum A_j}$$

which leads to

$$\kappa(e) = \frac{\alpha_e P_e \sum A_j}{P_g - P_e}. \tag{8-15}$$

Again $\kappa(e)$ has units of square centimeters per hydrogen particle. The stimulated emission factor should not be used for $\kappa(e)$.

Electron scattering increases in importance for low pressures. For main sequence stars $P_g \sim P_H \sim 10^3$ to 10^5 dynes/cm^2, giving $\kappa(e)/P_e \sim 10^{-27}$ to 10^{-29} cm^2 per H particle per unit electron pressure and so it is small compared to other absorbers. But for supergiants the pressures are 10^1 or 10^2 dynes/cm^2, making $\kappa(e)$ much larger.

Rayleigh scattering can be important in the UV in stars cool enough to have molecules in their atmospheres. The reader is referred to Vitense (1951), Gingerich (1964), Bode (1965), or Vardya (1970) for further details.

OTHER SOURCES OF OPACITY

Every treatment of absorption coefficients has to have a section such as this. We take this opportunity to give reference to the bound–free and free–free absorption of neutral helium (Mihalas 1970). Qualitatively the helium continuum has much in common with that of hydrogen. Ionization from the ground state corresponds to 504 Å radiation. The excitation energies are so large that helium need only be included in calculations of continua for O and early B stars.

Several molecular ions such as CN^-, C_2^-, and H_2O^- are likely to be important continuous absorbers in very late type stars. A general approximation for free–free absorption coefficients of many atomic and molecular species is presented by Dalgarno and Lane (1966). They explicitly consider H, He, N, O, Ne, H_2, N_2, and O_2. Linsky (1969) considers molecular hydrogen and other absorbers expected to be at work in late type stars. Water vapor opacity has been calculated by Auman (1966, 1967). A brief review of opacity in late type stars is given by Vardya (1970). A more detailed discussion of some absorbers is presented by Massey (1969).

Unidentified sources of opacity may exist in the ultraviolet for stars like the sun. The cause of such opacity has been sought by Bonnet (1968) and Linsky (1970), and more recently by Degl'innocenti and Noci (1973). Several suspects have been investigated without success. The latter authors suggest the cumulative effect of many absorption lines may be responsible, a possibility pursued previously by Holweger (1970) for the blue and near UV regions of the solar spectrum.

THE TOTAL ABSORPTION COEFFICIENT

We add the various absorption coefficients, reducing by the stimulate emission factor those terms not implicitly including it,

$$\kappa = \left(\left[\kappa(H_{bf} + \kappa(H_{ff}) + \kappa(H_{bf}^-)\right](1 - 10^{-\chi_\lambda \theta}) + \kappa(H_{ff}^-) + \cdots \right)\left(1 + \frac{\Phi(H)}{P_e} \right)^{-1}$$
$$+ \kappa(\text{metals}) + \kappa(He_{ff}^-) + \kappa(e) + \cdots \quad (8\text{-}16)$$

The factor $(1 + \Phi(H)/P_e)^{-1}$ converts the foregoing absorption coefficients from normalization per neutral hydrogen atom to per hydrogen particle, thus taking into account the ionization of hydrogen at high temperatures.

The absorption coefficient in square centimeters per H particle as given above is then converted to units of square centimeters per gram via division by the number of grams per H particle

$$\kappa_\nu = \frac{\kappa}{\sum A_j \mu_j}. \quad (8\text{-}17)$$

The mass, μ_j, of the jth element in the atmosphere is 1.6606×10^{-24} g times the element's atomic weight. A_j again denotes the number abundance relative to hydrogen. A typical value of $\sum A_j \mu_j$ is 2×10^{-24} g per H particle.

Some investigators prefer to interpolate in tables of precomputed absorption coefficients. One of the first comprehensive tabulations was done by Vitense (1951). A summary of these is given by Allen (1963). More recently Bode (1965) has computed tables of absorption coefficients, which are again summarized by Allen (1973).

REFERENCES

Allen, C. W. 1963. *Astrophysical Quantities*, 2nd ed., Athlone, University of London, London, p. 96.

Allen, C. W. 1973. *Astrophysical Quantities*, 3rd ed., Athlone, University of London, London, p. 99.

Auman, J. R. Jr. 1966. *Colloquium on Late Type Stars*, M. Hack, Ed., Osservatorio Astronomico, Trieste, p. 215.

Auman, J. R. Jr. 1967. *Astrophys. J. Suppl.* **14**, 171.

Bates, D. R. 1951. *Mon. Not. Roy. Astron. Soc.* **111**, 303.

Bates, D. R. 1952. *Mon. Not. Roy. Astron. Soc.* **112**, 40.

Bates, D. R., K. Ledsham, and A. L. Stewart 1953. *Phil. Trans. Roy. Soc. Lond. Ser. A* **246**, 215.

Bode, G. 1965. *Institutsbericht*, Institut für Theoretische Physik und Sternwarte der Universität Kiel, Kiel, Germany.

Bonnet, R. 1968. *Ann. Astrophys.* **31**, 597.

Buckingham, R. A., S. Reid, and R. Spence 1952. *Mon. Not. Roy. Astron. Soc.* **112**, 382.

Burgess, A. and M. J. Seaton 1960. *Mon. Not. Roy. Astron. Soc.* **120**, 121.

Dalgarno, A. and N. F. Lane 1966. *Astrophys. J.* **145**, 623.

Degl'innocenti, E. L. and G. Noci 1973. *Sol. Phys.* **29**, 287.

Doughty, N. A., P. A. Frazer 1966. *Mon. Not. Roy. Astron. Soc.* **132**, 267.

Doughty, N. A., P. A. Frazer, and R. P. McEachran 1966. *Mon. Not. Roy. Astron. Soc.* **132**, 255.

Doyle, R. O. 1968. *Astrophys. J.* **153**, 987.

Gaunt, J. A. 1930. *Phil. Trans. Roy. Soc. Lond. Ser. A* **229**, 163.

Geltman, S. 1962. *Astrophys. J.* **136**, 944.

Geltman, S. 1965. *Astrophys. J.* **141**, 376.

Gingerich, O. 1964. *Proceedings of the First Harvard–Smithsonian Conference on Stellar Atmospheres*, Smithsonian Astrophysical Observatory Special Report No. 167, Cambridge, Mass., p. 17.

Gingerich, O. and J. C. Rich 1966. *Astron. J.* **71**, 161.

Goldberg, L., E. A. Müller, and L. H. Aller 1960. *Astrophys. J. Suppl.* **5**, 1.

Holweger, H. 1970. *Astron. Astrophys.* **4**, 11.

John, T. L. 1966. *Mon. Not. Roy. Astron. Soc.* **131**, 315.

John, T. L. 1968. *Mon. Not. Roy. Astron. Soc.* **138**, 137.

Karzas, W. J., and R. Latter 1961. *Astrophys. J. Suppl.* **6**, 167.

Kramers, H. A. 1923. *Phil. Mag.* **46**, 836.

Krogdahl, W. S., and J. E. Miller 1967. *Astrophys. J.* **150**, 273.

Linsky, J. L. 1969. *Astrophys. J.* **156**, 989.

Linsky, J. L. 1970. *Sol. Phys.* **11**, 198.

Massey, H. S. W. 1969. *Electronic and Ionic Impact Phenomena*, Oxford University Press, p. 1255.

Matsushima, S. 1964. *Proceedings of the First Harvard–Smithsonian Conference on Stellar Atmospheres*, Smithsonian Astrophysical Observatory Special Report No. 167, Cambridge, Mass., p. 42.

Matsushima, S. 1968. *Astrophys. J.* **154**, 715.

Matsushima, S. and L. D. Travis 1973. *Astrophys. J.* **181**, 387.

McDowell, M. R. C., J. H. Williamson, and V. P. Myerscough 1966. *Astrophys. J.* **144**, 827.

Menzel, D. H. and C. L. Pekeris 1935. *Mon. Not. Roy. Astron. Soc.* **96**, 77.

Mihalas, D. 1967. *Astrophys. J.* **149**, 169.

Mihalas, D. 1970. *Stellar Atmospheres*, W. H. Freeman, San Francisco, p. 120.

Milne, E. A. 1924. *Phil. Mag.* **47**, 209.

Myerscough, V. P. 1968. *Astrophys. J.* **152**, 1115.

Peach, G. 1967. *Mem. Roy. Astron. Soc. (Lond.)* **71**, 1.

Pecker, J. C. 1965. *Annu. Rev. Astron. Astrophys.* **3**, 143.

Porter, J. R., S. G. Tilford, and K. G. Widing 1967. *Astrophys. J.* **147**, 179.

Rich, J. C. 1967. *Astrophys. J.* **148**, 275.

Somerville, W. B. 1964. *Astrophys. J.* **139**, 192.

Stilley, J. L. and J. Callaway 1970. *Astrophys. J.* **160**, 245.

Travis, L. D. and S. Matsushima 1968. *Astrophys. J.* **154**, 689.

Unsöld, A. 1955. *Physik der Sternatmosphären*, 2nd ed., Springer-Verlag, Berlin, p. 168.

Vardya, M. S. 1970. *Annu. Rev. Astron. Astrophys.* **8**, 87.

Vitense, E. 1951. *Z. Astrophys.*

Wildt, R. 1939. *Astrophys. J.* **89**, 295.

CHAPTER NINE

THE MODEL PHOTOSPHERE

What is a model photosphere and why build one? It would seem logical to take our stellar observations and deduce from them the physical conditions existing in the atmosphere of the star—somewhat like a parallax measurement yields the distance to a star. Alas, the formation of the stellar spectrum is not so simply related to the physical state of the atmosphere as distance is to parallax. There are many physical variables and a rigorously definitive interpretation cannot be made in most instances. We are led to hypothesize a model to organize and relate the details conveyed in the starlight. The model resembles a scientific theory in that it is constructed on the basis of our observations and known physical laws. We may then gather additional observations to test our model much like we would test a theory. The model is modified and improved at each step of such an idealized case.

When our model closely matches the observable information, we begin to feel that we "understand" the phenomenon being modeled, and, in fact, once we believe our model is pretty good and worthy of some trust, we may begin to deduce certain properties of the star such as its effective temperature, surface gravity, chemical composition, radius, and rate of rotation.

The model photosphere consists of a table of numbers giving the source function and the pressure as a function of optical depth. Additional columns may be added to the table depending on what the purpose is in constructing the model. For example, one might list the density, geometrical depth, absorption coefficient, and mean intensity.*

The material in this chapter is directed toward models that can be readily constructed. They are somewhat rudimentary compared to real stars. The

---

* An example can be found in Table 9-3.

reader should not jump to the conclusion that such models are useless—they are in fact indispensable tools for the analysis of stellar spectra. The major simplifications are as follows:

1. Plane parallel geometry, making all physical variables a function of only one space coordinate.
2. Hydrostatic equilibrium, meaning that the photosphere is not undergoing large scale accelerations comparable to the surface gravity; there is no significant mass loss.
3. Fine structure, such as granulation, sunspots (starspots?), and prominences, is negligible.
4. Magnetic fields are excluded.

It is also true that in the majority of cases we are thinking in terms of the thermodynamic equilibrium relations describing the excitation and ionization of the gas (eqs. 1-16 and 1-17) and in terms of the black body source function $B_\nu(T)$. The thermodynamic equilibrium concept is applied to relatively small volumes of the model's photosphere—volumes with dimensions of the order of unity in optical depth. For this reason the name local thermodynamic equilibrium, usually abbreviated LTE, is coined. The photosphere may then be characterized by one physical temperature at each depth. The excitation, ionization, source function, and thermal velocity distributions in the vicinity of one point are all described by this unique temperature. Progressing outward through the photosphere, each successive volume is assigned a lesser temperature so that in the LTE situation it is customary to replace the tabulation of the source function as it varies with depth by a tabulation of the temperature. The model then consists of temperature and pressure given as a function of optical depth. As we shall see, $B_\nu(T)$ makes a reasonable source function for many situations but can fail badly, especially for spectral lines.

On the practical side, we shall emphasize the use of the scaled solar temperature distribution in model construction. The justification of the scientific applicability of such a scheme is given below. The scaling procedure has its weaknesses, but because of the overwhelming simplicity it introduces, we use it here without hesitation or apology.

THE HYDROSTATIC EQUATION

The relation between pressure and optical depth is established by assuming hydrostatic equilibrium. Consider the small volume in Fig. 9-1. The difference in pressure between the top and bottom of the volume is the weight per unit area of the mass in the volume. The weight is the product of the density,

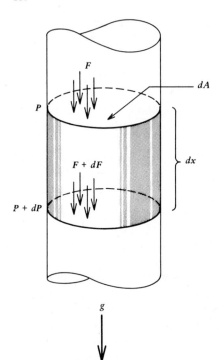

Fig. 9-1 The force is exerted by the overlying gas on the area dA to give a pressure P. The weight of the material in the volume $dx\, dA$ adds a force dF in the depth dx thereby increasing the pressure by dP.

volume, and surface gravity. We let dF represent this weight,

$$dF = \rho\, dA\, dx\, g.$$

Or, the increment in pressure over the distance dx is

$$dP = \frac{dF}{dA} = \rho g\, dx.$$

With x increasing inward as the pressure does, no negative sign appears in this relation. Now $d\tau_v = \kappa_v \rho\, dx$ so

$$\frac{dP}{d\tau_v} = \frac{g}{\kappa_v}, \tag{9-1}$$

which is the hydrostatic equation we need.

The pressure in eq. 9-1 is the total pressure supporting the small volume element. In most stars the gas pressure accounts for the bulk of the total pressure. In extreme cases other pressures may be significant compared to P_g. Examples are: (1) radiation pressure in very hot stars (refer to eq. 5-11) with

$$P_R = \frac{4}{3}\frac{\sigma}{c}T = 2.520 \times 10^{-15}T^4 \text{ dynes/cm}^2$$

(numerical values are given for main sequence stars in Table 9-1), (2) magnetic pressure given by $H^2/4\pi$ which we neglect, and (3) turbulence pressure often assumed to be $\frac{1}{2}\rho v^2$ where ρ and v are the density and root mean square velocity of the turbulent elements. This expression holds for random velocities that may or may not describe the aerodynamic behavior (see Chapter 18). In the following material only P_g is retained in the hydrostatic equation.

Table 9-1   Radiation Pressure Tabulation

T, °K	P_R, dyne/cm^2	P_g (typical), dyne/cm^2
4,000	0.6	10^5
8,000	10	10^4
12,000	52	3×10^3
16,000	165	3×10^3
20,000	403	5×10^3

Integration of the hydrostatic equation can be done in many ways, one of which is to write eq. 9-1 as

$$P_g^{1/2}\,dP_g = \frac{P_g^{1/2}g}{\kappa_0\,dt_0}$$

where κ_0 is κ_ν at some reference wavelength which we choose to be 5000 Å. Integration gives

$$P_g = \left[\frac{3}{2}g\int_0^{\tau_0}\frac{P_g^{1/2}}{\kappa_0\,dt_0}\right]^{2/3} = \left[\frac{3}{2}g\int_{-\infty}^{\log\tau_0}\frac{t_0 P_g^{1/2}}{\kappa_0\log e}\,d\log t_0\right]^{2/3}. \qquad (9\text{-}2)$$

Integration on a logarithmic optical depth scale gives better numerical precision—a detail that is obvious if you plot the integrand as a function of depth for a typical case. The procedure is to guess at $P_g(\tau_0)$ and do the numerical integration. The new value of $P_g(\tau_0)$ obtained from eq. 9-2 can be used in the next iteration until convergence is obtained. With a reasonable guess at $P_g(\tau_0)$, 2–4 iterations are needed. A poor guess will extend the iteration a cycle or two.

Before the computation in eq. 9-2 can proceed, we must know κ_0 as a function of τ_0 since κ_0 appears in the integrand. In Chapter 8 we saw that κ_ν was dependent on the temperature and the electron pressure. We conclude that we must know how these two variables, T and P_e, depend on τ_0. First let us consider $T(\tau_0)$ and then later $P_e(\tau_0)$.

THE TEMPERATURE DISTRIBUTION IN THE SUN

The solar temperature distribution can be obtained from limb darkening measurements of the type shown in Fig. 9-2. The reason for the limb darkening is that the continuum source function decreases outward. As we look toward the limb we see higher photospheric layers. In fact, from eq. 7-12,

$$I_\nu(0) = \int_0^\infty S_\nu e^{-\tau_\nu \sec \theta} \sec \theta \, d\tau_\nu, \qquad (9\text{-}3)$$

we see that the exponential extinction varies as $\tau_\nu \sec \theta$ so the position of optical depth along the line of sight moves upward, that is, to smaller τ_ν as the line of sight is shifted from the center to the limb. In the approximate grey case solution we found a linear source function, eq. 7-41, which we write as

$$S_\nu = a + b\tau_\nu.$$

Then from eq. 9-3 one gets

$$I_\nu(0) = a + b \cos \theta,$$

which says that at $\cos \theta = \tau_\nu$, the specific intensity on the surface at the position θ and the source function at the depth τ_ν are equal. This is referred to as the Eddington–Barbier relation. Measurements of I_ν across the solar disc can be used to map the depth dependence of S_ν in this way.

More complex analytical solutions for $S_\nu(\tau_\nu)$ have been conceived (e.g., Pierce and Waddell 1961 and Mitchell 1959). Numerical solutions can also be obtained by adjusting S_ν until the proper θ dependence is attained.

The concept of the depth of formation can be developed beyond the rudimentary relation described above by looking at the depth development of $I_\nu(\tau_\nu)$. More specifically, let us consider the integrand of eq. 9-3 rewritten on a logarithmic scale

$$I_\nu(0) = \int_{-\infty}^\infty S_\nu(\tau_\nu) e^{-\tau_\nu \sec \theta} \tau_\nu \frac{\sec \theta \, d \log \tau_\nu}{\log e}. \qquad (9\text{-}4)$$

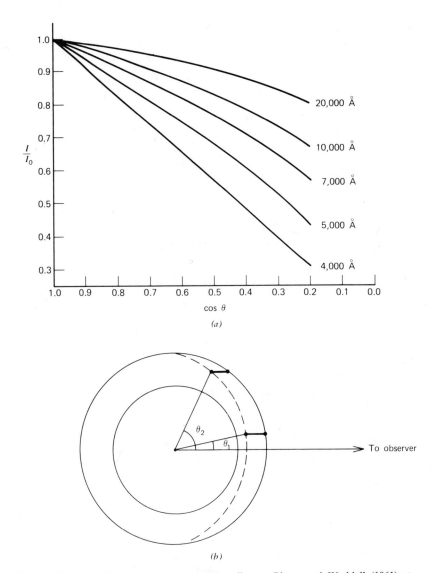

Fig. 9-2 (a) The limb darkening observation according to Pierce and Waddell (1961) at several wavelengths. The 4000 Å curve is nearly linear. (b) A schematic illustration of the cause of limb darkening. The bottom and top of the photosphere are indicated by the inner and outer circles. Penetration to unit optical depth, as shown by the heavy line segments, corresponds to different depths in the photosphere depending on θ. The dashed curve indicates the surface that is unit optical depth into the photosphere. The radiation seen at the disc position θ_2 is characteristic of the higher cooler layers than the radiation seen at the disc position θ_1.

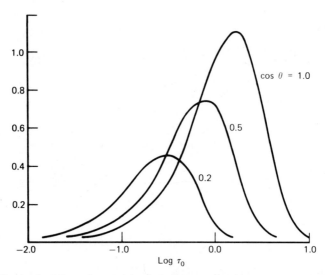

Fig. 9-3 The depth of formation at three limb distances showing quantitatively the decrease in intensity (area under curve) and shift to higher photospheric layers for lines of sight closer to the limb. (Adapted from Pierce and Waddell 1961.)

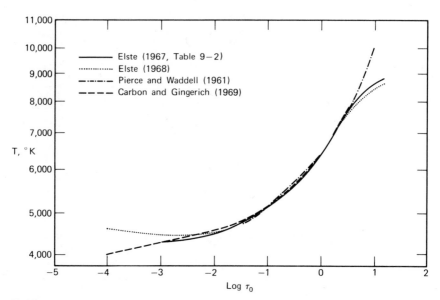

Fig. 9-4 The solar temperature distribution based on limb darkening measurements.

The depth dependence of this integrand is shown in Fig. 9-3, which is adapted from the work of Pierce and Waddell (1961). The figure shows how the surface intensity originates in higher layers for the light we see coming from positions on the solar disc closer to the limb.

The range in τ_v over which S_v is defined by a single limb darkening curve is relatively small and corresponds to those optical depths where the integrand of eq. 9-4 is large. A complete determination of the depth dependence of the source function requires measurements over many spectral regions, thereby making use of the wide range in the strength of κ_v and the corresponding change in the effective depth of emission. Measurements in the UV and IR, for example, can both be used to explore the source function higher than the temperature minimum because the continuous absorption in these spectral regions is large compared to the absorption in the visible. Details can be found in Lambert (1971) and Vernazza et al. (1973). We are more concerned

Table 9-2 Solar Temperature Distribution

Log τ_{5000}	T	Log τ_{5000}	T
-4.0	4440	0.9	5210
-3.0	4440	0.8	5294
-2.9	4447	0.7	5387
-2.8	4456	0.6	5490
-2.7	4466	0.5	5605
-2.6	4480	0.4	5734
-2.5	4494	0.3	5880
-2.4	4508	0.2	6043
-2.3	4528	0.1	6224
-2.2	4553	0.0	6429
-2.1	4579	0.1	6656
-2.0	4611	0.2	6904
-1.9	4644	0.3	7177
-1.8	4684	0.4	7467
-1.7	4724	0.5	7728
-1.6	4773	0.6	7962
-1.5	4822	0.7	8172
-1.4	4872	0.8	8358
-1.3	4931	0.9	8512
-1.2	4995	1.0	8630
-1.1	5061	0.1	8727
-1.0	5132	0.2	8811

Source. Adapted from Elste's Model 10.

here with the photospheric layers for which measurements in the visible window suffice to define the needed portion of $S_\nu(\tau_\nu)$.

On the practical side of things, the relative limb darkening I_ν/I_0 (where I_0 is the intensity at the center of the disc) is measured in one set of observations. Then eq. 9-4 is used in a normalized form by dividing by I_0, and $S_\nu(\tau_\nu)/I_0$ is obtained. The shape of the source function is found from the shape of the limb darkening. Another set of measurements is made to obtain I_0. The absolute scale of the source function is obtained from the absolute specific intensity. Finally the source function is set equal to the Planck function to obtain the temperature as a function of depth.

Several investigators have determined $T(\tau_0)$. A few of these results are shown in Fig. 9-4. We shall adopt the Elste 10 model temperature distribution as a standard of comparison (Elste 1967). The values of $T(\tau_0)$ are given in Table 9-2.

TEMPERATURE DISTRIBUTIONS IN OTHER STARS

No reliable measurements of limb darkening have been possible for stars. Instead, the run of temperature with depth is usually obtained in one of two ways. The first is theoretical. One requires flux constancy with depth according to eq. 7-24, 7-28, or 7-29. In Chapter 7 we saw how this could be done analytically for the grey case using the Eddington approximation. In the nongrey models, numerical integrations are performed. The procedures for obtaining flux constancy are all iterative, starting with some initial $S_\nu(\tau_0)$, computing $\mathcal{F}_\nu(\tau_0)$ by eq. 7-15, and finally correcting $S_\nu(\tau_0)$ in some way to meet the required condition.

The most useful method was developed by Avrett and Krook (1963a, 1963b) and Avrett (1964). A good example of its use is the work of Carbon and Gingerich (1969). Other methods are discussed in the 1st Harvard Smithsonian Conference (Gingerich 1964) and by Mihalas (1967, 1970). Usually the resulting $S_\nu(\tau_0)$ is related to $T(\tau_0)$ by Planck's equation. Flux constancy values of $0.1-1\%$ are typical.

The correctness of the derived solution depends upon our having the correct behavior of κ_ν with depth. Although we have neglected the bound–bound opacity in the construction of the models, in the present context this introduces an error that ultimately must be corrected. One approach to the problem has been made by Mihalas and Morton (1965), Mihalas (1966), and Van Citters and Morton (1970) who include the strongest lines in the calculation of early type spectra. In cooler stars where there are many more lines, various statistical approaches have been used, for example, by Allen (1970), Strom and Kurucz (1966), Carbon and Gingerich (1969), Athay (1970),

Fowler (1974), and Carbon (1974). The inclusion of lines causes a redistribu-
tion of flux in the spectrum and the temperature distribution responds to
bring this about under the flux constancy constraint. The blockage of flux
by the lines results in a heating of the deeper layers. The whole process is
referred to as line blanketing because of this. A comparison is made in Fig. 9-5
between models with and without lines. The effect on the temperature
gradient is quite significant when translated into the behavior of spectral
features having different depths of formation. Fowler (1974) has presented
evidence that indicates the importance of line blanketing in Sirius. A good
qualitative discussion is presented by Pecker (1973) and an earlier collection
of work can be found in the IAU colloquium edited by Böhm (1966). The
blanketing is also a function of the chemical composition.

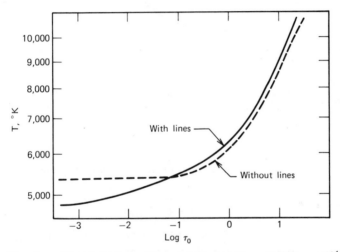

Fig. 9-5 The effects of line blanketing on $T(\tau_0)$ are illustrated. $T_{eff} = 6500°K, g = 10^4 \, cm/sec^2$,
solar abundances. (Data from Strom and Kurucz 1966.)

The second point of uncertainty with the flux constancy method arises in
the treatment of convection. According to the discussion in Chapter 7 a
convective flux given by eq. 7-44 should be included when $d \log T/d \log P_g$
exceeds the adiabatic gradient. The flux constancy condition then becomes
a constant radiative plus convective flux (eq. 7-24). The mixing length
remains a free parameter introducing a large uncertainty in the deep layers
as shown in Fig. 9-6. Fortunately so little surface flux arises in these layers
that the spectrum is hardly affected. In stars earlier than about F0 to F5,
convection probably does not exist in the photosphere.

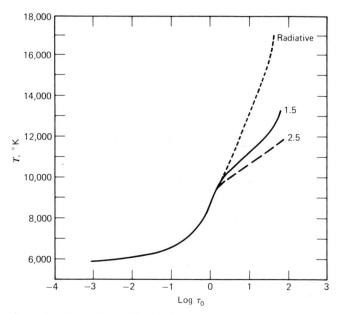

Fig. 9-6 Convection changes the model's $T(\tau_0)$ in the deep layers. The cases for no convection and mixing lengths of 1.5 and 2.5 pressure scale heights are shown. Very little light escapes from the layers below log $\tau_0 = 0.5$. (Data from Carbon and Gingerich 1969.)

The second method of obtaining $T(\tau_0)$ is to scale the solar temperature distribution,

$$T(\tau_0) = S_0\, T_\odot(\tau_0). \qquad (9\text{-}5)$$

The first indication we had that such a procedure might be possible was in the solution of the grey case where we found (eq. 7-43)

$$T(\tau) = \{\tfrac{3}{4}[\tau + q(\tau)]\}^{1/4} T_{\text{eff}}$$

or

$$T(\tau) = \frac{T_{\text{eff}}}{T_{\text{eff}}^\odot}\, T_\odot(\tau).$$

In the grey case the scaling factor is the ratio of effective temperatures.

The comparison of theoretical temperature distributions, computed on the basis of flux constancy, shows that a scaling relation holds, as seen in Figs. 9-7 and 9-8. The ratio of $T(\tau_0)/T_\odot(\tau_0)$ shown in Fig. 9-8 should be constant with τ_0 if eq. 9-5 is to hold exactly. A constant ratio describes the curves in Fig. 9-8 to within a few per cent, with the larger deviations occurring

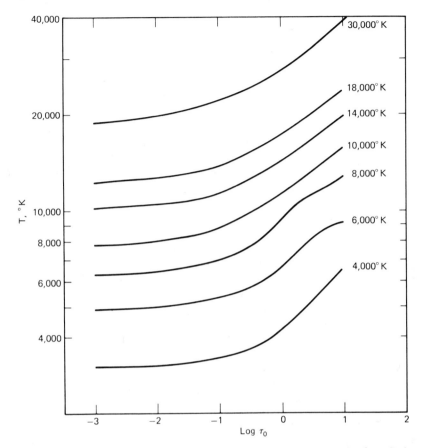

Fig. 9-7 Temperature distributions for a range in effective temperature. (Models from Carbon and Gingerich 1969.)

in the cool models. A similar comparison of models having different surface gravity shows $T(\tau_0)$ is independent of g to within a few per cent.

Suppose we desire $T(\tau)$ for a model having a specified T_{eff}. Can we scale $T_\odot(\tau_0)$ by the ratio of effective temperatures as we did in the grey case? The answer appears to be yes. We draw this conclusion from Fig. 9-9 where the S_0's are obtained from a match of theoretical $T(\tau_0)$'s to $T_\odot(\tau_0)$. The straight line represents the $S_0 = T_{eff}/T_{eff}^\odot$ relation. The points fall close to this line with some deviation to higher T_{eff} for the coolest and hottest models. A diagram similar to Fig. 9-9 was published by Gray (1967) and showed that the $T_{eff}/T_{eff}^\odot$ scaling did not hold. But in the years following this result, the flux constant models have been improved primarily through the inclusion of

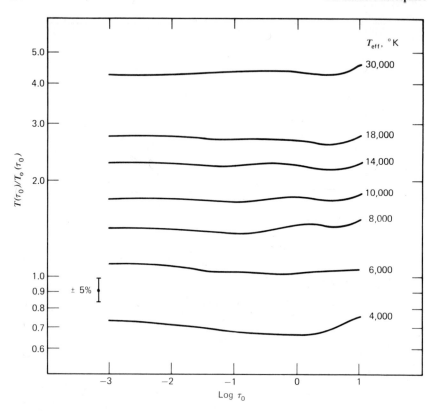

Fig. 9-8 The ratios of temperature distributions. $T(\tau_0)$ from Fig. 9-7 has been divided by the solar temperature distribution of Table 9-2. A constant ratio would indicate exact scaling.

better and more inclusive absorption coefficients. Those improvements have brought the flux constancy results closer to the scaled $T(\tau)$. Additional support for the validity of using the scaling approach is found in the work of Linsky and Ayres (1973). They find the temperature minimum scales by the ratio of effective temperatures for Procyon and the sun.

In many applications the scaling factor S_0 can be arbitrarily adjusted without asking what T_{eff} the model has. The effective temperature is an unneeded parameter in these cases. Examples are discussed in Chapter 15.

The observed solar $T(\tau_0)$ has, of course, the effects of line blanketing and convection in it. Scaling the solar distribution then includes some allowance for these effects. This is proper and advantageous when dealing with normal stars. For metal poor or metal rich stars, one must expect deviations in $T(\tau_0)$ in response to the different electron pressure and opacity. According

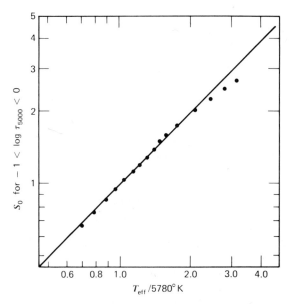

Fig. 9-9 The average value of $T(\tau_0)/T_\odot(\tau_0)$ in the range $-1 < \log \tau_0 < 0$ has been taken from Fig. 9-8 (along with additional models) and adopted as the scaling factor S_0. There is a very close relation between S_0 and T_{eff}.

to the calculations of Carbon and Gingerich (1969), at $T_{\text{eff}} = 6000°$K the effect is very small.

The method of scaling the solar temperature distribution has been used by many people including Cayrel and Jugaku (1963), Warner (1964), Cayrel and Cayrel (1963), and Gray (1967).

At this point we assume $T(\tau_0)$ has been determined by one of these two methods, and we turn our attention to the calculation of electron pressure.

THE P_g-P_e-T RELATION

When solving eq. 9-2 for the gas pressure, we start with an initial guess at $P_g(\tau_0)$. We then need $P_e(\tau_0)$ given $P_g(\tau_0)$ and not the other way around. The number of electrons generated in the photospheric material depends on the chemical composition. We let the number of ions per cubic centimeter of the jth element be N_{1j} and the number of neutrals per cubic centimeter, N_{0j}. Then the ionization eq. 1-17 is

$$\frac{N_{1j}}{N_{0j}} = \frac{\Phi_j(T)}{P_e}.$$

If double ionizations are neglected,* we can write $N_{1j} = N_{ej}$ = the number of electrons per cubic centimeter contributed by the jth element, and

$$\frac{\Phi_j(T)}{P_e} = \frac{N_{ej}}{N_{0j}}$$

$$= \frac{N_{ej}}{N_j - N_{ej}}.$$

The total number of jth element particles is $N_j = N_{1j} + N_{0j}$. We solve for N_{ej}.

$$N_{ej} = N_j \frac{\Phi_j(T)/P_e}{1 + \Phi_j(T)/P_e}.$$

The pressures are

$$P_e = \sum_j N_{ej} kT$$

$$P_g = \sum_j (N_{ej} + N_j) kT$$

which we ratio as follows

$$\frac{P_e}{P_g} = \frac{\sum N_{ej} kT}{\sum (N_{ej} + N_j) kT}$$

$$= \frac{\sum N_j \{\Phi_j(T)/P_e [1 + \Phi_j(T)/P_e]\}}{\sum N_j (1 + \{\Phi_j(T)/P_e [1 + \Phi_j(T)/P_e]\})}.$$

We introduce the number abundance of the element, $A_j \equiv N_j/N_H$ where N_H is the number of hydrogen particles per cubic centimeter. The relation we need is

$$P_e = P_g \frac{\sum A_j \{\Phi_j(T)/P_e [1 + \Phi_j(T)/P_e]\}}{\sum A_j (1 + \{\Phi_j(T)/P_e [1 + \Phi_j(T)/P_e]\})}. \tag{9-6}$$

The equation is transcendental in P_e and is solved by iteration. Notice that the $\Phi_j(T)$'s are constants for such an iteration.

The results for the solar chemical composition are shown in Fig. 9-10. Hydrogen controls the relation between P_e and P_g at high temperatures and,

---

* When double ionizations are important, the temperatures are so high that hydrogen is ionized and, for normal chemical compositions, will dominate P_e.

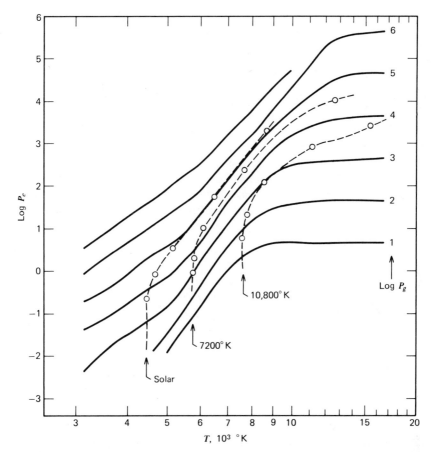

Fig. 9-10 The relation among P_g, P_e, and T. Lines of constant gas pressure are labeled with log P_g. The dashed lines show the relations for three models. The surface gravity is 10^4 cm/sec^2 for the two hotter models. Optical depth, τ_0, in decades is denoted by the circles with the lowest circle at $\tau_0 = 10^{-3}$ and the highest at 10. The cooler models show a nearly constant P_g behavior over a significant range in depth.

of course, $P_g = 2P_e$ in that case. The metals dominate at cooler temperatures where hydrogen is not ionized and then the P_e–P_g–T relation is sensitive to the chemical makeup of the model. It is necessary to include C, Si, Fe, Mg, Ni, Cr, Ca, Na, and K when dealing with the ionization equilibrium of models having normal chemical composition. We may find ourselves in a new iterative situation if the chemical composition of the star we are modeling comes out, after analysis, to be different from what we assume in the P_e calculation.

COMPLETION OF THE MODEL

Now we are in a position to compute eq. 9-2. Taking $T(\tau_0)$ and our guess at $P_g(\tau_0)$, we compute $P_e(\tau_0)$ and then $\kappa_0(\tau_0)$. Then eq. 9-2 gives $P_g(\tau_0)$ and we have completed the first cycle of the pressure iteration.* Integration is continued to convergence and we have a completed model for the specified $T(\tau_0)$. If we are using a scaled solar temperature distribution, we are finished. The model has been made and we are ready to calculate auxiliary information from the model (such as the geometrical depth and surface flux). If we are determining $T(\tau_0)$ by the flux constancy condition, we are not done but must calculate $\mathscr{F}_\nu(\tau_\nu)$ by eq. 7-15 for many frequencies across the spectrum so that $\int_0^\infty \mathscr{F}_\nu \, d\nu$ can be properly evaluated as a function of depth. The temperature distribution is then corrected and the model is recomputed until flux constancy is achieved.

THE GEOMETRICAL DEPTH SCALE

The geometrical depth can be computed from $dx = d\tau_0 / \kappa_0 \rho$,

$$
\begin{aligned}
x(\tau_0) &= \int_0^{\tau_0} \frac{dt_0}{\kappa_0 \rho} \\
&= \int_{-\infty}^{\log \tau_0} \frac{kT t_0}{\kappa_0 \sum A_j \mu_j P_g} \frac{d \log t_0}{\log e}
\end{aligned}
\tag{9-7}
$$

where ρ has been replaced by using the perfect gas law, eq. 1-6, and $\sum A_j \mu_j$ is the number of grams per hydrogen particle as in eq. 8-17. An alternative method is to integrate on a P_g scale starting with $dP_g = \rho g \, dx$. Then

$$
x(P_g) = \frac{1}{g} \int_0^{P_g} \frac{kT}{\sum A_j \mu_j} \frac{dp}{p}.
\tag{9-8}
$$

This is an interesting form because it shows that the thickness of the atmosphere is inversely proportional to the surface gravity since $T(P_g)$ is essentially independent of g. The run of geometrical depth with optical depth is nearly linear as shown in Fig. 9-11.

---

* At the outer boundary, when starting the integration, it is necessary to make some simple trapezoidal integrations for a step or two in $\log \tau_0$. After that, Simpson's rule can be used. Usually a few extra steps are used at the top of the photospheric model to ensure the pressure at the depths where we need it, is independent of how we start the integration.

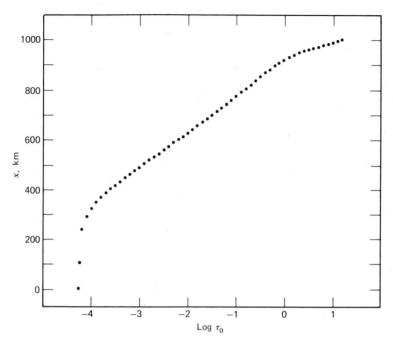

Fig. 9-11 Geometrical depth is nearly linear with logarithmic optical depth. The behavior higher than $\log \tau_0 = -4$ is a result of starting the integration and does not represent physical behavior.

COMPUTATION OF THE SPECTRUM

The spectrum is computed using eq. 7-16,

$$\mathscr{F}_v = 2\pi \int_{-\infty}^{\infty} S_v(\tau_0) E_2(\tau_v) \frac{\kappa_v \tau_0}{\kappa_0} \frac{d \log \tau_0}{\log e} \tag{9-9}$$

with $S_v(\tau_0) = B_v[T(\tau_0)]$ in the LTE case and where the optical depth scales are related by

$$dx = \frac{d\tau_v}{\kappa_v \rho} = \frac{d\tau_0}{\kappa_0 \rho}$$

or

$$\tau_v = \int_0^{\tau_0} \frac{\kappa_v(t_0)}{\kappa_0(t_0)} dt_0 = \int_{-\infty}^{\log \tau_0} \frac{\kappa_v(t_0)}{\kappa_0(t_0)} t_0 \frac{d \log t_0}{\log e}.$$

The amount of light contributed to the surface flux as a function of depth in the atmosphere (on a log τ_0 scale) is specified by the integrand of eq. 9-9. We call this integrand the flux contribution function. It is similar to the contribution function used with I_ν but now includes an integration over the stellar disc. In Fig. 9-12 are representative flux contribution functions. They show where in depth the surface flux originates and where, for example, $T(\tau_0)$ must be defined in order to compute the surface flux. The concept of a depth of formation has meaning in this regard. The contribution functions in Fig. 9-12 indicate that the radiation at 8000 Å is formed higher than the radiation at 5000 Å. This reflects in part the smaller absorption at 5000 Å. A good illustration of the depth of formation concept arises in discussing the behavior of the model flux above and below the Balmer discontinuity. As shown in Fig. 9-13, the flux at 3660 Å arises from significantly deeper layers than the flux at 3640 Å. The depth of formation is dependent upon

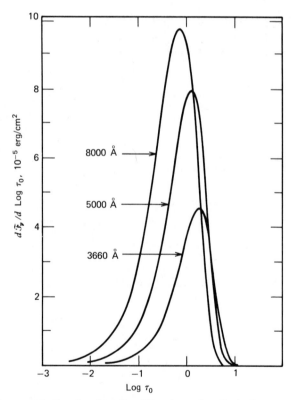

Fig. 9-12   Flux contribution functions for different wavelengths in the spectrum of a solar model.

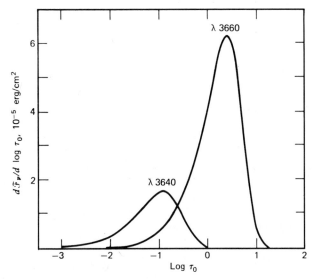

Fig. 9-13 The depth of formation is very different for flux on opposite sides of the Balmer discontinuity.

the temperature of the model. Changes such as those in Fig. 9-14 are typical. Surface gravity effects are much smaller, being barely noticeable in the hotter models and completely negligible in the cooler ones.

There are other techniques for computing the flux and they generally lead to other integrands. For example, suppose we integrate eq. 7-16 by parts to obtain

$$\mathcal{F}_\nu = \pi S_\nu(0) + 2\pi \int_0^\infty \frac{dS_\nu}{d\tau_\nu} E_3(\tau_\nu) \, d\tau_\nu. \tag{9-10}$$

Again, S_ν is replaced by $B_\nu(T)$ in LTE and $S_\nu(0) = B_\nu(T_0)$ where T_0 is the boundary temperature. The formulation 9-10 is interesting because it shows that the flux arises from the gradient of the source function. Greater contributions to the flux occur in models where $dS_\nu/d\tau_\nu$ is larger.

Notice, however, that in this formulation we have the curious situation of the flux depending on T_0 even though there may be no flux contributed from the boundary. Furthermore, the integrand $(dS_\nu/d\tau_\nu) E_3(\tau_\nu)$ is not an increment of flux. It follows that this integrand cannot be used as a contribution function in the usual sense of the term (see also Edmonds 1969).

Continuous spectra from these types of models agree reasonably well with the observations. The details of model continua as they relate to real stars are discussed in Chapter 10.

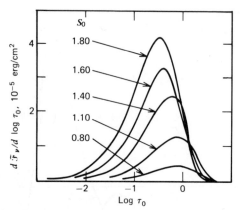

Fig. 9-14   Contribution functions at $\lambda7000$ as a function of model temperature. Hotter models show more flux and a shift toward higher layers.

PROPERTIES OF MODELS: PRESSURE RELATIONS

The relation between temperature and gas pressure for four models is shown in Fig. 9-15. It is clear that $T(P_g)$ does not scale the way $T(\tau_0)$ does. Although P_g could have been used as an independent variable in the integration of the hydrostatic equation, the scaling procedure would then be inapplicable. A surface gravity increase moves the relations of Fig. 9-15 to larger pressures as explained below, but the shape of each relation stays the same.

It is in the log T-log P_g plane where the slope exceeding $1 - 1/\gamma \sim 0.4$ implies instability to convection. Many models are convectively unstable by this criterion and, as indicated earlier in the chapter, convection should then be included in the energy transport. The $T(P_g)$ relation is altered by the convection, but the gradient does not normally reach the adiabatic value until below the photosphere because radiative transport is still competitive with the convection at the relatively low photospheric densities. The models of Fig. 9-6 are shown in Fig. 9-16. The radiative calculation gives the very high temperatures. The lower temperature relations depend on the value chosen for the mixing length.

The depth dependence of gas pressure and electron pressure are shown in Figs. 9-17 and 9-18. They are typical of cool star models. Both of these pressures increase with depth, but the electron pressure increases much more due to the enhanced level of ionization in deeper layers.

Both P_g and P_e increase at all optical depths when the surface gravity is increased. The inserts in Figs. 9-17 and 9-18 show the gravity dependence at fixed optical depths. The P_g dependence is very linear and can be written

$$P_g \doteq \text{const } g^p \tag{9-11}$$

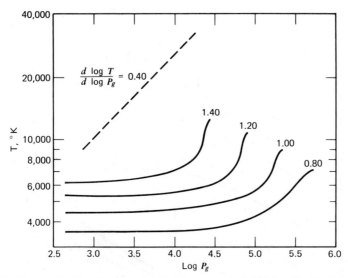

Fig. 9-15 The log T − log P_g plane showing four models (labeled with S_0). The $1 − 1/\gamma$ temperature gradient is indicated.

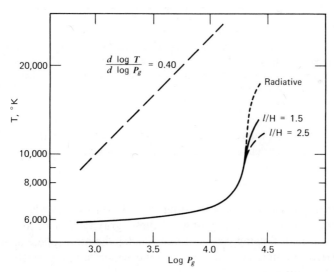

Fig. 9-16 The log T − log P_g relations for the models of Fig. 9-6.

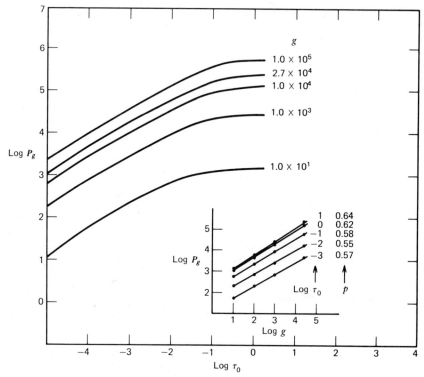

Fig. 9-17 The gas pressure shows a nearly linear relation with logarithmic optical depth but levels off in the deep layers. Each line represents a model with the solar temperature and the surface gravity shown to the right of the curve. The insert shows that log P_g at any given depth varies linearly with log g. The slope p is indicated.

where the power, p, ranges from 0.64 to 0.57 in going from deep to shallow layers. The P_e dependence is not quite so linear, but we may still approximate the relation by

$$P_e \doteq \text{const } g^p \qquad (9\text{-}12)$$

where p ranges from 0.30 to 0.45 going from deep to shallow layers.

An approximate understanding of these gravity dependences can be obtained in the following way. Let us consider the atmosphere as if it were composed of a single typical element (one which supplies electrons!) and reduce eq. 9-6 to

$$P_e = \frac{\Phi(T)/P_e}{1 + 2\Phi(T)/P_e} P_g$$

or

$$P_e^2 = \Phi(T)P_g - 2\Phi(T)P_e.$$

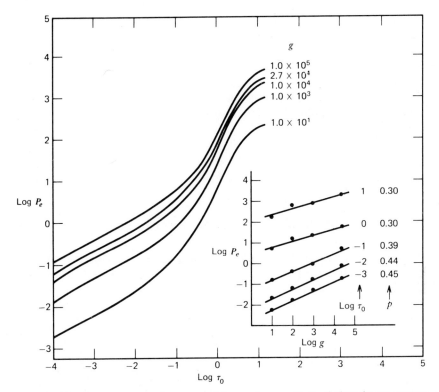

Fig. 9-18   The electron pressure shows a much larger change with depth than the gas pressure (Fig. 9-17). Each line represents a model with the solar temperature and the surface gravity shown next to each curve. The insert shows that log P_e at any given depth varies more or less linearly with log g. The slope p is indicated.

But, as we have seen, $P_e \ll P_g$, in solar-type stars, so the expression reduces to

$$P_e^2 \doteq \Phi(T)P_g. \tag{9-13}$$

For situations in which the temperature is constant, we expect the electron pressure to be proportional to the square root of the gas pressure.

The κ_0 in eq. 9-2 is proportional to P_e, and therefore to $P_g^{1/2}$, for stars in which the negative hydrogen ion dominates the opacity. The integrand of eq. 9-2 is then independent of pressure and

$$P_g = g^{2/3} \left[\frac{3}{2} \int_{-\infty}^{\log \tau_0} \frac{t_0 P_g^{1/2}}{\kappa_0 \log e} \, d \log t_0 \right]^{2/3}$$

$$= C(T)g^{2/3}.$$

That is, for a constant temperature, $C(T)$ is constant and P_g varies according to eq. 9-11 with $p = \frac{2}{3}$. Then by eq. 9-13, we must have

$$P_e \doteq \text{const } g^{1/3}$$

in agreement with eq. 9-12.

In hotter models hydrogen becomes ionized and $P_e \doteq \frac{1}{2}P_g$. The absorption coefficient goes with increasing temperature from being proportional to P_e for the negative hydrogen ion to being independent of P_e for neutral hydrogen to being proportional to P_e again for electron scattering. The exponent p in eq. 9-11 ranges from ~ 0.6 to ~ 0.9 and back again over this temperature sequence.

THE EFFECT OF CHEMICAL COMPOSITION

The pressure structure in a model is also dependent on the chemical composition. The nature of the dependence can be seen in Figs. 9-19 and 9-20. A cooler model is shown to enhance the effect of the different composition. Clearly, as one considers hotter and hotter models, hydrogen takes over

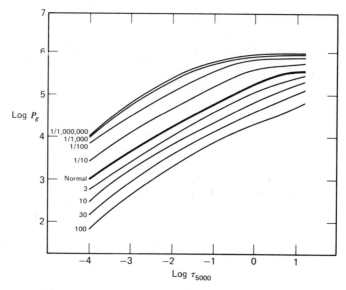

Fig. 9-19 An increase in metal abundance (factor shown) leads to a decrease in gas pressure at each optical depth for the cool models shown. This is a result of the increase in opacity caused by the increase in electron donors.

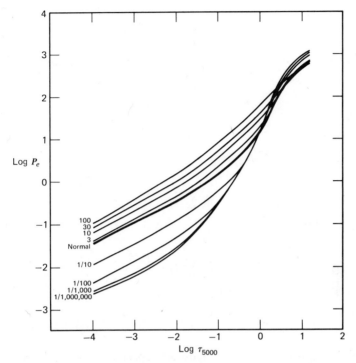

Fig. 9-20 An increase in metal abundance (factor shown) leads to an increase in the electron pressure at most depths in the cool models shown. The interesting crossover in the deeper layers arises because P_e is set there by the hydrogen ionization. If the metals are overabundant, a given τ_0 builds up sooner along the line of sight and we do not see as deeply into the ionized hydrogen region.

as the electron donor and the chemical composition enters only weakly. Increasing the metals increases the number of free electrons, which results in a larger continuous absorption coefficient and finally a shorter optical penetration into the atmosphere. The gas pressure at a given optical depth must decrease with increasing metal content. Suppose we write, using $\sum A_j$ for the sum of the metal abundances,

$$\frac{P_e}{P_g} \frac{1}{\sum A_j} = \frac{P_e}{P_g} \frac{N_H}{\sum N_j} \frac{kT}{kT} = \frac{P_e}{P_g} \frac{P_H}{\sum N_j kT} = \frac{P_e}{\sum N_j kT}$$

where P_H is the partial pressure of hydrogen. But the sum of the element numbers is comprised of the sum of both ions and neutrals

$$\sum N_j = \sum (N_1 + N_0)_j$$

and

$$P_e = N_e kT = \sum N_{1j} kT$$

assuming single ionizations. So finally the original expression becomes

$$\frac{P_e}{P_g} \frac{1}{\sum A_j} = \frac{\sum N_{1j}}{\sum (N_1 + N_0)_j}.$$

There are two obvious cases as follows.

1. Metals ionized; $\sum N_{1j} \gg \sum N_{0j}$ and

$$\frac{P_e}{P_g} \frac{1}{\sum A_j} = 1. \qquad (9\text{-}14)$$

2. Metals neutral; $\sum N_{1j} \ll \sum (N_1 + N_0)_j$ and

$$\frac{P_e}{P_g} \frac{1}{\sum A_j} \ll 1.$$

For the sun, case 1 is applicable and we write

$$dP_g = \frac{g}{\kappa_0} d\tau_0 = \frac{g}{P_e \kappa_0 / P_e} d\tau_0 = \frac{g}{P_g \sum A_j \kappa_0 / P_e} d\tau_0.$$

We assume κ_0 is dominated by hydrogen opacity and does not depend on the metal abundance except through the electron pressure. Upon integration,

$$\frac{1}{2} P_g^2 = \frac{1}{\sum A_j} \int_0^{\tau_0} \frac{g}{\kappa_0 / P_e} d\tau_0$$

and since κ_0 / P_e is independent of pressure and g is being held constant,

$$P_g = a(T)(\sum A_j)^{-1/2} \qquad (9\text{-}15)$$

where $a(T)$ is a constant when T is fixed. Then from eq. 9-14 it follows that

$$P_e = \text{const} \, (\sum A_j)^{1/2}. \qquad (9\text{-}16)$$

While these expressions are certainly not rigorous, they do give the proper qualitative behavior when hydrogen is not an electron donor. In reality hydrogen does take over in the deeper photospheric layers even for relatively cool models such as those shown in the Figs. 9-19 and 9-20. A similar derivation for case 2 above leads to

$$P_g = \text{const} \, (\sum A_j)^{-1/3}$$
$$P_e = \text{const} \, (\sum A_j)^{1/3}.$$

It is also instructive to compare the effects of g with the effects of $\sum A_j$ on the electron pressure as shown in Figs. 9-18 and 9-20. In the higher layers, $\log \tau_0 \sim -2$, decreasing either of these variables produces similar results. In the deeper layers where hydrogen is ionized, the chemical composition has only a small effect. We conclude that changes in chemical composition can mimic surface gravity effects (and vice versa) for those spectral features formed in the upper photosphere.

The helium abundance can alter the pressure structure by a small amount. The effect comes in through the sum of the atomic weights, $\sum A_j \mu_j$, used to reduce κ per hydrogen particle to κ_v per gram. By using eq. 8-17, we can write

$$dP_g = \frac{g}{\kappa_0} d\tau_0$$

$$= \frac{g}{\kappa} \sum A_j \mu_j \, d\tau_0$$

where the unsubscripted κ is the square centimeter per hydrogen particle absorption coefficient. We see that g and $\sum A_j \mu_j$ play the same role. Since by eq. 9-11 $P_g \doteq \text{const } g^{2/3}$, we expect

$$P_g = \text{const } (\sum A_j \mu_j)^{2/3}.$$

But

$$\sum A_j \mu_j = A(\text{H})\mu_{\text{H}} + A(\text{He})\mu_{\text{He}} + \cdots$$
$$\doteq [1 + 4A(\text{He})] \times 1.66 \times 10^{-24} \text{ g}$$

since the other elements contribute only a very small mass. Therefore

$$P_g \doteq \text{const } [1 + 4A(\text{He})]^{2/3}$$

$$P_e \doteq \text{const } [1 + 4A(\text{He})]^{1/3}$$

(9-17)

describes the dependence of the pressures on the helium abundance.

CHANGES WITH TEMPERATURE

Finally let us turn our attention to the temperature dependence of the pressure structure. The model data are presented in Figs. 9-21 and 9-22. The run of pressure at a given optical depth is a useful quantity, especially when we want to understand the behavior of spectral lines (Chapter 13). The gas pressure decreases at a given τ_0 as we consider hotter and hotter

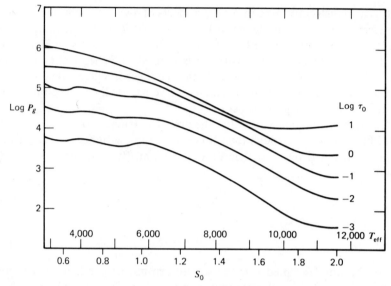

Fig. 9-21   The gas pressure decreases at each optical depth as the model temperature rises.

models. The electron pressure does the reverse. For the cooler stars we find it useful to write

$$P_e = \text{const } e^{\Omega T} \tag{9-18}$$

as an approximation to the curves in Fig. 9-22. The values of Ω are shown and are in the range of 0.001 deg^{-1}.

MODEL TABULATIONS

A small grid of models is given Table 9-3. These are scaled solar models based on the temperature distribution given in Table 9-2 [also the $T(\tau)$ column in the $S_0 = 1.00$ model of Table 9-3]. From these tabulations and the graphs on the preceding pages you can see what the numbers are like in model photospheres.

A great many grids of models have been published. Two of the most used are those of Carbon and Gingerich (1969) and Mihalas (1966). Models of hot stars can be found in the papers by Van Citters and Morton (1970), Underhill (1968), and Strom and Avrett (1965). Cool star models have been constructed by Tsuji (1966), Gingerich et al. (1966), and Myerscough (1968), for example. Models explicitly for giants have been computed by Parsons

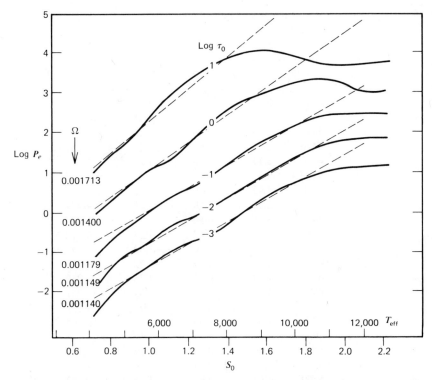

Fig. 9-22 The electron pressure increases with increasing model temperature until a leveling off occurs. During the increase, the behavior is approximately linear. The Ω of eq. 9-18 is shown on the left.

(1968), for supergiants by Groth (1961) and Osmer (1972), and for helium rich cases by Böhm-Vitense (1967) and Klinglesmith and Hunger (1969). White dwarf atmospheres are considered by Terashita and Matsushima (1966, 1969), by Weidemann (1968), and by Bues (1973). And the list goes on. But in many cases you may find it more useful to compute your own models.

REFLECTION

The models we have considered here are simple but moderately realistic. They are all determinate upon specification of the temperature parameter (S_0 or T_{eff}, for example), the pressure parameter (g), and the chemical composition.

Table 9-3 Model Photospheres*

$\log \tau_0$	T	$\log P_g$	$\log P_e$	$\log \kappa_0/P_e$	x
	$S_0 = 1.00$		$g = 2.74 \times 10^4$	$A(\text{He}) = 0.10$	
−4.4	4440	2.17	−1.81	−1.24	−918
−4.2	4440	2.67	−1.37	−1.25	−680
−4.0	4440	2.90	−1.16	−1.25	−595
−3.8	4440	3.06	−1.02	−1.25	−551
−3.6	4440	3.19	−0.90	−1.25	−516
−3.4	4440	3.31	−0.80	−1.26	−484
−3.2	4440	3.43	−0.70	−1.26	−455
−3.0	4440	3.54	−0.60	−1.26	−427
−2.8	4456	3.66	−0.50	−1.27	−400
−2.6	4480	3.77	−0.40	−1.28	−372
−2.4	4508	3.88	−0.30	−1.30	−345
−2.2	4552	3.99	−0.19	−1.32	−317
−2.0	4611	4.10	−0.07	−1.34	−289
−1.8	4684	4.22	0.05	−1.37	−261
−1.6	4773	4.33	0.17	−1.41	−233
−1.4	4874	4.44	0.30	−1.44	−204
−1.2	4995	4.55	0.44	−1.50	−175
−1.0	5132	4.66	0.58	−1.54	−145
−0.8	5294	4.76	0.74	−1.59	−114
−0.6	5490	4.87	0.92	−1.66	−83
−0.4	5733	4.97	1.14	−1.73	−53
−0.2	6043	5.05	1.43	−1.81	−24
0.0	6429	5.11	1.78	−1.90	0
0.2	6904	5.16	2.18	−2.01	20
0.4	7467	5.20	2.60	−2.12	35
0.6	7962	5.22	2.93	−2.19	47
0.8	8358	5.25	3.17	−2.24	58
1.0	8630	5.27	3.32	−2.26	68
1.2	8811	5.30	3.43	−2.27	80

(continued)

* Integrations were done with 10 steps/decade in τ_0 where τ_0 is the optical depth at 5000 A. Normal solar composition.

Table 9-3 (*continued*)

$\log \tau_0$	T	$\log P_g$	$\log P_e$	$\log \kappa_0/P_e$	x
	$S_0 = 1.5$	$g = 1.0 \times 10^4$		$A(\text{He}) = 0.10$	
−4.0	6660	1.41	0.06	−0.41	−2642
−3.0	6661	2.35	0.55	−0.86	−1607
−2.0	6917	3.04	1.10	−1.11	−895
−1.0	7699	3.51	1.90	−1.16	−358
0.0	9643	3.77	2.96	−0.99	0
1.0	12945	3.88	3.54	−1.12	287
	$S_0 = 0.70$	$g = 4.0 \times 10^4$		$A(\text{He}) = 0.10$	
−4.0	3108	3.24	−2.08	−0.54	−300
−3.0	3108	3.91	−1.52	−0.54	−214
−2.0	3227	4.47	−0.99	−0.63	−146
−1.0	3593	5.02	−0.30	−0.87	−74
0.0	4500	5.50	0.83	−1.31	0
1.0	6041	5.87	1.95	−1.82	73
	$S_0 = 1.00$	$g = 1.0 \times 10^3$		$A(\text{He}) = 0.10$	
−4.0	4440	2.13	−1.85	−1.24	−15081
−3.0	4440	2.78	−1.26	−1.25	−10604
−2.0	4611	3.32	−0.73	−1.33	−7082
−1.0	5132	3.86	−0.05	−1.53	−3311
0.0	6429	4.25	1.32	−1.89	0
1.0	8630	4.36	2.86	−2.12	1316

A reflection back on the original assumptions listed at the beginning of the chapter tells us how such models can be improved. Stars having extended atmospheres, as giants to supergiants do, may be better described in spherical rather than plane parallel geometry. When mass loss and other aerodynamic phenomena occur, we must question the validity of hydrostatic equilibrium, and so on down the list. In short, the reader need to extend these concepts of a model to suit his own research and understanding. We make some elementary extensions of these models in Chapters 17 (Stellar Rotation) and 18 (Turbulence).

REFERENCES

Allen, C. W. 1970. *Mon. Not. Roy. Astron. Soc.* **148**, 435.

Athay, R. G. 1970. *Astrophys. J.* **161**, 713.

Avrett, E. H., and M. Krook 1963a. *Astrophys. J.* **137**, 874.

Avrett, E. H., and M. Krook 1963b. *J. Quant. Spectrosc. Radiat. Transfer*, **3**, 107.

Avrett, E. H. 1964. *Proceedings First Harvard–Smithsonian Conference on Stellar Atmospheres*, Smithsonian Astrophysical Observatory Special Report No. 167, Cambridge, Mass., p. 83.

Böhm, K. H. 1966. *J. Quant. Spectrosc. Radiat. Transfer* **6**, 534.

Böhm-Vitense, E. 1967. *Astrophys. J.* **150**, 483.

Bues, I. 1973. *Astron. Astrophys.* **28**, 181.

Carbon, D. F., and O. Gingerich 1969. *Theory and Observation of Normal Stellar Atmospheres*, O. Gingerich, Ed., MIT Press, Cambridge, Mass., p. 377.

Carbon, D. F. 1974. *Astrophys. J.* **187**, 135.

Cayrel, R. and G. Cayrel 1963. *Astrophys. J.* **137**, 447.

Cayrel, R. and J. Jugaku 1963. *Ann. Astrophys.* **26**, 495.

Edmonds, F. N. Jr. 1969. *J. Quant. Spectrosc. Radiat. Transfer* **9**, 1427.

Elste, G. 1967. *Astrophys. J.* **148**, 857.

Elste, G. 1968. *Sol. Phys.* **3**, 106.

Fowler, J. W. 1974. *Astrophys. J.* **188**, 295.

Gingerich, O., D. W. Latham, J. Linsky, and S. S. Kumar 1966. *Colloquium on Late-Type Stars*, M. Hack, Ed., Osservatorio Astronomico, Trieste, p. 291.

Gingerich, O. 1964. *Proceedings First Harvard-Smithsonian Conference on Stellar Atmospheres*, Smithsonian Astrophysical Observatory Special Report No. 167, Cambridge, Mass.

Gray, D. F. 1967. *Astrophys. J.* **149**, 317.

Groth, H. G. 1961. *Z. Astrophys.* **51**, 231.

Klinglesmith, D. A. and K. Hunger 1969. *Theory and Observation of Normal Stellar Atmospheres*, O. Gingerich, Ed., MIT Press, Cambridge, Mass., p. 161.

Lambert, D. L. 1971. *Phil. Trans. Roy. Soc. Lond. Ser. A* **270**, 3.

Linsky, J. L. and T. Ayres, 1973. *Astrophys. J.* **180**, 473.

Mihalas, D. 1966. *Astrophys. J. Suppl.* **13**, 1.

Mihalas, D. 1967. *Methods in Computational Physics*, Vol. 7, B. Alder, S. Fernback, and M. Rotenberg, Eds., Academic, New York, p. 1.

Mihalas, D. 1970. *Stellar Atmospheres*, W. H. Freeman, San Francisco.

Mihalas, D. and D. C. Morton 1965. *Astrophys. J.* **142**, 253.

Mitchell, W. E. Jr. 1959. *Astrophys. J.* **129**, 93.

Myerscough, V. P. 1968. *Astrophys. J.* **153**, 421.

Osmer, P. S. 1972. *Astrophys. J. Suppl.* **24**, 255.

Parsons, S. B. 1968. *Astrophys. J. Suppl.* **18**, 127.

Pecker, J. C. 1973. *Problems of Calibration of Absolute Magnitudes and Temperature of Stars*, IAU Symposium No. 54, B. Hauck and B. E. Westerlund, Eds., Reidel, Dordrecht, p. 173.

Pierce, A. K. and J. H. Waddell 1961. *Mem. Roy. Astron. Soc.* **68**, 89.

Strom, S. E. and E. H. Avrett. 1965. *Astrophys. J. Suppl.* **12**, 1.

Strom, S. E. and R. L. Kurucz 1966. *J. Quant. Spectrosc. Radiat. Transfer* **6**, 591.

Terashita, Y. and S. Matsushima 1966. *Astrophys. J. Suppl.* **13**, 461.

Terashita, Y. and S. Matsushima 1969. *Astrophys. J.* **156**, 203.

Tsuji, T. 1966. *Colloquium on Late-Type Stars*, M. Hack, Ed. Osservatorio Astronomico, Trieste, p. 260.

Underhill, A. B. 1968. *Bull. Astron. Inst. Neth.* **19**, 500.

Van Citters, G. W. and D. C. Morton 1970. *Astrophys. J.* **161**, 695.

Vernazza, J. E., E. H. Avrett, and R. Loeser 1973. *Astrophys. J.* **184**, 605.

Warner, B. 1964. *Observatory* **84**, 258.

Wiedemann, V. 1968. *Annu. Rev. Astron. Astrophys.* **6**, 351.

CHAPTER TEN

THE MEASUREMENT AND BEHAVIOR OF STELLAR CONTINUA

Through the visible window of the earth's atmosphere, 3300 to $\sim$ 12,000 Å, we measure portions of a stellar continuum. The primary aim is to obtain the photospheric temperature scale. In the hotter stars the shape of the continuum is molded by the bound–free absorption of neutral hydrogen. We can measure pieces of the Balmer and Brackett continua and the complete Paschen continuum. The Balmer and Paschen discontinuities are pressure sensitive in the temperature range of late A and F stars. In cooler stars where the negative hydrogen ion dominates, we are interested in the continuum from about 4000 Å and longer for establishing the temperature. For these cool stars there are no significant pressure effects in the continuum.

Our equipment must be capable of measuring radiation over a wide wavelength range either simultaneously or sequentially. A star of interest is normally measured in two or three steps: (1) comparison of the star's continuum to a standard star, (2) calibration of the shape of the standard star's continuum, and, in some cases, (3) calibration of the absolute flux at one or more points in the continuum of the standard star. We are most concerned with the execution of step 1 and the various types of stellar continua that are observed or calculated.

SCANNERS AND POLYCHROMATORS

Photoelectric photometers having low, but precisely defined, resolution are suitable for measuring the continuum. The optical layouts of three different scanning monochromators or scanners are shown in Figs. 10-1, 10-2, and 10-3. These instruments are used at the Cassegrain focus. The first system is a modified Czerny–Turner configuration and is representative of the majority of scanners. The spectrum is scanned over the exit slit by rotating the grating. A reference photomultiplier monitors the seeing fluctuations. By rationing the signal from the spectrum measuring photomultiplier with the reference channel signal, these fluctuations are largely eliminated. In the Gillieson (Gillieson 1949, Schroeder 1966) system shown in Fig. 10-2, the grating is illuminated in convergent light. Some coma is introduced by doing this (Murty 1962), but it is tolerable in low resolution work such as this. A still

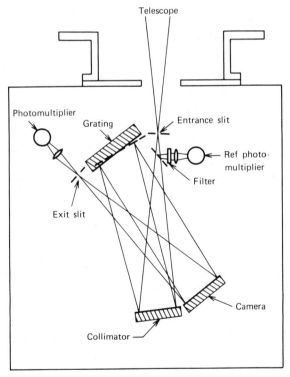

Fig. 10-1   A Cassegrain spectrum scanner using a separate collimator and camera and a plane diffraction grating. The spectrum is moved across the exit slit by rotating the grating. (Adapted from Willstrop 1965.)

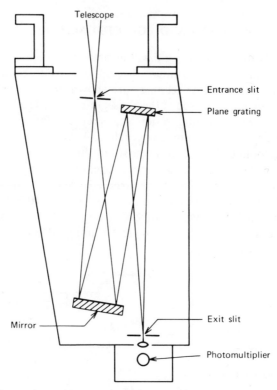

Fig. 10-2 A Cassegrain spectrum scanner using one mirror and a plane diffraction grating. The spectrum is scanned by grating rotation about an axis displaced from the grating.

simpler system (Fig. 10-3), and probably the only astronomical scanner of its kind, has been built by Liller (1963). A single optical element, the concave grating, is used. The system is extremely compact having a 600 lines/mm grating of 25 mm × 30 mm ruled area and a 11.5 cm radius of curvature. Other scanners have been built by Liller and Aller (1954), Tull (1963), Oke and Greenstein (1961), and Wampler (1966) to cite a few. A novel scanning filter photometer has been put into use by Trodahl et al. (1973).

The entrance aperture of a scanner is made small enough to isolate the image of the star and to exclude as much sky background as possible. But it is large enough to include the full seeing disc of the star.

The spectral purity for the conventional spectrograph having an entrance slit W' is given by eq. 3-21,

$$\Delta\lambda = -W' \frac{\cos\alpha}{f_{\text{coll}}} \frac{d}{n}$$

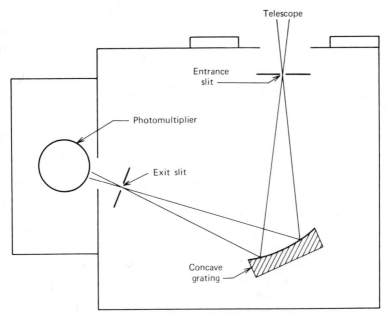

Fig. 10-3 A Cassegrain spectrum scanner using a concave grating. The spectrum is scanned by rotating the grating.

where α is the angle of incidence on the grating, f_{coll} is the collimator focal length, and d is the grating ruling spacing for the nth order spectrum. In this case W' is replaced by the seeing disc of the star. Let us suppose the seeing disc has an angular size $\Delta\phi$ and the telescope effective focal length is F_{eff}. Then $W' = \Delta\phi F$ and the spectral purity is

$$\Delta\lambda = -\cos\alpha\,\frac{F_{eff}}{f_{coll}}\frac{d}{n}\,\Delta\phi. \tag{10-1}$$

We see that seeing fluctuations result in spectral purity fluctuations, but the effect can be reduced by making the ratio F_{eff}/f_{coll} as small as possible. This is equivalent to making the ratio of telescope aperture to collimated beam size as small as possible (since the same f ratio is involved for both). It is to accomplish this, that some scanners employ an inverted Cassegrain configuration for a collimator, namely, a large effective collimator focal length can be incorporated into a small space.

In other than the usual mirror configurations, these same conclusions hold, but the $\cos\alpha$ term may be more complicated. For example, in a Gillieson system one has

$$\Delta\lambda = \frac{-\cos\alpha}{(1 + 4\tan^2\beta)^{1/2}} \frac{F_{eff}}{f_{coll}} \frac{d}{n} \Delta\phi \tag{10-2}$$

where β is the diffraction angle of the grating and f_{coll} is the focal length of the only mirror. Typically $\Delta\lambda$ is an angstrom or two.

The use of a narrow entrance slit of the type used in conventional spectroscopy could be used to improve the resolution in some cases but, due to seeing fluctuations, a slit would introduce a large noise into the output signal. Ratioing (such as shown in Fig. 10-1) then becomes a necessity. Usually, however, much lower resolution is used. The exit slit, defining the portion of the spectrum to be recorded, may be set at 50 Å. Values of 10–200 Å have been used. So usually $\Delta\lambda$ given in the above equations is small or very small compared to the actual resolution.

The exit slit width, imaged seeing disc, and aberrations combine together to make the instrumental profile.* Diffraction is negligible. The continuum is a smoothly varying function, however, and even the low resolution of a scanner is more than adequate. It is not normally necessary to consider corrections for the instrumental profile of a scanner used to measure the continuum. Detailed consideration of the instrumental profile is appropriate when dealing with spectral lines. This is provided in Chapter 12.

The scanning is accomplished by rotating the grating. The angle between the collimator axis and the camera axis is a constant, let us say θ_0. The grating equation can then be written

$$\frac{n\lambda}{d} = \sin\alpha + \sin\beta$$

$$= \sin(\theta_0 - \beta) + \sin\beta$$

$$= \sin\theta_0\cos\beta + (1 - \cos\theta_0)\sin\beta.$$

In the design by Tull (1963), for example, the grating normal is toward the camera, making β small. Then $\sin\beta \doteq \beta$, $\cos\beta \doteq 1$, and λ is linearly related to β as long as β remains small. The position in the spectrum can be set to about 1 Å using standard machine shop construction.

Observation in two or sometimes three grating orders is necessary to cover the full visible window. Filters are used to remove the light from overlapping orders.

---

* The linear dispersion is 10–20 Å/mm in most working instruments. A linear dispersion of this size is convenient to work with. Notice that the resolution and linear dispersion are not related in a photoelectric system unless one makes the dispersion so small that it becomes impossible to make the exit slit small enough. See Chapter 12 for additional discussion.

The choice of photomultiplier may present a problem. The S-20 cathode response is excellent over wavelengths less than 7500 Å, although bialkali tubes have higher quantum efficiency in the UV. The Ga–As cathode may be most suitable in the red and up to 8000 Å. At the longer wavelengths the S-1, with its low quantum efficiency, may be the only choice. Generally then, it is necessary to switch photomultipliers to measure across the whole visible window.

With relatively small telescope apertures ($\sim$ 50–100 cm) it is possible to measure stars as faint as 8–10th magnitude using a scanning monochromator. The efficiency of the operation can be increased by using a polychromator (or multichannel monochromator). The monochromator is

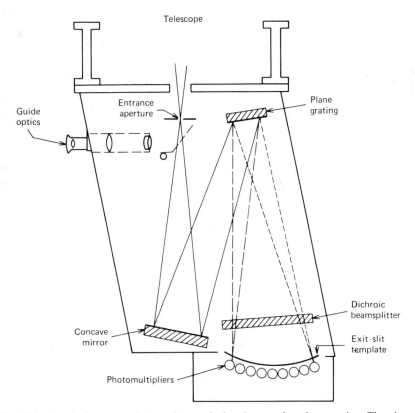

Fig. 10-4 A polychromator that employs a single mirror and a plane grating. The photomultipliers simultaneously measure many bands of light as defined by the exit slit template. A second array of photomultipliers lies below the figure and measures light directed to it by the dichroic beamsplitter. In this way simultaneous measurements are made in both the first and the second grating orders.

converted to a polychromator by placing an array of detectors in the spectrum. These detectors may be photomultipliers, photodiodes, image tube scanners, and so on. Since all channels are recorded simultaneously, a polychromator has all the advantages of a rationing system in addition to an increase in efficiency proportional to the number of channels. Polychromators using photomultipliers have been built by Oke (1969), Rodgers et al. (1973), Gray, and Willstrop (1965a). Oke's machine uses a plane grating and a folded collimator of approximately twice the focal length of the camera. Thirty-three photomultipliers are spread across the 3100–11,000 Å region. A pulse counting system is used with each photomultiplier. The instrument is designed to measure very faint stars. Gray uses the Gillieson optical scheme. Eighteen photomultipliers are spread across the 3300–11,000 Å window as shown in Fig. 10-4, and each channel has a DC integrator for measuring the output current. The instrument is designed to measure relatively bright stars.

One of the most advanced polychromator detector systems is the image tube–image dissector combination (Robinson and Wampler 1972a, 1972b). With it, measurements in approximately 4000 channels have been made on faint sources (see Butler et al. 1973, for example). Other detectors, such as those mentioned at the end of Chapter 4, should prove very useful for polychromators.

The spectral response of a monochromator or a polychromator shows a large and sometimes discontinuous change with wavelength. The strongest factors affecting the response are the grating blaze distribution and detector response. Lesser factors are contributed by filters, reflecting surfaces, beam splitters, and lenses. The nonuniform spectral response does not cause a serious problem in the differential comparison of stars. But as the faint star limit of the instrument is reached the low sensitivity regions are lost first— usually the 8000—11,000 Å section.

THE OBSERVATION

The linearity of the photodetectors and their attendant circuitry are depended on throughout. Any star of interest is compared with some standard star for which the continuum energy distribution has already been measured. The ratio of the output signals, after correction for sky plus dark signal and atmospheric extinction, is the ratio of fluxes for the two stars. The sky correction is obtained by interspersing measurements of the sky (next to each star) between the stellar measurements. The sky measurements include the dark signal, so when the sky reading is subtracted from the stellar measurement, the dark signal is automatically subtracted out along with it.

One of the stars must be observed for a sufficiently long time to define the slope of the extinction curve of the type shown in Fig. 10-5. As long as the plane parallel approximation* can be applied to the terrestrial atmosphere we expect the observed flux to vary with angle Z from the zenith according to

$$F_\nu = F_\nu^0 e^{-\tau_\nu \sec Z}$$

or, as more commonly written,

$$\log F_\nu = \log F_\nu^0 - \tau_\nu \log e \sec Z \qquad (10\text{-}3)$$

where F_ν^0 is the flux outside the atmosphere and τ_ν is the optical depth through the atmosphere toward the zenith. The coefficient of $\sec Z$, $\tau_\nu \log e$, is often called an extinction coefficient and, of course, is the slope of the line in Fig. 10-5. For comparative measurements it is not necessary to know $\tau_\nu \log e$ or

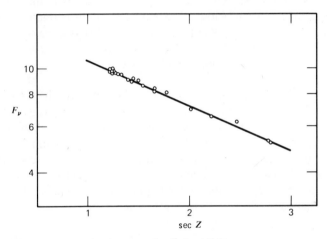

Fig. 10-5 An extinction curve for ξ^2 Cet $\lambda 4167$.

F_ν^0 explicitly, although they can be calculated if desired. Suppose we compare two stars and write, using eq. 10-3,

$$\log \frac{F_\nu(1)}{F_\nu(2)} = \log \frac{F_\nu^0(1)}{F_\nu^0(2)} - \tau_\nu \log e(\sec Z_1 - \sec Z_2). \qquad (10\text{-}4)$$

The left side of the equation is equal to the ratio of data signals after allowance for sky and any amplifier gain differences. The flux ratio on the right side of the equation is the answer we are after. It is desirable to have measurements

* Curvature corrections can easily be applied if necessary. Then $\sec Z$ in eq. 10-3 is replaced with a more complicated quantity called the air mass. See Hardie (1962) for details.

of both stars near the same sec Z so the last term in eq. 10-4 is small. Then eq. 10-4 can be viewed as a simple interpolation along the extinction curve from which we find $F_v(1)$ at sec Z_2 or vice versa. The solution can be arrived at graphically or algebraically.

There are several assumptions incorporated in the discussion: the sky is uniform and stable so that the extinction coefficient is the same for both stars and constant for the duration of the observations; the star used to define the extinction curve is not a variable star; and the effective wavelength of the spectral band is not a function of sec Z. Difficulties with the first two assumptions are reasonably obvious if the observations are carefully made. A change in effective wavelength with sec Z is a well known phenomena in broadband photometry (see, e.g., Hardie 1962). It results from the change in the energy distribution within the spectral band caused by the wavelength dependent extinction. Rayleigh scattering is the most pronounced extinction mechanism (the λ^{-4} term below). One can describe the extinction coefficient by (see Hayes and Latham 1975 for a more complete discussion)

$$\tau_v \log e = A + \frac{B}{\lambda} + \frac{C}{\lambda^4}.$$

The steep wavelength dependence can cause the V band of the UBV system to change its effective wavelength by tens of angstroms when measurements are taken at large zenith angles. A very similar effect occurs when comparing stars of different intrinsic energy distributions such as a B and a K star. But in the situation where the stellar continuum is to be compared with model atmosphere calculations, we generally use much higher spectral resolution than afforded by the broadband UBV photometry. Over a 50 Å spectral band, the change in effective wavelength is negligible.

ABSOLUTE CALIBRATION

Vega (A0 V) has been adopted as the primary photometric standard. Secondary standards have been set up as described in the next section. The absolute calibration of the energy distribution of Vega is done in two parts. First the shape is measured, referring the flux at various points in the continuum to the flux at $\lambda 5556$. Then the actual number of photons comprising the flux at $\lambda 5556$ is measured. Absolute measurements are more vulnerable to poor sky conditions because eq. 10-3 rather than eq. 10-4 must be used.

The shape of Vega's continuum has been measured by several astronomers. It is not simply a matter of taking a spectrum scanner, measuring the star, and correcting for extinction because the sensitivity of the instrument is a strong function of wavelength. The effect of the spectral response can be removed by using the scanner to make a differential measurement between the star and a standard light source. The standard sources that have been used are standard lamps, black bodies, and synchrotron radiators. Standard lamps are intermediate standards and require previous calibration, usually against a black body. They are ribbon filament light bulbs operating at about 2800°K and have only weak radiation in the UV. Calibration and use of standard lamps are described by Bless et al. (1968). Black bodies are usually operated at the melting temperature of some metal. Platinum black bodies operate at 2045°K, copper at 1358°K, and gold at 1338°K (see Chapter 6). The latter two are too cool for blue and UV observations. The synchrotron radiation is more suitably distributed for UV calibration. Each of these standard sources is subject to fundamental uncertainties like those discussed in Chapter 6 for the black body. The result is increased uncertainty in the UV and IR flux of Vega.

Absolute calibrations of Vega have been given in recent years by Hayes (1970), Oke and Schild (1970), Hayes and Latham (1975), and Hayes et al. (1975), and in the UV by Code and Bless (1972). Earlier measurements by Willstrop (1965b), Glushneva (1964), Oke (1964, 1965), Kharitonov (1963), and Code (1960) should probably be viewed as superseded by the more recent results

In Fig. 10-6 some of the results are compared. Not shown in the figure are the calibrations of Hayes (1970) and Oke and Schild (1970). These two calibrations showed reasonable agreement over the Paschen continuum with only a small systematic difference in the gradient. But in the Balmer and Brackett continua, Hayes' points lie substantially lower than those of Oke and Schild. The new measurements in the infrared by Hayes et al. (1975) and the rediscussion by Hayes and Latham (1975) reconcile some of these differences. The result is the adopted calibration given in Table 10-1 and shown by the connected points in Fig. 10-6.

The absolute flux at 5556 Å is more difficult to measure. Code (1960) reviews and explains some of these difficulties and gives a value measured in collaboration with R. Bless, 3.9×10^{-9} erg/sec cm^2 Å, for a star having a visual magnitude* $V = 0.00$. A similar calibration was performed by Willstrop (1965b, 1960) who found 3.84×10^{-9} erg/sec cm^2 Å at 5438 Å, which according to his data for the A0 star 58 Aql translates into 3.7×10^{-9} erg/sec cm^2 Å at 5556 Å for $V = 0.00$. Kharitonov (1963) obtained the considerably smaller value of 3.49×10^{-9} erg/sec cm^2 Å. More recent measurements by Oke and Schild (1970) give 3.48×10^{-9} and those by Hayes and Latham (1975) give 3.56×10^{-9}, both in agreement with Kharitonov. A value of 3.50×10^{-9} erg/sec cm^2 Å is probably the best value to adopt, being uncertain by an estimated $\pm 2\%$.

---

* The V magnitude of Vega is usually taken to be 0.03 in these calculations.

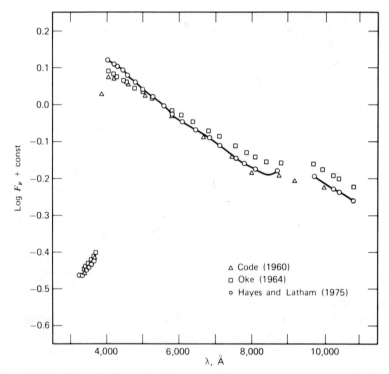

Fig. 10-6 Absolute calibrations of the shape of Vega's energy distribution are shown. The calibrations by Hayes (1970) and Oke and Schild (1970) are not shown because of crowding. The adopted calibration (Table 10-1) is shown by the solid line.

Although no one has measured the relation between the narrow band used for the absolute flux calibration and the V magnitude band, it is generally assumed that the apparent visual magnitude of the star can be used to scale the absolute flux at $\lambda5556$ for any star of known visual magnitude. In this way we write

$$\log F_\lambda = -0.400V - 8.456 \text{ erg/sec cm}^2 \text{ Å}$$

$$\log F_\nu = -0.400V - 19.443 \text{ erg/sec cm}^2\text{c/sec.} \qquad (10\text{-}5)$$

A small but systematic error is introduced when these expressions are used for stars of different temperatures because the effective wavelength of the V band is 5480 Å, that is, 76 Å shortward of the standard reference wavelength. From consideration of stellar energy distributions it can be shown that a

Table 10-1   The Shape Calibration for Vega

λ	$\log F_\nu + \text{const}$	λ	$\log F_\nu + \text{const}$
3300	−0.455	6056	−0.044
3400	−0.452	6436	−0.067
3500	−0.439	6800	−0.088
3571	−0.431	7100	−0.109
3636	−0.424	7550	−0.144
4036	+0.120	7780	−0.159
4167	+0.111	8090	−0.172
4255	+0.105	8400	−0.190*
4464	+0.095	8708	−0.180
4566	+0.080	9700	−0.194
4785	+0.061	9950	−0.209*
5000	+0.041	10250	−0.228
5263	+0.020	10400	−0.234
5556(ref)	0.000	10800	−0.260
5840	−0.025		

Source.   Adapted from Hayes and Latham (1975).
* Interpolated from graph in Fig. 10-6.

color term of the form

$$\log \frac{F(\lambda 5556)}{F(\lambda 5480)} = -0.006 + 0.018\,(B - V)$$

describes the error. We see that the flux ratio differs only a small amount from unity except for extreme values of color index and that the error committed is less than the uncertainty in the calibration.

The solar continuum has been subject to extensive measurements. We cite the work of Labs and Neckel (1971, 1972), the results of which are shown in Fig. 10-7, and of Arvesen et al. (1969). As Code (1973) has pointed out, the UV and IR radiation comprise a small portion of the total solar radiation. Code in this same article also compared the average energy distribution of a solar type star $(B - V = 0.63)$ based on the calibrations of Hayes (1970) and of Oke and Schild (1970) with the solar measurements. There is much better agreement now than there was with earlier stellar calibrations.

The absolute flux at 5556 Å for a star having $V = 0.00$, as derived from the solar measurements, is 3.6×10^{-9} erg/sec cm^2 Å according to Code (1973) and is uncertain by about $\pm 3\%$. The main uncertainty in this value arises from error in the apparent visual magnitude of the sun.

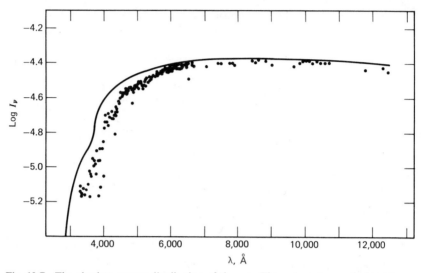

Fig. 10-7 The absolute energy distribution of the sun. The apparent smoothed continuum is indicated by the solid line. The mean flux in 20.5 Å bands is shown by the points. (Data from Labs and Neckel 1967, 1971, 1972.)

PHOTOMETRIC STANDARD STARS

Early type stars are chosen as photometric standards because their spectra are the least cluttered by absorption lines. The particular wavelength regions to be measured are selected to be as free from lines as possible. These regions are tabulated in the first column of Table 10-1.

The grid of standards chosen by Oke (1964) is well distributed in right ascension and has generally been used. Both Oke (1964) and Hayes (1970) have given tabulations of the energy distribution of secondary standards relative to Vega. The relative errors are in the $1-2\%$ range. These secondary standards are listed in Table 10-2.

MEASURED CONTINUA

The continuum or energy distribution is a strong function of temperature. Figure 10-8 displays a range of energy distributions as a function of spectral type and $B - V$ color index. For stars hotter than about A0, the slope of the Paschen continuum becomes less sensitive to the temperature of the star. Most of the flux is emitted in the Balmer continuum and ultraviolet measurements are called for. The same happens in reverse for the cool stars, leading one to make IR measurements.

Table 10-2   Photometric Standard Stars

		HR	α(1980)	δ(1980)	V	B − V	Sp
o^2	Cet	718	$2^h\ 27^m\ \ 5^s$	$+8°\ 23'$	4.28	−0.06	B9 III
ε	Ori	1903	$5^h\ 35^m\ 11^s$	$-1°\ 11'$	1.70	−0.19	B0 Ia
γ	Gem	2421	$6^h\ 36^m\ 34^s$	$+16°\ 25'$	1.93	0.00	A0 IV
η	Hya	3454	$8^h\ 42^m\ 11^s$	$+3°\ 27'$	4.30	−0.20	B3 V
α	Leo	3982	$10^h\ 07^m\ 18^s$	$+12°\ 04'$	1.36	−0.11	B7 V
θ	Crt	4468	$11^h\ 35^m\ 40^s$	$-9°\ 41'$	4.70	−0.08	B9 V
θ	Vir	4963	$13^h\ 08^m\ 55^s$	$-5°\ 26'$	4.37	−0.02	A1 V
α	Lyr	7001	$18^h\ 36^m\ 15^s$	$+38°\ 46'$	0.03	0.00	A0 V
κ	Aql	7446	$19^h\ 35^m\ 49^s$	$-7°\ 05'$	4.96	−0.01	B0.5 III
58	Aql	7596	$19^h\ 53^m\ 43^s$	$+0°\ 14'$	5.62	+0.11	A0 V
ε	Aqr	7950	$20^h\ 46^m\ 36^s$	$-9°\ 34'$	3.77	+0.01	A1 V
29	Psc	9087	$0^h\ \ 0^m\ 48^s$	$-3°\ \ 9'$	5.10	−0.14	B8 III

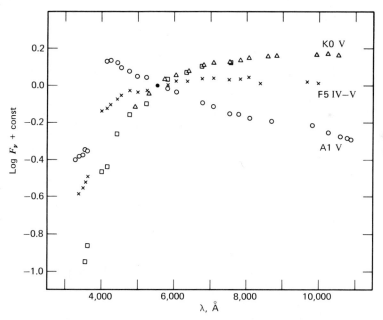

Fig. 10-8   Measured continua for stars ranging from early A to early K in spectral type. Normalization to λ5556 is shown. Absorption by lines cause the irregularities in the Paschen continua. (Data from Bessell 1967, Whiteoak 1967, Gray 1967, and Hayes 1968 by private communication.)

Schild et al. (1971), Wolff et al. (1968), and Kodaira (1970) have published energy distributions for B and A stars and use them to determine temperatures. Willstrop (1965b) gives an extensive catalog of measurements. Oke and Conti (1966) have measured several Hyades stars. Jones (1966) and Whiteoak (1967) give energy distributions for many cooler stars. Breger and Kuhi (1970) used their energy distribution data for surface gravity studies. Infrared measurements of the energy distributions of very cool stars have been made by Wing (1966), Danielson (1966), and Gillett et al. (1968). Mean energy distributions as a function of spectral type have been compiled by O'Connell (1973) and Spinrad and Taylor (1971).

CONTINUA FROM PHOTOSPHERIC MODELS

The temperature and pressure sensitivity of the continuum can easily be explored using models of the type described in Chapter 9. The resulting continua are seen in Figs. 10-9 and 10-10. Notice in Fig. 10-9 how the absolute flux increases and the slope changes in going to hotter models. The Paschen continuum slope and Balmer discontinuity strength are plotted in Fig. 10-11. Both parameters show a strong temperature dependence. The rise

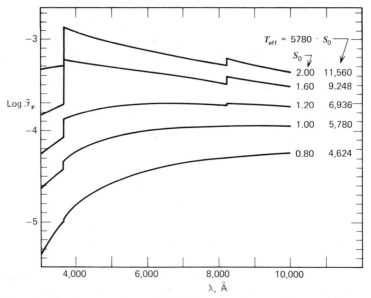

Fig. 10-9 Theoretical continua for scaled solar $T(\tau_0)$ models as a function of temperature plotted on an absolute flux scale. Surface gravity is constant at 2.7×10^4 cm/sec².

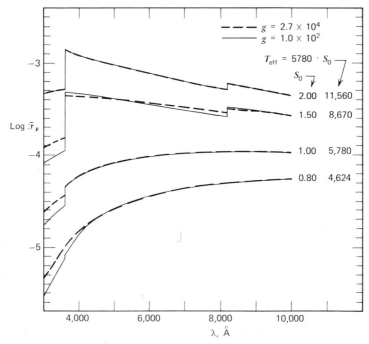

Fig. 10-10 Theoretical continua for scaled solar $T(\tau_0)$ models showing the surface gravity dependence at several temperatures.

and fall of the size of the Balmer jump shows the temperature range over which neutral hydrogen dominates the continuous opacity. At lower temperatures the negative hydrogen ion grows, while the neutral hydrogen diminishes from lack of excitation. The ratio of Balmer to Paschen absorption (i.e., the Balmer jump) increases with decreasing temperature (see eq. 8-5), but has little effect on the spectrum because of the strong H^- absorption. At the higher temperatures the Balmer jump is reduced as the ratio of Balmer to Paschen absorption decreases and as the hydrogen absorption fades with increased hydrogen ionization.

The observed Balmer jump according to Barbier (1958) is superimposed onto Fig. 10-11. The excellent agreement shown in the figure is probably spurious,* but it does show that the models behave somewhat like real stars.

---

* As discussed in Chapter 15, the scaled solar $T(\tau_0)$ probably gives too small a Balmer jump. Furthermore, the calculated spectra have no hydrogen lines in them so that D is measured differently for the spectra of real stars and these models. And, finally, the temperature scale used to assign T_{eff} to the observed stars is not definitive.

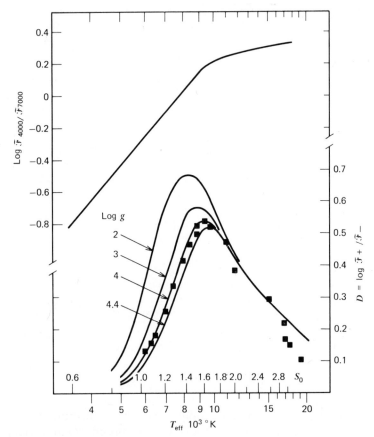

Fig. 10-11 (*Top*) The Paschen continuum slope varies linearly with temperature for the cooler stars. (*Bottom*) The Balmer jump is shown as a function of temperature for several values of surface gravity. The models use a scaled solar $T(\tau_0)$. Smaller gravity enhances the Balmer jump. The squares show observations tabulated by Barbier (1958).

Prior to fitting the model to the observations it is necessary to consider the absorption caused by the spectral lines that makes the measured energy distribution differ from the true continuum as emitted by the star.

LINE ABSORPTION

When a spectrum scanner measures the radiation in a portion of a stellar spectrum, the lines are, of course, included. Necessarily then the measured value lies below the true continuum. Ideally the model should include all the

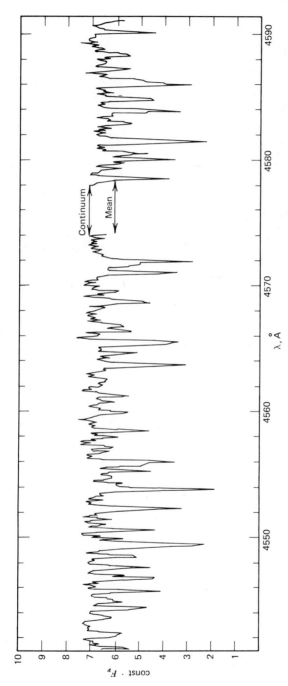

Fig. 10-12 A 50 Å portion of a microdensitometer tracing for the photographically recorded spectrum of η Cas A (G0V). The central wavelength is 4566 Å. The continuum and the mean flux levels are shown. Their ratio is 0.847.

lines in each measured spectral band. Then the average flux as observed and computed could be directly compared. As yet we are not able to make such a computation routinely and the expense of the computation may not be justifiable (but see Holweger 1970). Allowance for the line absorption can be made empirically by use of relatively high resolution spectrograms to obtain the ratio of average flux to continuum flux. Figure 10-12 is an example of such a measurement. The line absorption amounts to 15.3% in this 50 Å bandpass. The measured value is then divided by 0.847 to give the height of the stellar continuum.

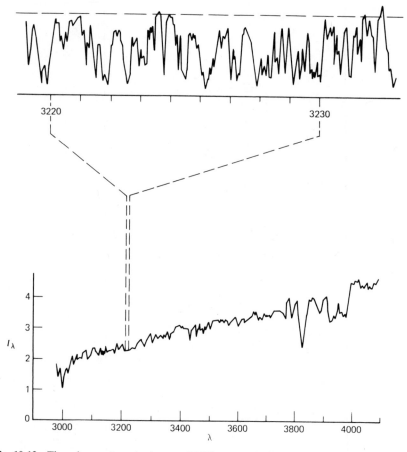

Fig. 10-13   The solar spectrum in the near UV illustrates the fallacy of choosing an apparent continuum over too limited a spectral region. (Adapted from Houtgast 1966. Courtesy of the International Astronomical Union.)

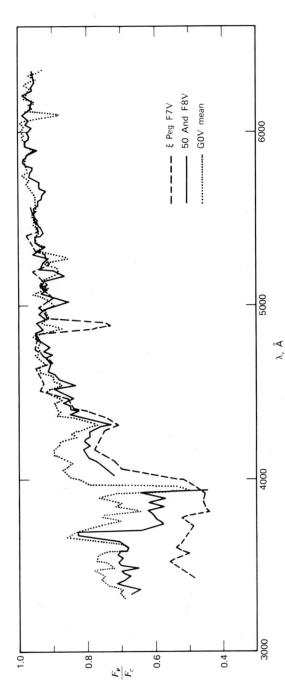

Fig. 10-14 The wavelength dependence in line absorption for three stars. The absorption increases strongly toward shorter wavelengths. The data for ξ Peg and 50 And are from Wildey et al. (1962) and for the G0V curve from Oke and Conti (1966).

Should the crowding of the stellar lines be so great that no continuum remains, the method cannot be used in any rigorous way. It is not always trivial to decide such a question and care must be exercised. For example, in Fig. 10-13 we see a situation that, while badly cut up with lines, looks as though a continuum can be set when viewed over a small portion of the spectrum. But on the broader view we see the whole spectral region to be depressed.

The amount of line absorption in any spectrum is greater for cooler stars, is greater for shorter wavelengths, and is greater for metal rich stars. It is a good practice to measure the line absorption for each individual star being studied. Typical changes can be seen in Fig. 10-14 and 10-15. The wavelength

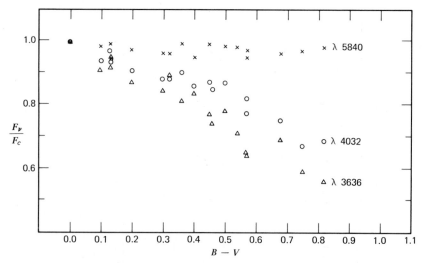

Fig. 10-15 The line absorption in three 50 Å spectral bands becomes larger in cooler stars. This is typical behavior. (Data from Baschek and Oke 1965, Gray 1967, Melbourne 1960, Oke and Conti 1966, Wildey et al. 1962.)

dependence is illustrated in Fig. 10-14 and the change with temperature is found in Fig. 10-15. A comparison of line absorption data taken from different publications may show significant differences.

The correction to the observed continua are quite significant, as shown in Fig. 10-16. When the corrections become large, the standard photographic measurement of line absorption is probably inadequate because of the relatively large photometric errors.

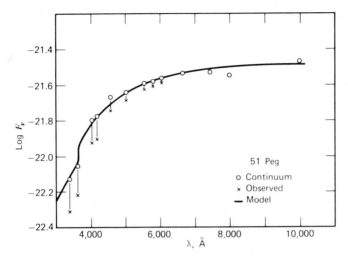

Fig. 10-16 ($\times$) The observed energy distribution in 51 Peg (G5V) according to Melbourne (1960) but corrected to the calibration of Vega. ($\bigcirc$) The deduced continuum after line absorption corrections are applied to the observations. (——) The continuum calculated from a model having $S_0 = 0.98$.

A COMPARISON OF MODEL TO STELLAR CONTINUA

An estimate of the proper model for a star can be made by comparing a grid of continua like the one shown in Figs. 10-9 and 10-10 with the observed energy distribution after correction for line absorption. New models can be calculated or interpolation can be done within a grid of models until the observations are matched within the errors. Such a fit is shown in Fig. 10-16. The star is relatively cool and the continuum is insensitive to surface gravity. The temperature remains as the lone adjustable parameter. For the F star in Fig. 10-17 the Paschen continuum is fit by temperature adjustment, but the Balmer jump then remains to be matched by adjustment of the model's surface gravity. In still hotter stars like the early A stars shown in Fig. 10-18, the Balmer jump has become nearly independent of surface gravity again.

In most cases the stellar continua can be fit sufficiently well by the calculations. Not all models fit perfectly, as shown in Fig. 10-18. Sources of uncertainty rest in (1) the accuracy of the absolute flux calibration, that is, the shape of the observed continuum, (2) the realism of the model, and (3) interstellar reddening.

The interstellar reddening is a problem for stars more distant than $\sim 10^2$ pc. If the star is reddened, use of the energy distribution to choose a model

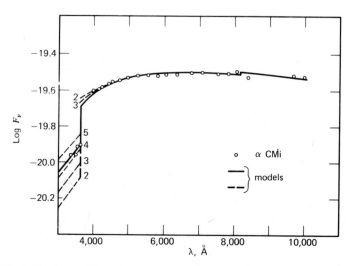

Fig. 10-17 A fit of a model to the continuum of Procyon (F5 IV–V): (———) model with $S_0 = 1.20$ and $g = 2.7 \times 10^4$ cm/sec^2; (---) model results for other gravities. The observations are adapted from Bessell (1967).

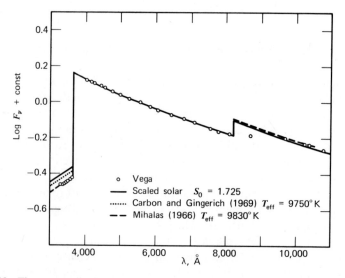

Fig. 10-18 The energy distribution of Vega from Table 10-1 is shown fitted with several models. The Paschen continuum is adequately matched by all models. The Balmer jump, which is only slightly gravity dependent for a star this hot, is fit more precisely with models having steeper temperature gradients.

becomes a greatly weakened procedure. Spectral lines and the Balmer jump must then be used to extract the necessary information.

With regard to point 2 above, the LTE models differ in essentially only one one way—the temperature gradient. Line blanketing remains the most uncertain influence on the gradient. The more heavily line blanketed models (steeper gradient) give a better fit in the A stars as shown in Fig. 10-18.

The question of whether or not the continuum source function is sufficiently approximated by $B_\nu(T)$ has been considered by several people, including Mihalas and Stone (1968), Mihalas (1967), Strom and Kalkofen (1966), Smith and Strom (1969), and Mihalas and Auer (1970). The Paschen continuum is shown by these investigations to be adequately described by LTE models. The Balmer and Paschen jumps may show non-LTE effects of a few per cent and we would expect the error to increase with decreasing pressure, a result shown quantitatively in the calculations of the cited references. A definitive observational test may be possible with the newest absolute calibrations of the Balmer and Brackett continua of Vega.

BOLOMETRIC FLUX

Through the visible window we measure part of the flux emitted by the star. The bolometric flux or luminosity refers to the total flux emitted by the star of radius R,

$$L \equiv 4\pi R^2 \int_0^\infty \mathcal{F}_\nu \, d\nu.$$

It is this quantity that is important for studies of the effective temperature scale (Chapter 15) and in linking stellar atmosphere to stellar interior calculations.

The direct method of determining the nonvisible window flux is to observe from rockets, satellites, and high altitude balloons. In recent years a large number of UV observations have been made in this way (see the review articles by Code 1973 and Bless and Code 1972, and the papers by Davis and Webb 1970 and Beeckmans et al. 1974). A much less certain method is to calculate the flux in the nonvisible spectrum using a model photosphere (e.g., Oke and Conti 1966). Either way one obtains the ratio, a, of the bolometric flux to the flux in some spectral window

$$a \equiv \frac{\int_0^\infty F_\nu \, d\nu}{\int_0^\infty W(\nu) F_\nu \, d\nu} = \frac{\int_0^\infty \mathcal{F}_\nu \, d\nu}{\int_0^\infty W(\nu) \mathcal{F}_\nu \, d\nu} \tag{10-6}$$

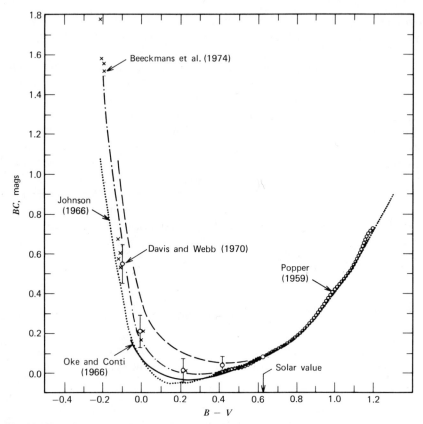

Fig. 10-19 Bolometric corrections (in magnitudes) for main sequence stars according to several investigators shown as a function of color index. Normalization is to the nominal solar value $B - V = 0.63$, $BC = 0.07$.

where F_v denotes flux measured at the earth and $\mathcal{F}_v$ denotes flux emitted at the star. The window through which we record the ground based data has the response $W(v)$.

In certain applications the bolometric flux is compared to the flux seen in the visual band of the UBV system. Then $W(v)$ is the response function for V given in Fig. 1-4. The bolometric correction* is defined to be

$$BC = 2.5 \log a + \text{const}$$
$$= m_V - m_{\text{bol}}$$
$$= M_V - M_{\text{bol}}.$$

* It is customary to use magnitude units. The bolometric correction is sometimes defined with the opposite sign.

The additive constant arises because the zero point of the magnitude scale is arbitrary and set by adoption. It was customary to set equal to zero the minimum in the relation between bolometric correction and spectral type or color index. According to Popper (1959) the sun has a bolometric correction of 0.07 when this is done. But because the bolometric correction for the sun has been well determined, the value of 0.07 is now adopted as the fundamental point to which modern bolometric correction scales are normalized. The solar color index is 0.628 according to Fernie et al. (1971) and Croft et al. (1972).

The bolometric flux of a star can be ratioed to the same quantity for the sun giving

$$M_{bol} - M^{\odot}_{bol} = M_V - M^{\odot}_V - (BC - BC_{\odot})$$

or

$$-2.5 \log \left(\frac{L}{L_{\odot}} \right) = M_V - M^{\odot}_V - (BC - BC_{\odot}).$$

The luminosity of the sun follows from the solar constant of 1.36 kW/m^2 (Neckel and Labs 1973) or $L_{\odot} = 3.825 \times 10^{33}$ erg/sec. A recent review of the apparent visual magnitude of the sun by Johnson (1965) results in -26.74 or $M^{\odot}_V = 4.83$.

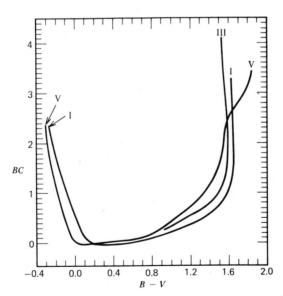

Fig. 10-20 Bolometric corrections for luminosity classes I, III, and V according to Johnson (1966). The V curve is also shown in Fig. 10-19.

The large change in bolometric correction with color index is shown in Figs. 10-19 and 10-20. Unfortunately the scatter is uncomfortably large. The bolometric corrections for non-main sequence stars according to Johnson (1966) can be seen in Fig. 10-20. Earlier estimates are tabulated by Wildey (1963) and a weighted summary is given by Schlesinger (1969).

REFERENCES

Arvesen, J. C., R. N. Griffin, Jr., and B. D. Pearson 1969. *Appl. Opt.* **8**, 2215.

Barbier, D. 1958. *Handb. Phys.* **50**, 322.

Baschek, B., and J. B. Oke 1965. *Astrophys. J.* **141**, 1404.

Beeckmans, F., D. Macau, and D. Malaise 1974. *Astron. Astrophys.* **33**, 93.

Bessell, M. S. 1967. *Astrophys. J. Lett.* **149**, L67.

Bless, R. C., and A. D. Code 1972. *Annu. Rev. Astron. Astrophys.* **10**, 197.

Bless, R. C., A. D. Code, and D. J. Schroeder 1968. *Astrophys. J.* **153**, 545.

Breger, M., and L. V. Kuhi 1970. *Astrophys. J.* **160**, 1129.

Butler, D., R. P. Kraft, J. S. Miller and L. B. Robinson 1973. *Astrophys. J. Lett.* **179**, L73.

Carbon, D., and O. Gingerich 1969. *Theory and Observation of Normal Stellar Atmospheres*, O. Gingerich, Ed., MIT Press, Cambridge, Mass., p. 377.

Code, A. D. 1960. *Stellar Atmospheres*, J. L. Greenstein, Ed., University of Chicago Press, Chicago, Chapter 2.

Code, A. D. 1973. *Problems of Calibration of Absolute Magnitudes and Temperatures of Stars*, IAU Symposium No. 54, B. Hauck and B. E. Westerlund, Eds., Reidel, Dordrecht, p. 131.

Croft, S. K., D. H. McNamara, and K. A. Feltz, Jr. 1972. *Publ. Astron. Soc. Pac.* **84**, 515.

Danielson, R. E. 1966. *Colloquium on Late-Type Stars*, M. Hack, Ed., Osservatorio Astronomico, Trieste, p. 198.

Davis, J., and R. J. Webb 1970. *Astrophys. J.* **159**, 551.

Fernie, J. D., J. P. Hagen, Jr., G. L. Hagen, and L. McClure 1971. *Publ. Astron. Soc. Pac.* **83**, 79.

Gillett, F. C., F. J. Low, and W. A. Stein 1968. *Astrophys. J.* **154**, 677.

Gillieson, A. H. C. P. 1949. *J. Sci. Instrum.* **26**, 335.

Glushneva, I. N. 1964. *Sov. Astron. AJ* **8**, 163.

Gray, D. F. 1967. *Astrophys. J.* **149**, 317.

Hardie, R. H. 1962. *Astronomical Techniques*, W. A. Hiltner, Ed., University of Chicago Press, Chicago, p. 178.

Harris, D. L. 1963. *Basic Astronomical Data*, K. Aa. Strand, Ed., University of Chicago Press, Chicago, p. 263.

Hayes, D. S. 1970. *Astrophys. J.* **159**, 165.

Hayes, D. S., and D. Latham 1975. *Astrophys. J.* **197**, 593.

Hayes, D. S., D. Latham, and S. H. Hayes 1975. *Astrophys. J.* **197**, 587.

Holweger, H. 1970. *Astron. Astrophys.* **4**, 11.

Houtgast, J. 1966. *Abundance Determinations in Stellar Spectra*, IAU Symposium No. 26, H. Hubenet, Ed., Academic, London, p. 38.

Johnson, H. L. 1965. *Commun. Lunar Planet. Lab.* **3**, 67.

Johnson, H. L. 1966. *Annu Rev. Astron. Astrophys.* **4**, 193.

Jones, D. H. P. 1966. *Spectral Classification and Multicolour Photometry*, IAU Symposium No. 24, K. Loden, L. O. Loden, U. Sinnerstad, Eds., Academic, London, p. 141.

Kharitonov, A. V. 1963. *Sov. Astron. AJ* **7**, 258.

Kodaira, K. 1970. *Astrophys. J.* **159**, 931.

Labs, D., and H. Neckel 1967. *Z. Astrophys.* **65**, 133.

Labs, D., and H. Neckel 1971. *Sol. Phys.* **19**, 3.

Labs, D., and H. Neckel 1972. *Sol. Phys.* **22**, 64.

Liller, W., and L. H. Aller 1954. *Astrophys. J.* **120**, 48.

Liller, W. 1963. *Appl. Opt.* **2**, 187.

Melbourne, W. G. 1960. *Astrophys. J.* **132**, 101.

Mihalas, D. 1966. *Astrophys. J. Suppl.* **13**, 1.

Mihalas, D. 1967. *Astrophys. J.* **149**, 169.

Mihalas, D., and L. H. Auer 1970. *Astrophys. J.* **160**, 1161.

Mihalas, D., and M. E. Stone 1968. *Astrophys. J.* **151**, 293.

Morton, D. C., and T. F. Adams 1968. *Astrophys. J.* **151**, 611.

Murty, M. V. R. K. 1962. *J. Opt. Soc. Am.* **52**, 768.

Neckel, H., and D. Labs 1973. *Problems of Calibration of Absolute Magnitudes and Temperatures of Stars*, IAU Symposium No. 54, B. Hauck and B. E. Westerlund, Eds., Reidel, Dordrecht, p. 149.

O'Connell, R. W. 1973. *Astrophys. J.* **78**, 1074.

Oke, J. B. 1964. *Astrophys. J.* **140**, 689.

Oke, J. B. 1965. *Annu. Rev. Astron. Astrophys.* **3**, 23.

Oke, J. B. 1969. *Publ. Astron. Soc. Pac.* **81**, 11.

Oke, J. B., and P. S. Conti 1966. *Astrophys. J.* **143**, 134.

Oke, J. B., and J. L. Greenstein 1961. *Astrophys. J.* **133**, 349.

Oke, J. B., and R. E. Schild 1970. *Astrophys. J.* **161**, 1015.

Popper, D. M. 1959. *Astrophys. J.* **129**, 647.

Robinson, L. B., and E. J. Wampler 1972a. *Astrophys. J. Lett.* **171**, L83.

Robinson, L. B., and E. J. Wampler 1972b. *Publ. Astron. Soc. Pac.* **84**, 161.

Rodgers, A. W., R. Roberts, P. T. Rudge, and T. Stapinski 1973. *Publ. Astron. Soc. Pac.* **85**, 268.

Schild, R., D. M. Peterson, and J. B. Oke 1971. *Astrophys. J.* **166**, 95.

Schlesinger, B. M. 1969. *Astrophys. J.* **157**, 533.

Schroeder, D. J. 1966. *Appl. Opt.* **5**, 245.

Smith, M. A., and S. E. Strom 1969. *Astrophys. J.* **158**, 1161.

Spinrad, H. and B. Taylor 1971. *Astrophys. J. Suppl.* **22**, 445.

Strom, S. E. and W. Kalkofen 1966. *Astrophys. J.* **144**, 76.

Trodahl, H. J., D. J. Sullivan, and D. Beaglenhole 1973. *Publ. Astron. Soc. Pac.* **85**, 608.

Tull, R. G. 1963. *Photoelectric Spectrophotometry of M31*, doctoral dissertation, University of Michigan, Ann Arbor.

Wampler, E. J. 1966. *Astrophys. J.* **144**, 921.

Whiteoak, J. B. 1967. *Astrophys. J.* **150**, 524.

Wildey, R. L. 1963. *Nature* **199**, 988.

Wildey, R. L., E. M. Burbidge, A. R. Sandage, and G. R. Burbidge 1962. *Astrophys. J.* **135**, 94.

Willstrop, R. V. 1960. *Mon. Not. Roy. Astron. Soc.* **121**, 17.

Willstrop, R. V. 1965a. *Mon. Not. Roy. Astron. Soc.* **130**, 233.

Willstrop, R. V. 1965b. *Mem. Roy. Astron. Soc.* (*Lond.*) **69**, 83.

Wing, R. F. 1966. *Colloquium on Late-Type Stars*, M. Hack, Ed., Osservatorio Astronomico, Trieste, p. 231.

Wolff, S. C., L. V. Kuhi, and D. Hayes 1968. *Astrophys. J.* **152**, 871.

CHAPTER ELEVEN

THE LINE ABSORPTION COEFFICIENT

The line absorption coefficient plays a fundamental role in determining the shape of a spectral line. The situation in this regard is similar to the effect the continuous absorption coefficient has on the shape of the continuum. Lines are more interesting, however, because several different physical effects can enter the structuring of the final absorption coefficient as used in the solution of the transfer equation. Each of these effects has associated with it a function describing the variation in strength of the absorption with wavelength across the line; that is, each one of these processes has its own absorption coefficient. The processes we consider in this chapter are (1) natural atomic absorption, (2) pressure broadenings, of which there are several, and (3) the thermal broadening.

One of the remarkable results of these studies is that the natural atomic absorption and all the significant pressure broadenings have the same wavelength dependence in their individual absorption coefficients (with the notable exception of the hydrogen lines). This leads to a tremendous computational simplification as we shall see. The thermal broadening reflects the Maxwellian velocity distribution of the absorbing atoms and ions via the Doppler effect. The perpetual presence of thermal broadening complicates the discussion and, while it is not overly difficult to handle, it does require additional computation.

The atomic line absorption coefficient with units of square centimeters per absorber we call α (this is the same notation used in Chapter 8). It is well to remember, as explained in the second section of Chapter 8, that α is not a distribution, but $\alpha\,dv$ is and refers to the amount of power abstracted from a unit I_ν beam. Most physical processes deal with energy, thereby making frequency units appropriate. The linear Stark effect in hydrogen is an important exception and there the equations are expressed in wavelength units with explicit conversion by λ^2/c when necessary.

More advanced and/or comprehensive treatments of the line absorption coefficient can be found in Mihalas (1970), Cowley (1970), and Traving (1968), and for discussions of helium lines that we do not consider here at all see Mihalas et al. (1974), Barnard et al. (1974), and Griem (1968).

THE NATURAL ATOMIC ABSORPTION

The simplest classical model for the interaction of light with atoms is that of a plane electromagnetic wave interacting with dipoles. The absorption or attenuation of the wave, which is a consequence of the interaction, we now consider.

Imagine an incident electromagnetic wave, E, as it interacts with an ensemble of dipole charge oscillators representing the absorbing atoms in the gas. We choose the coordinates so E is traveling in the y direction. E must satisfy the wave equation

$$\frac{\partial^2 E}{\partial t^2} = v^2 \frac{\partial^2 E}{\partial y^2} \tag{11-1}$$

(where v is the wave velocity), but this is easily done by any field having $y - vt$ as its argument, that is,

$$E = f(y - vt) \tag{11-2}$$

is a solution of eq. 11-1. It is easy to show (see Stone 1963) that for light the wave velocity is related to the electrical and magnetic properties of the medium through which it is passing by

$$v = \left(\frac{\varepsilon_0 \mu_0}{\varepsilon \mu}\right)^{1/2} c.$$

Here ε is the dielectric constant, μ is the magnetic permeability, ε_0 is the permitivity, and μ_0 is the permeability of free space. Since we are dealing with gases and they have negligible magnetic permeability, $\mu = \mu_0$. Equation 11-2 can now be written

$$E = f\left(y - \sqrt{\frac{\varepsilon_0}{\varepsilon}}\, ct\right).$$

We can allow E to remain general by considering it to be a Fourier composition of sinusoidals. Because the electric field can be linearly decomposed, the manipulations performed on any one component can be performed as well on the others—the final answer being the sum of the manipulated components. This one Fourier component we write as

$$E = E_0 e^{i\omega(t - \sqrt{\varepsilon/\varepsilon_0}\, y/c)} \tag{11-3}$$

where ω denotes the angular frequency. We now choose E to be in the x direction, so then too the oscillation of the dipoles will be in the x direction.

The next step is to evaluate $\sqrt{\varepsilon/\varepsilon_0}$. We go back to the basic reason why ε differs from ε_0, which is the charge separation of the dipoles induced by the imposed field, E. The total electrical field is the sum of E and the field of the separated charges. The ratio of this total field to the field of the wave is $\varepsilon/\varepsilon_0$. If we let there be N dipoles per unit volume, then the ratio of the total field to E is

$$\frac{E + 4\pi Nqx}{E} = 1 + \frac{4\pi Nqx}{E} = \frac{\varepsilon}{\varepsilon_0} \tag{11-4}$$

where q is the size of the electrical charges of the dipole and x is their induced separation. In most stellar spectra we deal with electronic transitions so that we replace q by $e = 4.803 \times 10^{-10}$ esu.

The displacement, x, can be obtained by considering a single oscillator at $y = 0$ as it is forced into oscillation by E. Adding the acceleration, damping, and restoring force terms in that order we get the usual harmonic oscillator equation with the driving term on the right,

$$\frac{d^2x}{dt^2} + \gamma \frac{dx}{dt} + \omega_0^2 x = \frac{e}{m} E_0 e^{i\omega t} \tag{11-5}$$

where m (9.1095×10^{-28} g) is the mass of the electron. The damping constant, γ, is of some importance and is discussed in detail below. The driving wave on the right hand side of the equation is the same E of eq. 11-3 but with $y = 0$. The solution is found by choosing $x = x_0 e^{i\omega t}$. Then $dx/dt = i\omega x$ and $d^2x/dt^2 = -\omega^2 x$, all of which, when placed back into eq. 11-5, gives

$$-\omega^2 x + i\gamma\omega x + \omega_0^2 x = \frac{e}{m} E_0 e^{i\omega t}$$

or

$$x = \frac{e}{m} \frac{E_0 e^{i\omega t}}{\omega_0^2 - \omega^2 + i\gamma\omega} = \frac{e}{m} \frac{E}{\omega_0^2 - \omega^2 + i\gamma\omega}.$$

And so for eq. 11-4 where we need x/E, we obtain

$$\frac{\varepsilon}{\varepsilon_0} = 1 + \frac{4\pi Ne^2}{m} \frac{1}{\omega_0^2 - \omega^2 + i\gamma\omega}. \qquad (11\text{-}6)$$

In the wave of eq. 11-3 we need the square root of eq. 11-6, but it is clear that $\varepsilon \sim \varepsilon_0$ in a gas so the second term must be small compared to unity. We use the approximation that $(1 + \delta)^{1/2} \doteq 1 + \frac{1}{2}\delta$ for $\delta \ll 1$ and write

$$\sqrt{\frac{\varepsilon}{\varepsilon_0}} \doteq 1 + 2\pi \frac{Ne^2}{m} \frac{1}{\omega_0^2 - \omega^2 + i\gamma\omega}$$

which is the same as

$$\sqrt{\frac{\varepsilon}{\varepsilon_0}} = 1 + 2\pi \frac{Ne^2}{m} \frac{1}{(\omega_0^2 - \omega^2)^2 + \gamma^2\omega^2} - i2\pi \frac{Ne^2}{m} \frac{\gamma\omega}{(\omega_0^2 - \omega^2)^2 + \gamma^2\omega^2}. \qquad (11\text{-}7)$$

We define this to be, for convenience,

$$\sqrt{\frac{\varepsilon}{\varepsilon_0}} \equiv n - ik. \qquad (11\text{-}8)$$

Then eq. 11-3 can be written

$$E = E_0 e^{i\omega[t - (n - ik)y/c]}$$

or

$$E = E_0 e^{i\omega(t - ny/c) - k\omega y/c}. \qquad (11\text{-}9)$$

The result expressed in eq. 11-9 shows a real term in the exponent so there is an exponential extinction in the wave amplitude.† The intensity is proportional to EE^* where the asterisk means complex conjugate, giving

$$I = I_0 e^{-2k\omega y/c}$$

which is to be compared to the simple extinction law $I = I_0 e^{-l_\nu \rho y}$. We have called the absorption coefficient per unit mass l_ν. In this way we find

$$l_\nu \rho = \frac{4\pi Ne^2}{mc} \frac{\gamma\omega^2}{(\omega_0^2 - \omega^2)^2 + \gamma^2\omega^2} \qquad (11\text{-}10)$$

where k is taken from eqs. 11-7 and 11-8.

---

† While not a logical link in the derivation, notice that eq. 11-8 is a complex index of refraction. The quantity n is the usual index used in geometrical optics. For a prism the dispersive parameter $dn/d\lambda$ is of interest. Equation 11-7 can be used to show that $dn/d\lambda$ becomes larger as $\omega \to \omega_0$ but so does the absorption term. As a result a prism with high dispersive power is also plagued with high absorption.

Now this function peaks sharply so that nonzero values occur only when $\omega \sim \omega_0$. In this case then, we can write

$$\omega_0^2 - \omega^2 = (\omega_0 - \omega)(\omega_0 + \omega)$$
$$\doteq (\omega_0 - \omega)2\omega.$$

This brings eq. 11-10 into the form

$$l_\nu \rho = N \frac{\pi e^2}{mc} \frac{\gamma}{\Delta\omega^2 + (\gamma/2)^2}. \tag{11-11}$$

The basic form of this absorption coefficient is that of the dispersion profile (refer to eq. 2-7). It is also called a damping profile, a Lorentzian profile, a Cauchy curve, and, by some, the Witch of Agnesi. We shall continue to call it a dispersion profile.

If we now consider the absorption coefficient per atom, α, we write for eq. 11-11

$$l_\nu \rho = N\alpha. \tag{11-12}$$

Thus

$$\left.\begin{aligned}
\alpha &= \frac{2\pi e^2}{mc} \frac{\gamma/2}{\Delta\omega^2 + (\gamma/2)^2} \\[2mm]
&= \frac{e^2}{mc} \frac{\gamma/4\pi}{\Delta\nu^2 + (\gamma/2\pi)^2} \\[2mm]
&= \frac{e^2}{mc} \frac{\lambda^2}{c} \frac{\gamma\lambda^2/4\pi c}{\Delta\lambda^2 + (\gamma\lambda^2/4\pi c)^2}
\end{aligned}\right\} \tag{11-13}$$

We can show by direct integration of the dispersion profile in eq. 11-13 or by comparison with the unit area form in eq. 2-7 that

$$\int_0^\infty \alpha \, d\nu = \int_{-\infty}^\infty \alpha \, d\Delta\nu = \frac{\pi e^2}{mc}. \tag{11-14}$$

This is the energy/sec atom rad^2 c/sec absorbed by the total line from the unit I_ν beam. The total energy taken from an I_λ beam is

$$\int_0^\infty \alpha \, d\lambda = \int_{-\infty}^\infty \alpha \, d\Delta\lambda = \frac{\pi e^2}{mc} \frac{\lambda^2}{c}. \tag{11-15}$$

Actual measurements show $\int_0^\infty \alpha \, d\lambda$ to be smaller than indicated by eq. 11-14 or 11-15 and by an amount that differs from one spectral line to another.

A quantum mechanical treatment shows that it is useful to introduce a quantity called the oscillator strength, f, such that

$$\int_0^\infty \alpha \, dv = \frac{\pi e^2}{mc} f. \tag{11-16}$$

The oscillator strength, also called the f value, is different for each atomic level and is related to the atomic transition probability B_{lu} (defined in Chapter 5) as follows. The integral in eq. 11-16 must be equivalent to the probability of absorption times the unit energy for the transition according to eq. 5-20, or

$$\int_0^\infty \alpha \, dv = h v B_{lu}$$

which means

$$f = \frac{mc}{\pi e^2} h v B_{lu} = 7.484 \times 10^{-7} \frac{B_{lu}}{\lambda} \tag{11-17}$$

for λ in angstroms. Using eq. 6-7b we also see that

$$f = \frac{mc^3}{2\pi e^2 v^2} \frac{g_u}{g_l} A_{ul} = 1.884 \times 10^{-15} \lambda^2 \frac{g_u}{g_l} A_{ul} \tag{11-18}$$

where g_u and g_l are the statistical weights of the upper and lower levels, and again λ is in angstroms. Note in passing that the probability coefficients defined in eqs. 5-19 and 5-20 are per unit solid angle and not over the full 4π rad^2 of a unit sphere.

It is also possible to define an f value for emission. These f values are related by

$$g_u f_{em} = g_l f_{abs}.$$

Tabulations of measured f values give gf so that no ambiguity arises in use. Most f values are determined from empirical laboratory measurements, although in the less complicated cases calculations can be done. The hydrogen expression is

$$f = \frac{2^5}{3\sqrt{3}\pi} \frac{g}{l^5 u^3} \left[\frac{1}{l^2} - \frac{1}{u^2} \right]^{-3}$$

where u and l are the quantum numbers of the upper and lower levels as usual and g is the bound–bound Gaunt factor. Numerically $f = 0.6407$ for Hα, 0.1193 for Hβ, 0.0447 for Hγ, and so on (see Table 11-5). A few references to published f values are given in Chapter 14.

THE DAMPING CONSTANT FOR NATURAL BROADENING

The damping constant appearing in eq. 11-13 can be calculated from the classical dipole emission theory by taking a time average of the acceleration (Menzel 1961). An equation of the form

$$\frac{dW}{dt} = -\frac{2}{3}\frac{e^2}{m}\frac{\omega^2}{c^3}W \qquad (11\text{-}19)$$

results. The solution is $W = W_0 e^{-\gamma t}$ where $\gamma = 2e^2\omega^2/(3mc^3) = 0.22/\lambda^2$ for λ in centimeters. The half half width of α is $\gamma/2$, $\gamma/4\pi$, or $\gamma\lambda^2/4\pi c$, depending on which of the forms in eq. 11-13 are used. In the $\Delta\lambda$ form, the half half width is 0.59×10^{-4} Å for all lines. The classical damping constant is usually smaller by an order of magnitude compared to the observations and again a quantum–mechanical interpretation must be made.

A simple phenemological way to introduce the quantum aspects is to view W in eq. 11-19 as quantized and we write

$$W = N_u h\nu$$

where N_u is the population of the upper level for the line. Then eq. 11-19 can be rewritten

$$\frac{dN_u}{dt} = -\gamma N_u. \qquad (11\text{-}20)$$

But we already know from Chapters 5 and 6 that $dN_{ul}/dt = -A_{ul}N_u$ for each transition starting from the level u so that the total downward rate due to spontaneous emission is

$$\frac{dN_u}{dt} = \sum_{l<u}\frac{dN_{ul}}{dt} = -N_u\sum_{l<u}A_{ul}. \qquad (11\text{-}21)$$

A comparison of eqs. 11-20 and 11-21 leads us to write

$$\gamma_u = \sum_{l<u}A_{ul}. \qquad (11\text{-}22)$$

The physical meaning of this γ can be found by viewing it in terms of the lifetime of the level. Since A_{ul} has units of reciprocal seconds and specifies the probability that the electron will come down from the upper level, u, in 1 sec, the quantity

$$\Delta t = \frac{1}{\sum_{l<u}A_{ul}}$$

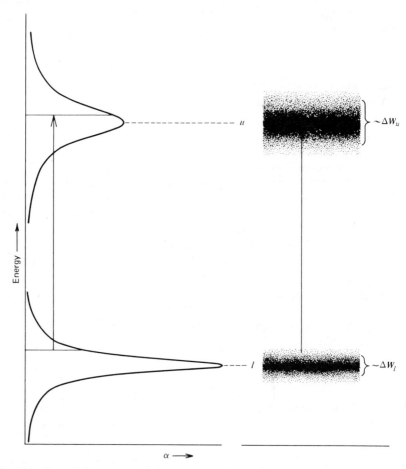

Fig. 11-1 A symbolic energy level diagram illustrating the energy width of the atomic levels. The transition starts somewhere in the lower level with a probability of a given energy coordinate given by α_l. The transition ends in the upper level with the a terminal energy whose probability is α_u.

corresponds to the probable time interval the electron will, if left alone, stay in the upper level.

This concept in turn can be coupled to the Heisenberg uncertainty principle. The uncertainty of the electron energy in the level u we define to be ΔW. Then

$$\Delta W_u \, \Delta t \gtrsim \frac{h}{2\pi}$$

or

$$\Delta W_u \gtrsim \frac{h}{2\pi} \sum_{l<u} A_{ul}.$$

In other words, the energy level u has an energy width set by nature at ΔW.

The lower level, l, can be expected to show a similar broadening. We therefore have a ΔW_l and γ_l for that level as well. Consider the absorption process depicted in Fig. 11-1. The electron leaves the level l from the energy range ΔW_l. The probability of the electron being at some point in this energy band is α_l. By a similar token the probability of the electron being at some energy within ΔW_u in the final state is α_u. Since any starting energy within the lower level can correspond to the full range in energy in the upper level we see that the full absorption coefficient is a convolution between α_u and α_l. In Chapter 2 we saw that a convolution between dispersion profiles results in a new dispersion profile having a larger γ given by

$$\gamma = \gamma_u + \gamma_l.$$

It is this expression that is used for the natural broadening component in the computation of $\kappa_\nu \rho$, although in the case of a strong radiation field it is sometimes expanded to

$$\gamma_u = \sum_{l<u} A_{ul} + \sum_{l<u} I_\nu B_{ul} + \sum_{k>u} I_\nu B_{uk}.$$

The second term allows for stimulate emission and the third for electrons leaving the level u for higher levels by absorption of a photon. A similar expression is written for γ_l.

The observed γ is still larger in many stars than this natural width would imply. We go on to discuss some of these other broadening mechanisms.

PRESSURE BROADENING

The term pressure broadening implies a collisional interaction between the atoms absorbing the light and other particles. The other particles can be ions, electrons, atoms of the same element as the absorbers or another type, or in cool stars they may be molecules. One imagines the atomic levels of the transition of interest to be disturbed such that their energy is altered. The distortion is a function of the separation, R, between the absorber and the perturbing particle. We expect the upper level, u, to be more strongly altered than the lower level, l. Figure 11-2 shows a schematic presentation of this situation where the potential energy of the level is plotted against the distance

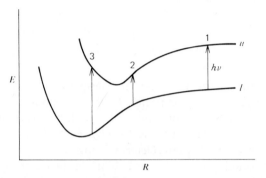

Fig. 11-2   The energies associated with two atomic levels u and l depend on the position R of the perturber. The transition between the levels u and l may be more or less energetic than the unperturbed case depending on the perturber distance.

to the perturber. Transitions of type 1 have the undisturbed energy. Transitions of type 2 have lower energy, while type 3 shows an increased energy. Depending on the distribution of encounter separations, R, and the shape of the energy curves, the net effect of all the absorbers along a line through the stellar photosphere may be a line shift and asymmetry as well as broadening. Line asymmetries and shifts due to collisions have been studied very little in stellar spectra. In keeping with the level of treatment, we shall concentrate on the broadening and refer the reader to the references for discussion of shifts and asymmetries.

The change in energy induced by the collision when plotted against R can often be approximated by a power law of the form

$$\Delta E = \frac{a}{R^n} \qquad a = \text{constant} \qquad (11\text{-}23)$$

where the integer n depends on the type of interaction. For example, the dipole coupling force between pairs of atoms of the same kind is proportional to R^{-4}, making ΔE proportional to R^{-3}, so $n = 3$ is used for this interaction. Dipole coupling perturbations are called resonance broadening. As a second example we cite the result of London (1930a, 1930b) who found theoretically that when the perturber is of a different species, the perturbation energy is proportional to R^{-6}, so $n = 6$. Table 11-1 gives a summary of the types of interactions that appear to be important in stars.*

---

* Resonance broadening, $n = 3$, may be important in some cases (see Cayrel and Traving 1960).

Table 11-1   Types of Pressure Broadening

n	Type	Lines Affected	Typical Perturber
2	Linear Stark	Hydrogen	Protons and electrons
4	Quadratic Stark	Most lines especially in hot stars	Electrons
6	van der Waals	Most lines especially in cooler stars	Neutral H

We can convert the energy change given in eq. 11-23 to a change in frequency in the spectrum by subtracting the equation for the lower level from the equation for the upper level, $\Delta E_u - \Delta E_l = h \, \Delta v$ or,

$$\Delta v = \frac{C_n}{R^n} \tag{11-24}$$

The interaction constant, C_n, must be measured or calculated for each transition and type of interaction. It is known for only a few lines.

In most of what we shall consider, it is assumed that no atomic transitions result from the collision itself. This assumption is called the adiabatic assumption. Baranger (1958) has given a generalized theory in which this limitation is not imposed. A review of the progress made in understanding pressure broadening has been given by Van Regemorter (1965). Additional material and references can be obtained from Griem (1974), Fuhr et al. (1973), and Van Regemorter (1973).

THE IMPACT APPROXIMATION

The temperatures of most stellar atmospheres lead to relatively high particle velocities. This is especially true for hydrogen and helium atoms and for electrons. Furthermore, for stars on or above the main sequence, the atmospheric pressures are modest ($\lesssim 1$ atm). Under these conditions one expects the duration of the collision to be small compared to the time between collisions. In this way one is led to consider the impact approximation as the starting point for a theoretical treatment. The impact approximation is also referred to as interruption and phase shift approximations.

A simple form of the impact treatment was proposed by Lorentz (1906) in which he assumed the electromagnetic wave was terminated by the impact, with the remaining photon energy being converted to kinetic energy. This simple treatment was refined under the hands of Lenz (1924), Kallman and

London (1929), and Weisskopf (1932). In place of a termination of the electro-magnetic wave, they showed how a phase change in the oscillation could occur and could account for the broadening.

Suppose, as shown in Fig. 11-3, we have a photon of duration W (typically about 10^{-9} sec) that we view as a product of an infinite sinusoid and a box. The spectrum of the electric field E is then a sinc $[\pi(v - v_0)W]$ function centered on the frequency of the sinusoid and of characteristic width $\Delta v = 1/W$. (Of course the intensity spectrum is EE^* or sinc^2 $\pi(v - v_0)W$.) If the same sinusoid is being absorbed or emitted by an atom that suffers several impacts from perturbers, one can view the wave as shown in Fig. 11-4. Several successive segments of the sinusoid result. The phase between segments has been changed. This can then be viewed as a sum of boxes of length W_j multiplied with sinusoidals. The original photon has been broken down into two or more shorter pieces. The Fourier transform of such a sum is the sum of the transforms. But since each W_j is smaller than the original W, the line is broadened with each $\Delta v_j = 1/W_j$ being greater than $\Delta v = 1/W$.

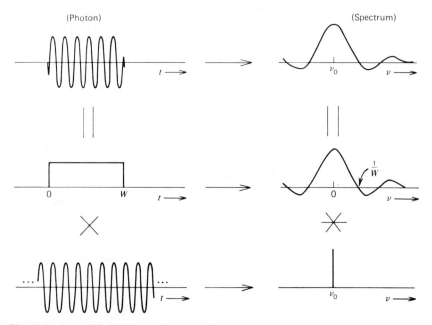

Fig. 11-3 A simplified way of viewing a photon and its spectrum. The sinusoidal oscillation lasts $\sim 10^{-9}$ sec in time or ~ 30 cm in length, but the duration of the photon determines the width of the spectral line.

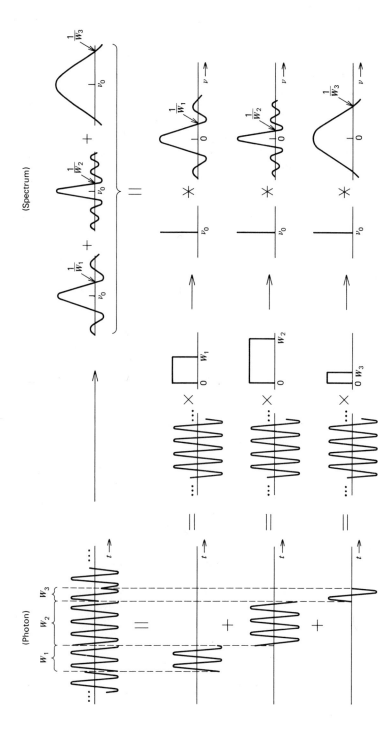

(Photon)

(Spectrum)

Fig. 11-4 The photon wave at the upper left corner is disturbed by collisions that produce phase shifts. Each of these segments can be thought of a product of an infinite sinusoidal with a box (center). In the transform domain (right) the sinc functions of the segmented photon sum together to give the spectrum. Each box is necessarily shorter than the full photon, making the spectral line broader as a result of the collisions.

233

The distribution of the W_j's along with the sinc function determines the shape of the absorption coefficient. The distribution, P, of W_j is*

$$dP(W_j) = e^{-W_j/W_0} \frac{dW_j}{W_0} \tag{11-25}$$

where W_0 is the average length of uninterrupted segment. The atomic line absorption coefficient is then proportional to the weighted average,

$$\int_0^\infty \left[\frac{\sin \pi(v - v_0)W}{\pi(v - v_0)W} \right]^2 e^{-W/W_0} \frac{dW}{W_0}.$$

This integral can be evaluated (see Petit Bois 1961, for example). The result is

$$\alpha = b \frac{2}{4\pi^2(v - v_0)^2 + (1/W_0)^2}$$

where b is a constant of proportionality. This can be rewritten

$$\alpha = \text{const} \frac{\gamma_n/4\pi}{(v - v_0)^2 + (\gamma_n/4\pi)^2}. \tag{11-26}$$

The quantity $\gamma_n = 2/W_0$ is defined to be the collisional damping constant.

This is a remarkable result because we have derived a dispersion profile similar to that for natural damping. Furthermore, the impact approach has led to this result quite independent of the type of collisional interaction.

In order to use eq. 11-26 in a line profile calculation, it is necessary to know γ_n, including its pressure and temperature dependences. Unlike radiation damping, collisional damping is a function of depth in the stellar photosphere.

THEORETICAL EVALUATION OF γ_n

The simplest approach is to assume all encounters can be divided into two groups, depending on the strength of the encounter. We measure this strength by the size of the phase shift. If we find the phase is shifted by, say, 1 radian

---

* The derivation of this distribution is as follows. Let $P(W)$ be the probability that there is, and $p(W)$ that there is not, an impact collision in the interval of time W. Then $p(W) = 1 - P(W)$. Now in the interval of time between W and $W + dW$ the probability of collision is dW/W_0 where W_0 is the average collision time for the gas. Therefore the probability of no collision in dW is $p(dW) = 1 - dW/W_0$. Probability theory tells us that $p(W + dW) = p(W)p(dW) = p(W)(1 - dW/W_0)$ or $p(W + dW) = p(W) - p(W) dW/W_0$. However, from a Taylor expansion we have $p(W + dW) = p(W) + (dp/dW) dW + \dots$, so that we are led to equate dp/dW with $-p/W_0$, which upon integration gives $p(W) = e^{-W/W_0}$. Finally, $P(W) = 1 - p(W) = 1 - e^{-W/W_0}$ and $dP(W) = e^{-W/W_0} dW/W_0$.

or more, we count it, but if the shift is less we neglect it. Clearly this is an order of magnitude estimate for there is no reason to choose 1 radian or π radians and so on. The chosen value simply represents what investigators in the area consider a significant phase shift. The frequency change is given by eq. 11-24. The cumulative effect of this change in frequency is the phase shift we seek,

$$\phi = 2\pi \int_0^\infty \Delta v \, dt = 2\pi \int_0^\infty C_n R^{-n} \, dt. \qquad (11\text{-}27)$$

We can evaluate this integral with a precision consistent with the situation by assuming the perturber moves past the atom on a straight line trajectory

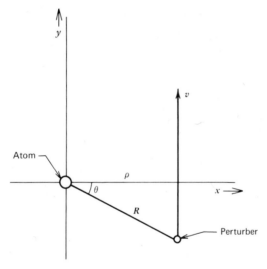

Fig. 11-5 The perturber passed the atom at a distance ρ with a velocity v. The y axis is chosen parallel to v.

and maintains a constant velocity v. Figure 11-5 shows the geometry including the impact parameter ρ and the angle θ. We can write

$$\rho = R \cos \theta$$

so that eq. 11-27 becomes

$$\phi = \int_0^\infty C_n \cos^n \theta \, \frac{dt}{\rho^n}.$$

Using the fact that $v = dy/dt = (\rho/\cos^2 \theta)(d\theta/dt)$ gives $dt = (\rho/v)(d\theta/\cos^2 \theta)$ and

$$\phi = \frac{C_n}{v\rho^{n-1}} \int_{-\pi/2}^{\pi/2} \cos^{n-2} \theta \, d\theta. \tag{11-28}$$

The value of the integral is given in Table 11-2 for the values of n in which we have interest. At this point we use our definition of a significant phase shift and set $\phi = 1$ radian, and thereby define a limiting impact parameter

$$\rho_0 \equiv \left[\frac{C_n}{v} \int_{-\pi/2}^{\pi/2} \cos^{n-2} \theta \, d\theta \right]^{1/(n-1)} \tag{11-29}$$

As stated at the beginning of the section, we count only those collisions that have $\rho < \rho_0$. The number of such collisions is given by $\pi\rho_0^2 vNT$ where

Table 11-2

n	$\int_{-\pi/2}^{\pi/2} \cos^{n-2} \theta \, d\theta$
2	π
3	2
4	$\pi/2$
6	$3\pi/8$

$\pi\rho_0^2 v$ is the volume swept out per second, T is the time interval over which we count the collisions and N is the number of perturbers per unit volume. If we set $T = W_0$ (from eq. 11-25) then on the average the number of collisions must be 1 and $\pi\rho_0^2 vNW_0 = 1$. In this way.

$$\gamma_n \equiv \frac{2}{W_0} = 2\pi\rho_0^2 vN \tag{11-30}$$

with ρ_0 given by eq. 11-29. The average relative velocity, v, for the atom of mass m_A and the perturber of mass m_p is, according to eq. 1-12,

$$v = \left[\frac{8kT}{\pi} \left(\frac{1}{m_A} + \frac{1}{m_p} \right) \right]^{1/2}. \tag{11-31}$$

NUMERICAL CALCULATION OF γ_n

We now obtain full numerical expressions for the quadratic Stark and van der Waals broadening. The quadratic Stark effect arises from perturbations by charged particles. In a stellar atmosphere these are electrons and ions. From Table 11-2 and eq. 11-29 we have $\rho_0 = (\pi C_4/2v)^{1/3}$ from which we deduce

$$\gamma_4 = 2\pi v N \left(\frac{\pi C_4}{2v}\right)^{2/3}$$

$$= 28.8 v^{1/3} C_4^{2/3} N.$$

A slightly more complicated theory (e.g., Lindholm 1945, Foley 1946, Unsöld 1955) gives

$$\gamma_4 = 38.8 v^{1/3} C_4^{2/3} N \qquad (11\text{-}32)$$

which, considering the approximate nature of the calculation, is the same as the previous equation. Nevertheless, to be numerically consistent with material in the literature, we adopt eq. 11-32. Now we take eq. 11-32 and write

$$\gamma_4 = 38.8 C_4^{2/3} \left[\frac{8kT}{\pi}\left(\frac{1}{m_A} + \frac{1}{m_e}\right)\right]^{1/6} N_e$$

$$+ 38.8 C_4^{2/3} \sum_i \left[\frac{8kT}{\pi}\left(\frac{1}{m_A} + \frac{1}{m_i}\right)\right]^{1/6} N_i.$$

The value of γ_4 is so insensitive to the values of m_i and m_A that the mass of *any* real ion and atom can be adopted. Further, we can take $\sum N_i = N_e$ without serious error, so the summation reduces to

$$\gamma_4 = 38.8 C_4^{2/3} N_e \left(\frac{8kT}{\pi}\right)^{1/6}\left[\left(\frac{1}{m_A} + \frac{1}{m_e}\right)^{1/6} + \left(\frac{1}{m_A} + \frac{1}{m_i}\right)^{1/6}\right]$$

or

$$\log \gamma_4 = 19.4 + \tfrac{2}{3}\log C_4 + \log P_e - \tfrac{5}{6}\log T \qquad (11\text{-}33)$$

where we must remember that the constant 19.4 is uncertain.

Values of the interaction constant C_4 are known from laboratory measurements for a few lines. A few examples are cited in Table 11-3 and additional values can be found in Allen (1963), Griem (1964), and Landolt–Börnstein (1950). Calculation of C_4 can also be done. Mugglestone and O'Mara (1966) and Hunger (1960) give examples of such computations.

Table 11-3 Stark Interaction Constants

Line	$\log C_4$
Na $\lambda 5890$	-15.17
$\lambda 5896$	-15.33
Mg $\lambda 5172$	-14.52
$\lambda 5183$	-14.52
$\lambda 5528$	-13.12

We now turn our attention to the van der Waals case. It is sufficient for almost all stars to consider neutral hydrogen and helium as the only perturbers. This approximation can be made because these two elements overwhelmingly dominate the chemical composition (refer to Chapter 14).

As with the quadratic Stark case, we start with eq. 11-30 using ρ_0 given by eq. 11-29 and Table 11-2 where $n = 6$. We find

$$\gamma_6 = 17.0 v^{3/5} C_6^{2/5} N. \tag{11-34}$$

We let v_{H} and v_{He} be the relative velocity for hydrogen and helium, respectively, so that for the case being considered,

$$\gamma_6 = 17.0 v_{\mathrm{H}}^{3/5} C_6^{2/5}(\mathrm{H}) \left\{ \left(1 + \frac{\Phi(\mathrm{H})}{P_e} \right)^{-1} \right.$$
$$\left. + \left(\frac{v_{\mathrm{He}}}{v_{\mathrm{H}}} \right)^{3/5} \left[\frac{C_6(\mathrm{He})}{C_6(\mathrm{H})} \right]^{2/5} \frac{N(\mathrm{He})}{N(\mathrm{H})} \right\} \frac{N(\mathrm{H}) P_g}{\sum NkT}.$$

The $\Phi(\mathrm{H})$ used here is the same as $\Phi(\mathrm{H})$ used in eq. 8-16 and the factor $(1 - \Phi(\mathrm{H})/P_e)^{-1}$ equals the number of neutral divided by total hydrogen particles, thereby taking into account the ionization of hydrogen. $N(\mathrm{H})$ represents the total number of hydrogen particles, neutrals and ions, per cubic centimeter even though γ_6 is due only to the neutrals. The last factor $P_g/\sum NkT$ is unity by definition since $\sum N$ represents the total number of particles per cubic centimeter and it is used in the next step.

The velocity ratio $(v_{\mathrm{He}}/v_{\mathrm{H}})^{3/5}$ is 0.684, and it is found by experiment that $[C_6(\mathrm{He})/C_6(\mathrm{H})]^{2/5} = 0.619$. Using in addition eq. 11-31 for v_{H}, we reduce eq. 11-34 to

$$\gamma_6 = 3.90 \times 10^{19} \left(0.992 + \frac{1}{\mu} \right)^{3/10} C_6^{2/5}(\mathrm{H})$$
$$\times \left\{ \left(1 + \frac{\Phi(\mathrm{H})}{P_e} \right)^{-1} + 0.422 A(\mathrm{He}) \right\} \frac{P_g T^{-7/10}}{\sum A}$$

or

$$\log \gamma_6 = 19.6 + \frac{3}{10} \log \left(0.992 + \frac{1}{\mu}\right) + \frac{2}{5} \log C_6(\mathrm{H}) + \log \left[\left(1 + \frac{\Phi(\mathrm{H})}{P_e}\right)^{-1}\right.$$

$$\left. + 0.422 A(\mathrm{He})\right] - \log \sum A + \log P_g - \frac{7}{10} \log T \qquad (11\text{-}35)$$

in which $\sum A$ is written for the sum of the element abundances expressed as a fraction of hydrogen by number and μ is the atomic mass (in atomic mass units) of the atom being perturbed. The second, fourth, and fifth terms in eq. 11-35 are often small enough to be negligible, especially in view of the approximate nature of the discussion leading to the expression. The essential behavior of γ_6 can be seen by neglecting these small terms,

$$\log \gamma_6 \cong 19.6 + \tfrac{2}{5} \log C_6(\mathrm{H}) + \log P_g - \tfrac{7}{10} \log T.$$

Unsöld (1955) has shown how the perturbation energy can be evaluated in an approximate way. The difference between the energies for the two levels of the transition leads to the expression

$$C_6(\mathrm{H}) = 0.3 \times 10^{-30}\left[\frac{1}{(I - \chi - \chi_\lambda)^2} - \frac{1}{(I - \chi)^2}\right] \qquad (11\text{-}36)$$

where I is the ionization potential and χ the excitation potential of the lower level in electron volts for the atom of interest. χ_λ denotes $1.24 \times 10^4/\lambda$ with λ in angstroms and is the energy of a photon in the line.

An illustration of the van der Waals and quadratic Stark damping is given for the Na D lines in Fig. 11-6. Log $C_6(\mathrm{H})$ has a value of -31.7. The damping constants are shown as a function of optical depth in a solar type model and it can be seen how the Stark broadening is much less than the van der Waals broadening over most of the atmosphere. For other lines, such as the Mg b lines, the difference is less and the Stark broadening dominates in the deeper layers.

Equations 11-35 and 11-36 have been used in a great many line profile calculations. The validity of γ_6 computed in this way from the R^{-6} potential has been seriously questioned by Kusch (1958), Roueff and Van Regemorter (1969, 1971), Blackwell et al. (1972), and others. Theoretical and observational studies have in some cases produced scaling or enhancement factors to be multiplied with the values given by eq. 11-35. Work of this sort has been reported by Fullerton and Cowley (1971), Cowley et al. (1968), Burgess and Grindley (1970), and Holweger (1971, 1972). More complete revisions have also been made, such as those by Lewis et al. (1971, 1972), and Brueckner (1971). Even in this case the dependence of γ_6 on pressure and temperature

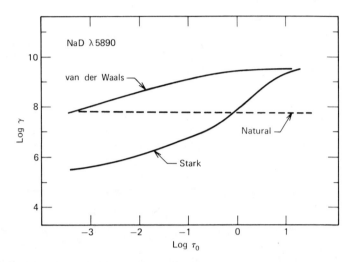

Fig. 11-6 The damping constants are shown as a function of depth in a solar model. The van der Waals constant is computed using eqs. 11-35 and 11-36. The Stark constant comes from eq. 11-33. For comparison, the natural radiation damping according to eq. 11-19 is shown.

is nearly the same as given by eq. 11-35, with the major difference being in the absolute size of γ_6. Additional references can be found in Van Regemorter (1973).

LIMITATION OF THE IMPACT MODEL

We based the previous calculations on the impact model, noting that the pressures and temperatures in most stellar atmospheres favor the assumptions of the concept. The Δv spectral broadening corresponds to $1/W$ where W is the time interval between collisions. In order to class a collision as an impact, W must be large compared to the duration of the collision. This implies small Δv or a shift not too far from the line center. A more quantitative description of the validity of the impact assumption is as follows.

We define $\Delta t_b = \rho/v$ where ρ is the average impact parameter and v the average impact velocity. Δt_b is a measure of the duration of the collisions and for the impact approximation to hold, we need $W \gg \Delta t_b$. Or in terms of frequency $\Delta v \ll \Delta v_b$ where

$$\Delta v_b = \frac{1}{\Delta t_b} = \frac{v}{\rho}$$

or

$$\rho = \frac{v}{\Delta v_b}.$$

Equation 11-24 gives $\Delta v = C_n/R^n$ from which we can write $\Delta v_b = C_n/\rho^n$. Then using $\rho = v/\Delta v_b$ one gets

$$\Delta v_b = \left(\frac{v^n}{C_n}\right)^{1/(n-1)} \tag{11-37}$$

We now calculate Δv_b for a typical case to see if we are justified in applying the impact treatment. Consider the Na D_2 line for which we have values of C_4 and C_6. In the Stark case $\log C_4 = -15.2$, and for a perturbing ion having an atomic mass of 20 the thermal velocity in a solar type star is about 5×10^5 cm/sec, giving $\Delta v_b = 5 \times 10^{12}$ sec^{-1} or $\Delta \lambda_b \cong 60$ Å. This is large compared to the observed line widths, and perturbation by electrons or protons give a $\Delta \lambda_b$ still larger. Switching to the van der Waals case, $\log C_6 = -31.7$, leading to a $\Delta \lambda_b \sim 10^2$ Å. We conclude that the impact approximation is appropriate in the stellar atmosphere context for the metal lines.

The same type of calculation done for the hydrogen Balmer lines (where the linear Stark effect is applicable) shows a different situation. The $\Delta \lambda_b$ for electrons is of the same order as the width of the lines. For protons the $\Delta \lambda_b$ is very much *smaller*. Here is a situation where, at least for the proton and other ion perturbations, a new approach is called for. When $\Delta \lambda \gg \Delta \lambda_b$ the duration of the collision is long compared to the time between collisions (the inverse of the impact case). The logical limiting approximation is to treat the distribution of perturbers as stationary. This case is called the static approximation (also quasistatic or statistical).

THE IONIC BROADENING OF HYDROGEN LINES

Struve (1929) gave convincing arguments that the great widths of the hydrogen lines are due to the linear Stark effect. That portion of the linear Stark effect induced by *ions* near the hydrogen atom is the subject of this section. The contribution of the electrons is dealt with in the next section.

Holtsmark (1919) began the studies of theoretical distribution of ions in a plasma and the subsequent broadening caused in removing the degeneracy of the hydrogen energy levels. The splitting of the energy levels can be expressed as the wavelength shift of the spectral components,

$$\Delta \lambda_j = c_j E \tag{11-38}$$

in which E is the electric field at the hydrogen atom and c_j the constant of proportionality for the jth component. The electric field, E, depends on the distribution of ions. We imagine the situation to look like Fig. 11-7. If we let R be the separation of some ion from the hydrogen atom, then the field from that ion is

$$E = \frac{e}{R^2} \tag{11-39}$$

where we have assumed the perturbing ion is singly ionized.*

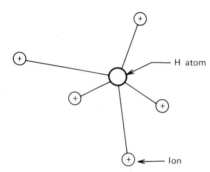

Fig. 11-7 The distribution of ions in space gives a nonzero electric field at the position of the atom. This field causes a perturbation in the energy levels of the atom.

The strongest interaction occurs between the atom and the closest perturber. The lowest order approximation giving the distribution of electric fields is begun by assuming pure binary encounters and completely neglecting the lesser fields contributed by the other ions slightly farther away. (This approach is also known as the nearest neighbor approximation.) We take N to be the number density of perturbers. The probability of finding an ion in the shell bounded by R and $R + dR$ (see Fig. 11-8) is proportional to the number of ions expected to be in that size volume, namely, $4\pi R^2 \, dRN$. We seek that ion that is closest to the atom so we combine the probability that the ion is *not* in the sphere of radius R but *is* in the shell of thickness dR. The probability, $P(R)$, of both these conditions occurring at the same time is the product of their individual probabilities,

$$P(R) \, dR = P_n(R)4\pi R^2 \, dRN \tag{11-40}$$

where $P_n(R)$ denotes the probability of no ion being in the sphere. We know that as R increases, $P_n(R)$ must diminish according to the increased likelihood

* From eqs. 11-38 and 11-39 we see that the spectral shift can be written in the form of eq. 11-24 with $n = 2$.

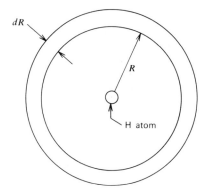

Fig. 11-8 The shell of radius R and thickness dR around the hydrogen atom has a volume $4\pi R^2\, dR$.

of encompassing some ion. This leads us to write

$$P_n(R + dR) = P_n(R)(1 - 4\pi R^2\, dR N)$$
$$= P_n(R) - 4\pi R^2 N P_n(R)\, dR.$$

But since quite generally

$$P_n(R + dR) = P_n(R) + \frac{dP_n}{dR}\, dR,$$

we write

$$\frac{dP_n}{dR} = -4\pi R^2 N P_n$$

which upon integration gives

$$P_n(R) = P_n(0)e^{-(4\pi/3)R^3 N}.$$

When $R = 0$, $P_n = 1$ so the integration constant need not be written again. It is convenient to use the mean distance between ions, $R_0 = (4\pi N/3)^{-1/3}$, as the unit of length. We then have

$$P_n(R) = e^{-(R/R_0)^3}$$

and finally, from eq. 11-40

$$P(R)\, dR = e^{-(R/R_0)^3} \frac{3R^2\, dR}{R_0^3}$$

$$= 3\left(\frac{R}{R_0}\right)^2 e^{-(R/R_0)^3}\, d\left(\frac{R}{R_0}\right).$$

This distribution in R can be translated into the distribution of electric fields by using $E = e/R^2$, $E_0 \equiv e/R_0^2$, and $|d(R/R_0)| = \frac{1}{2}(E_0/E)^{3/2} d(E/E_0)$. One finds

$$P(E) \, dE = \frac{3}{2} \left(\frac{E_0}{E}\right)^{5/2} e^{-(E_0/E)^{3/2}} \frac{dE}{E_0}$$

or

$$P(\beta) \, d\beta = \tfrac{3}{2}\beta^{-5/2}e^{-\beta^{-3/2}} \, d\beta \tag{11-41}$$

where $\beta \equiv E/E_0$ is the field strength in units of E_0. Numerically $E_0 = 1.25 \times 10^{-9}N^{2/3}$.

This is the simplified distribution based on the nearest neighbor approximation which explicitly avoids the vector addition of E from several perturbers. More elaborate calculations take into account the full distribution of perturbers and so one obtains a somewhat more accurate result for $P(\beta)$ called the Holtsmark distribution. For example, the Holtsmark distribution as given by Underhill and Waddell (1959) is

$$P(\beta) \, d\beta = \frac{4}{3\pi} \sum_{j=0}^{\infty} (-1)^j \frac{\Gamma[(4j + 6)/3]}{(2j + 1)} \beta^{2j+2} \tag{11-42}$$

in which Γ is the gamma function. Figure 11-9 shows the perturber distribution according to eqs. 11-41 and 11-42. The two distributions give the same result for large fields as one would expect based on the simple argument that strong fields arise from close encounters, in which case one ion dominates and the nearest neighbor approximations are fulfilled. Still more detailed calculations have been done (Ecker 1957, Mozer and Baranger 1960, and Cooper 1968), but because of the low densities encountered in most stellar atmospheres these refinements have marginal effect.

In all these cases one can derive an asymptotic expression for the line wing. Large $\Delta\lambda$ implies large β, in which case eq. 11-41 gives

$$P(\Delta\lambda) \, d\Delta\lambda = \text{const } \Delta\lambda^{-5/2}$$

where we have used $\Delta\lambda = c_j E_0$ (eq. 11-38 combined with the definition of β). Aller (9163) gives this wing absorption coefficient per hydrogen atom as

$$\alpha = 3.21 \times 10^2 C_u \frac{P_e}{T} \Delta\lambda^{-5/2} \tag{11-43}$$

where it is assumed that $N_e = N$ and for which the coefficients C_u are tabulated in Table 11-4 for a few lines assuming $\Delta\lambda$ is in angstroms. With modern methods of computing, one is usually not satisfied with using the asymptotic form in eq. 11-43.

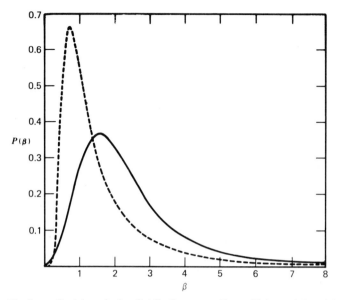

Fig. 11-9 The linear Stark broadening distributions according to Holtsmark (——) (eq. 11-42) and the nearest neighbor approximation (----) (eq. 11-41).

Table 11-4 Coefficients for Eq. 11-43

Line	C_u
Hα	3.13×10^{-16}
Hβ	0.88×10^{-16}
Hγ	0.44×10^{-16}
Hδ	0.32×10^{-16}
Hε	0.26×10^{-16}

The complete hydrogen Stark broadening, arising from many atoms but normalized to the f value for the line, is the weighted sum of the individual Stark components with each showing a $P(\beta)$ or $P(\Delta\lambda)$ distribution according to eq. 11-42. Following the customary definition (Underhill 1951) we denote the f values of the unshifted Stark components as f_0 and define the shifted components' f value as

$$f_{\pm} \equiv \frac{1}{l^2} \sum_j f_j,$$

the summation being carried over all shifted components. The principal quantum number of the lower level has been called l. The Stark profile is then defined as the weighted sum

$$S\left(\frac{\Delta\lambda}{E_0}\right)\frac{d\Delta\lambda}{E_0} \equiv \sum \frac{f_j}{\sum f_j} P(\beta)\,d\beta = \sum \frac{f_j}{l^2 f_\pm} P\left(\frac{\Delta\lambda}{c_j E_0}\right)\frac{d\Delta\lambda}{c_j E_0}. \qquad (11\text{-}44)$$

This equation defines the Stark profile $S(\Delta\lambda/E_0)$. The sum S is normalized to unity according to

$$\int_{-\infty}^{\infty} S\left(\frac{\Delta\lambda}{E_0}\right)\frac{d\Delta\lambda}{E_0} = 1.$$

In Fig. 11-10, $S(\Delta\lambda/E_0)$ is shown for the first few Balmer lines. The oscillator strengths are given in Table 11-5.

The energy subtracted from a unit strength I_λ beam can now be written

$$\alpha\,d\Delta\lambda = \frac{\pi e^2}{mc}\frac{\lambda^2}{c}\left[\frac{f_\pm\,S(\Delta\lambda/E_0)}{E_0} + f_0\delta(\Delta\lambda)\right]d\Delta\lambda. \qquad (11\text{-}45)$$

This expression for α is evaluated numerically using tables of $S(\Delta\lambda/E_0)$ such as those given by Underhill and Waddell (1959). $\delta(\Delta\lambda)$ is a delta function at the line center.

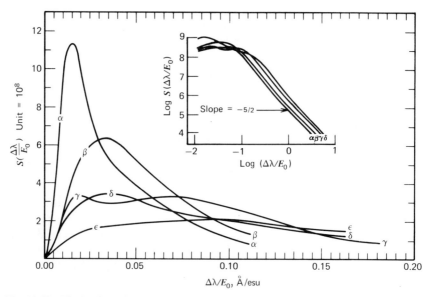

Fig. 11-10   The Stark profiles $S(\Delta\lambda/E_0)$ are shown for the first few Balmer lines. Only half the profiles are shown since they are symmetric. The insert shows the same functions in logarithmic coordinates where the $\Delta\lambda^{-5/2}$ wing is apparent. (Data from Underhill and Waddell 1959.)

Table 11-5   Hydrogen Oscillator Strengths

Line	u	f_0	$f_\pm$	f
Hα	3	0.248689	0.392058	0.640742
Hβ	4	0.00	0.119321	0.119321
Hγ	5	0.007014	0.037656	0.044670
Hδ	6	0.00	0.022093	0.022093
Hε	7	0.001310	0.011394	0.012704
Hζ	8	0.00	0.008036	0.008036

THE ADDITION OF ELECTRON BROADENING IN HYDROGEN LINES

Until about 1954 (see Griem 1954) it was thought that broadening due to electrons could be neglected. This proved not to be the case and several formulations have been advanced that include the electron contribution.

The impact approximation for electron broadening gives a dispersion profile that is added to the ionic contribution. Consider, for example, eq. 11-43. In such a wing formulation we can use $\text{const}/\Delta\lambda^2$ for the dispersion profile, giving

$$\alpha = 321 \, C_u \frac{P_e}{T} \frac{1}{\Delta\lambda^{5/2}} + \frac{\text{const}}{\Delta\lambda^2}$$

or

$$\alpha = 321 \, C_u \frac{P_e}{T} \Delta\lambda^{-5/2}[1 + R \, \Delta\lambda^{1/2}] \tag{11-46}$$

in place of eq. 11-43. A tabulation of R has been given by Griem et al. (1959). An approximate analytical expression is

$$R = \frac{4.6 Y(u)}{T^{1/2}} \left[\log\left(\frac{4 \times 10^6 T}{u^2 \, N_e^{1/2}}\right) - 0.125 \right] \frac{u^5 + l^5}{u^2 l^2 (u^2 - l^2)^{1/2}}$$

which is a slight modification of the formula given by Griem (1962) and in which u and l are the quantum numbers of the upper and lower levels of the transition. The constant $Y(u)$ is 1.5 for Hα, 1.12 for Hβ, 1.06 for Hγ, and 0.96 for Hδ.

A more general formulation uses the full Stark broadening expressed in eq. 11-45 rather than the asymptotic eq. 11-43. Several recipes have been advanced for α. See, for example, the work of Griem (1962, 1967) with application by Baschek (1964), Stienon (1964), Peterson and Shipman (1973), and Fowler (1974).

As mentioned earlier, $\Delta\lambda_b$ for electrons is of the same order as the line width. It is not entirely clear that the impact approximation is suitable. Edmonds et al. (1967) have worked out a semiempirical formulation in which both electrons and ions are treated with the static approximation. They do this by an empirical manipulation of the perturber density as a function of $\Delta\lambda$. Since $N_{ion} = N_e$ to a reasonable approximation, the static eq. 11-45 can be used with E_0 calculated using $2N_e$ as the number of perturbers per cubic centimeter when both electrons and ions cause broadening. As $\Delta\lambda$ becomes less than $\Delta\lambda_b$ one expects a transition from static to impact regimes. But laboratory empirical measurements show that the electron contribution decreases as $\Delta\lambda$ becomes small and goes to practically zero as $\Delta\lambda$ decreases to $\lesssim \Delta\lambda_b/25$. Empirically then, this small $\Delta\lambda$ case is given using a perturber density of N_e rather than $2N_e$. The smooth transition needed between these cases is made by letting N be a function of position from line center according to

$$N = N_e\left[1.5 + 0.5\left(\frac{7\,\Delta v/\Delta v_b - 1}{7\,\Delta v/\Delta v_b + 1}\right)\right]$$

where Δv_b is given by eq. 11-37. Their formulation also includes the effects of charge correlation and shielding. The value of α is tabulated for the first 18 series members in the Lyman, Balmer, Paschen, and Brackett series. There have been several applications of this formulation, for example, Olson (1968) and Strom and Peterson (1968). A more complete theoretical approach can be found in the work of Vidal, et al. (1971, 1973).

All the pressure broadening is a result of shifts in the individual atomic spectral lines due to energy perturbation. Other wavelength shifts result from the Doppler effect. We now turn our attention to this line broadening mechanism.

THERMAL BROADENING

Each atom has a component of velocity along our line of sight to the star due to its thermal motion. We denote this velocity, the radial velocity, by v_r. Then the Doppler shift of the line emitted or absorbed by the atom is given by

$$\frac{\Delta\lambda}{\lambda} = \frac{\Delta v}{v} = \frac{v_r}{c} \tag{11-47}$$

where c is the velocity of light. The distribution of $\Delta\lambda$'s gives us the shape of the absorption coefficient, but since $\Delta\lambda$ and v_r are proportional we may simply use the distribution of atomic velocities according to the v_r form of eq. 1-9,

$$\frac{dN}{N} = \frac{1}{v_0\sqrt{\pi}} e^{-v_r^2/v_0^2} \, dv_r \tag{11-48}$$

where the variance, v_0^2, is related to the temperature by

$$v_0^2 = \frac{2kT}{m}. \tag{11-49}$$

Here m is the mass of the atom and k is Boltzmann's constant. The Doppler wavelength shift associated with v_0 is

$$\Delta\lambda_D \equiv \frac{v_0}{c}\lambda = \frac{\lambda}{c}\sqrt{\frac{2kT}{m}} = 4.301 \times 10^{-7}\lambda\sqrt{\frac{T}{\mu}} \tag{11-50}$$

in which μ is the atomic weight in atomic mass units. The corresponding frequency shift is

$$\Delta v_D \equiv \frac{v_0}{c}v = \frac{v}{c}\sqrt{\frac{2kT}{m}} = 4.301 \times 10^{-7}v\sqrt{\frac{T}{\mu}}. \tag{11-51}$$

The distribution of $\Delta\lambda$'s is then

$$\frac{dN}{N} = \frac{1}{\sqrt{\pi}} e^{-(\Delta\lambda/\Delta\lambda_D)^2} \, d\left(\frac{\Delta\lambda}{\Delta\lambda_D}\right). \tag{11-52}$$

The energies removed from unit intensity beams are $(\pi e^2 f/mc)(\lambda^2/c)$, and $(\pi e^2 f/mc)$ times dN/N or

$$\alpha \, d\lambda = \frac{\sqrt{\pi}e^2}{mc} f \frac{\lambda^2}{c} \frac{1}{\Delta\lambda_D} e^{-(\Delta\lambda/\Delta\lambda_D)^2} \, d\lambda \tag{11-53}$$

and

$$\alpha \, dv = \frac{\sqrt{\pi}e^2}{mv} f \frac{1}{\Delta v_D} e^{-(\Delta v/\Delta v_D)^2} \, dv. \tag{11-54}$$

These expressions define α for the thermally broadened absorption coefficient.

COMBINING ABSORPTION COEFFICIENTS

In the preceding pages we have seen that there are many processes: natural broadening, Stark broadening, van der Waal's broadening, and thermal broadening, all of which can take place simultaneously. Consider any two of these absorption coefficients and imagine the first one to be synthesized as a series of δ functions. Each of these δ functions experiences the broadening of the second absorption coefficient. In short, the two distributions must be convolved together to obtain the combined result. If we expand this to include all of the above mentioned processes, we get

$$\alpha(\text{total}) = \alpha(\text{nat}) * \alpha(\text{Stark}) * \alpha(\text{v.d.W.}) * \alpha(\text{thermal}) \qquad (11\text{-}55)$$

and $\int_0^\infty \alpha(\text{total})\,dv$ is normalized to $\pi e^2 f/mc$. The first three coefficients have a dispersion profile according to the impact approximation (hydrogen lines are treated separately below). In Chapter 2 we saw that such convolutions result in a new dispersion profile with $\beta = \sum \beta_j$ where the β_j's are the half half widths of the individual profiles.

It is immediately clear that the first three coefficients in eq. 11-55 can be written as a single dispersion profile in which the damping constant is

$$\gamma = \gamma_{\text{nat}} + \gamma_4 + \gamma_6 \qquad (11\text{-}56)$$

This is an extremely powerful result because of the great computational simplification.

The last convolution in eq. 11-55 (the thermal profile) cannot be handled so easily. We have by eq. 11-54

$$\alpha = \frac{\pi e^2}{mc} f\left(\frac{\gamma/4\pi^2}{(\Delta v)^2 + (\gamma/4\pi)^2}\right) * \left(\frac{1}{\sqrt{\pi}\,\Delta v_D} e^{-(\Delta v/\Delta v_D)^2}\right) \qquad (11\text{-}57)$$

where γ is now given by eq. 11-56. The convolution of the dispersion and Gaussian profiles is the Voigt function first mentioned in Chapter 2. The Voigt function is normalized to unit area and formally defined by

$$V(\Delta v, \Delta v_D, \gamma) \equiv \left(\frac{\gamma/4\pi^2}{(\Delta v)^2 + (\gamma/4\pi)^2}\right) * \left(\frac{1}{\sqrt{\pi}\,\Delta v_D} e^{-(\Delta v/\Delta v_D)^2}\right)$$

$$= \int_{-\infty}^{\infty} \left(\frac{\gamma/4\pi^2}{(\Delta v - \Delta v_1)^2 + (\gamma/4\pi)^2}\right)\left(\frac{1}{\sqrt{\pi}\,\Delta v_D} e^{-(\Delta v_1/\Delta v_D)^2}\right) d\Delta v_1.$$

$$(11\text{-}58)$$

If we choose to express Δv in units of Δv_D, $u \equiv \Delta v/\Delta v_D$, as suggested by the exponential term, and if in addition we use a for $(\gamma/4\pi)/\Delta v_D$, then we can write the Voigt function as

$$V(u, a) = \frac{1}{\sqrt{\pi}\,\Delta v_D} \frac{a}{\pi} \int_{-\infty}^{\infty} \frac{e^{-u_1^2}}{(u - u_1)^2 + a^2}\,du_1. \qquad (11\text{-}59)$$

The H function or Hjerting function $H(u, a)$ is a closely related function (Hjerting 1938) and is in common use along with $V(u, a)$. It is defined by

$$V(u, a) \equiv \frac{1}{\sqrt{\pi} \, \Delta v_D} \, H(u, a) = \frac{\lambda^2/c}{\sqrt{\pi} \, \Delta \lambda_D} \, H(u, a).$$

In the λ form (the right side of the above equation), $u \equiv \Delta \lambda / \Delta \lambda_D$ and $a \equiv (\lambda^2 \gamma / 4\pi c)/\Delta \lambda_D$. The H function is normalized to an area of $\sqrt{\pi}$.

The atomic absorption coefficient of eq. 11-55 we now write as

$$\alpha = \frac{\pi e^2}{mc} \, f V(u, a)$$

$$= \frac{\sqrt{\pi} e^2}{mc} \frac{f}{\Delta v_D} \, H(u, a) = \frac{\sqrt{\pi} e^2}{mc} \frac{\lambda^2}{c} \frac{f}{\Delta \lambda_D} \, H(u, a). \qquad (11\text{-}60)$$

The absorption coefficient can be calculated using the H function tabulations given in Table 11-6. Harris (1948) has given these functions which relate to $H(u, a)$ according to

$$H(u, a) = H_0(u) + a H_1(u) + a^2 H_2(u) + a^3 H_3(u) + a^4 H_4(u) + \cdots \qquad (11\text{-}61)$$

where the damping parameter a is a small number. In many applications the function is needed to only three or four place precision and then interpolation in Table 11-6 followed by application of eq. 11-61 is a suitable method for computing $H(u, a)$. In the far wing the asymptotic expressions

$$H_1(u) = \frac{0.56419}{u^2} + \frac{0.846}{u^4}$$

and

$$H_3(u) = \frac{-0.56}{u^4}$$

are useful for extending the values given in the table. When higher precision is needed, the tables of Finn and Mugglestone (1965), who give $H(u, a)$ to six figures, for $0 \le a \le 1$ and $0 \le u \le 22$, or the tables of Hummer (1965), who gives $H(u, a)/\pi^{1/2}$ to eight places for $0 \le a \le 0.5$ and $0 \le u \le 10$, are useful.

In some contexts it is better to renormalize the Voigt function to unit height and unit half-half width instead of unit area. Such a normalization is particularly useful for fitting Voigt functions to profiles and other data. Details are given by Elste (1953).

The absorption coefficient for hydrogen is given by eq. 11-55 with the Stark broadening, if we neglect the electron contribution, according to

Table 11-6 **The Functions** H_0, H_1, H_2, H_3, **and** H_4

u	$H_0(u)$	$H_1(u)$	$H_2(u)$	$H_3(u)$	$H_4(u)$
0.0	$+1.000\ 000$	$-1.128\ 38$	$+1.000\ 0$	-0.752	$+0.50$
0.1	$+0.990\ 050$	$-1.105\ 96$	$+0.970\ 2$	-0.722	$+0.48$
0.2	$+0.960\ 789$	$-1.040\ 48$	$+0.883\ 9$	-0.637	$+0.40$
0.3	$+0.913\ 931$	$-0.937\ 03$	$+0.749\ 4$	-0.505	$+0.30$
0.4	$+0.852\ 144$	$-0.803\ 46$	$+0.579\ 5$	-0.342	$+0.17$
0.5	$+0.778\ 801$	$-0.649\ 45$	$+0.389\ 4$	-0.165	$+0.03$
0.6	$+0.697\ 676$	$-0.485\ 52$	$+0.195\ 3$	$+0.007$	-0.09
0.7	$+0.612\ 626$	$-0.321\ 92$	$+0.012\ 3$	$+0.159$	-0.20
0.8	$+0.527\ 292$	$-0.167\ 72$	$-0.147\ 6$	$+0.280$	-0.27
0.9	$+0.444\ 858$	$-0.030\ 12$	$-0.275\ 8$	$+0.362$	-0.30
1.0	$+0.367\ 879$	$+0.085\ 94$	$-0.367\ 9$	$+0.405$	-0.31
1.1	$+0.298\ 197$	$+0.177\ 89$	$-0.423\ 4$	$+0.411$	-0.28
1.2	$+0.236\ 928$	$+0.245\ 37$	$-0.445\ 4$	$+0.386$	-0.24
1.3	$+0.184\ 520$	$+0.289\ 81$	$-0.439\ 2$	$+0.339$	-0.18
1.4	$+0.140\ 858$	$+0.313\ 94$	$-0.411\ 3$	$+0.280$	-0.12
1.5	$+0.105\ 399$	$+0.321\ 30$	$-0.368\ 9$	$+0.215$	-0.07
1.6	$+0.077\ 305$	$+0.315\ 73$	$-0.318\ 5$	$+0.153$	-0.02
1.7	$+0.055\ 576$	$+0.300\ 94$	$-0.265\ 7$	$+0.097$	$+0.02$
1.8	$+0.039\ 164$	$+0.280\ 27$	$-0.214\ 6$	$+0.051$	$+0.04$
1.9	$+0.027\ 052$	$+0.256\ 48$	$-0.168\ 3$	$+0.015$	$+0.05$
2.0	$+0.018\ 3156$	$+0.231\ 726$	$-0.128\ 21$	$-0.010\ 1$	$+0.058$
2.1	$+0.012\ 1552$	$+0.207\ 528$	$-0.095\ 05$	$-0.026\ 5$	$+0.056$
2.2	$+0.007\ 9071$	$+0.184\ 882$	$-0.068\ 63$	$-0.035\ 5$	$+0.051$
2.3	$+0.005\ 0418$	$+0.164\ 341$	$-0.048\ 30$	$-0.039\ 1$	$+0.043$
2.4	$+0.003\ 1511$	$+0.146\ 128$	$-0.033\ 15$	$-0.038\ 9$	$+0.035$
2.5	$+0.001\ 9305$	$+0.130\ 236$	$-0.022\ 20$	$-0.036\ 3$	$+0.027$
2.6	$+0.001\ 1592$	$+0.116\ 515$	$-0.014\ 51$	$-0.032\ 5$	$+0.020$
2.7	$+0.000\ 6823$	$+0.104\ 739$	$-0.009\ 27$	$-0.028\ 2$	$+0.015$
2.8	$+0.000\ 3937$	$+0.094\ 653$	$-0.005\ 78$	$-0.023\ 9$	$+0.010$
2.9	$+0.000\ 2226$	$+0.086\ 005$	$-0.003\ 52$	$-0.020\ 1$	$+0.007$
3.0	$+0.000\ 1234$	$+0.078\ 565$	$-0.002\ 10$	$-0.016\ 7$	$+0.005$
3.1	$+0.000\ 0671$	$+0.072\ 129$	$-0.001\ 22$	$-0.013\ 8$	$+0.003$
3.2	$+0.000\ 0357$	$+0.066\ 526$	$-0.000\ 70$	$-0.011\ 5$	$+0.002$
3.3	$+0.000\ 0186$	$+0.061\ 615$	$-0.000\ 39$	$-0.009\ 6$	$+0.001$
3.4	$+0.000\ 0095$	$+0.057\ 281$	$-0.000\ 21$	$-0.008\ 0$	$+0.001$

(Continued)

Table 11-6 (*Continued*)

u	$H_0(u)$	$H_1(u)$	$H_2(u)$	$H_3(u)$	$H_4(u)$
3.5	+0.000 0048	+0.053 430	−0.000 11	−0.006 8	0.000
3.6	+0.000 0024	+0.049 988	−0.000 06	−0.005 8	0.000
3.7	+0.000 0011	+0.046 894	−0.000 03	−0.005 0	0.000
3.8	+0.000 0005	+0.044 098	−0.000 01	−0.004 3	0.000
3.9	+0.000 0002	+0.041 561	−0.000 01	−0.003 7	0.000
4.0	+0.000 0001	+0.039 250	0.000 00	−0.003 3	0.000

u	$H_1(u)$	$H_3(u)$	u	$H_1(u)$	$H_3(u)$
4.0	+0.039 250	−0.003 29	8.0	+0.009 0306	−0.000 15
4.2	+0.035 195	−0.002 57	8.2	+0.008 5852	−0.000 13
4.4	+0.031 762	−0.002 05	8.4	+0.008 1722	−0.000 12
4.6	+0.028 824	−0.001 66	8.6	+0.007 7885	−0.000 11
4.8	+0.026 288	−0.001 37	8.8	+0.007 4314	−0.000 10
5.0	+0.024 081	−0.001 13	9.0	+0.007 0985	−0.000 09
5.2	+0.022 146	−0.000 95	9.2	+0.006 7875	−0.000 08
5.4	+0.020 441	−0.000 80	9.4	+0.006 4967	−0.000 08
5.6	+0.018 929	−0.000 68	9.6	+0.006 2243	−0.000 07
5.8	+0.017 582	−0.000 59	9.8	+0.005 9688	−0.000 07
6.0	+0.016 375	−0.000 51	10.0	+0.005 7287	−0.000 06
6.2	+0.015 291	−0.000 44	10.2	+0.005 5030	−0.000 06
6.4	+0.014 312	−0.000 38	10.4	+0.005 2903	−0.000 05
6.6	+0.013 426	−0.000 34	10.6	+0.005 0898	−0.000 05
6.8	+0.012 620	−0.000 30	10.8	+0.004 9006	−0.000 04
7.0	+0.011 8860	−0.000 26	11.0	+0.004 7217	−0.000 04
7.2	+0.011 2145	−0.000 23	11.2	+0.004 5526	−0.000 04
7.4	+0.010 5990	−0.000 21	11.4	+0.004 3924	−0.000 03
7.6	+0.010 0332	−0.000 19	11.6	+0.004 2405	−0.000 03
7.8	+0.009 5119	−0.000 17	11.8	+0.004 0964	−0.000 03
8.0	+0.009 0306	−0.000 15	12.0	+0.003 9595	−0.000 03

eq. 11-45. If we denote the convolution of all the α's except the Stark term by the Voigt function $V(u, a)$, eq. 11-55 can be written

$$\alpha = \frac{\pi e^2}{mc} \left[\frac{f_\pm S(\Delta\lambda/E_0)}{E_0} + f_0 \delta(\Delta\lambda) \right] * V(u, a)$$

$$= \frac{\pi e^2}{mc} \left[f_\pm S(\Delta\lambda/E_0) * \frac{V(u, a)}{E_0} + f_0 V(u, a) \right].$$ (11-62)

In terms of the Hjerting function, eq. 11-62 becomes

$$\alpha = \frac{\pi e^2}{mc} \frac{\lambda^2}{c} \left[f_\pm S\!\left(\frac{\Delta\lambda}{E_0}\right) * \frac{H(u, a)}{\sqrt{\pi}\,\Delta\lambda_D} \frac{1}{E_0} + f_0 \frac{H(u, a)}{\sqrt{\pi}\,\Delta\lambda_D} \right].$$ (11-63)

Since $S(\Delta\lambda/E_0)$ is a distribution in $\Delta\lambda$ we have expressed $V(u, a)$ in the $\Delta\lambda$ coordinate rather than using Δv.

In stars of early spectral type, the Stark broadening is so much greater than the width of $V(u, a)$ that $V(u, a)$ looks like a δ function by comparison. In such a case the convolution in eq. 11-63 gives $S(\Delta\lambda/E_0)$ back again and one can write

$$\alpha = \frac{\pi e^2}{mc} \frac{\lambda^2}{c} \left[\frac{f_\pm S(\Delta\lambda/E_0)}{E_0} + f_0 \frac{H(u, a)}{\sqrt{\pi}\,\Delta\lambda_D} \right].$$ (11-64)

This simplified form is usually used in model atmosphere computations for the ionic broadening of hydrogen line profiles. Inclusion of the electron broadening can take the form of a modified S as in the Edmonds, Schlüter, Wells (1967) formulation or an additional term may be added to eq. 11-64 such as that given by Griem (1962, 1967).

THE MASS ABSORPTION COEFFICIENT FOR LINES

Once we have established α, the absorption coefficient per atom, we can calculate the mass absorption coefficient needed in the radiative transfer equation and its solutions. To do this we write (eq. 11-12)

$$l_v = N\alpha/\rho$$ (11-65)

where N/ρ is the number of absorbers per gram. N can be broken down in the following way,

$$\frac{N}{\rho} = \left(\frac{N}{N_E}\right)\left(\frac{N_E}{N_H}\right)\frac{N_H}{\rho}$$ (11-66)

where N/N_E is the fraction of the element, E, that is capable of absorbing in the line of interest. This term takes into account the ionization and excitation of the element E. The ratio N_E/N_H is the number abundance, A, of the element E and N_H/ρ is the number of hydrogen atoms per gram of stellar material. It is more convenient usually to compute the number of grams of stellar material per hydrogen atom,

$$\frac{\rho}{N_H} = \sum_j \frac{N_j}{N_H} \mu_j = \sum_j A_j \mu_j. \tag{11-67}$$

The summation is carried over all the elements in the stellar material—each having an abundance $A_j = N_j/N_H$ and an atomic mass in grams of μ_j. The summations in eqs. 11-67 and eq. 8-17 are the same.

The mass absorption coefficient in square centimeters per gram for a typical metal line can now be written

$$
\begin{aligned}
l_v &= \frac{\sqrt{\pi}e^2}{mc} f \, \frac{H(u, a)}{\Delta v_D} \frac{N}{N_E} \frac{A}{\sum A_j \mu_j} (1 - 10^{-\chi\lambda\theta}) \\[2mm]
&= 1.4974 \times 10^{-2} \, \frac{Af}{\sum A_j \mu_j} \frac{N}{N_E} \frac{H(u, a)}{\Delta v_D} (1 - 10^{-\chi\lambda\theta}) \\[2mm]
&= 4.9947 \times 10^{-21} \, \frac{\lambda^2 Af}{\sum A_j \mu_j} \frac{N}{N_E} \frac{H(u, a)}{\Delta \lambda_D} (1 - 10^{-\chi\lambda\theta}) \\[2mm]
&= 1.4974 \times 10^{-2} \, \frac{Af\lambda}{\sum A_j \mu_j} \frac{N}{N_E} \frac{H(u, a)}{(2kT/m_E)^{1/2}} (1 - 10^{-\chi\lambda\theta})
\end{aligned}
\right\} \tag{11-68}
$$

for $\Delta \lambda_D$ and λ in angstroms. The stimulated emission factor, $1 - 10^{-\chi\lambda\theta}$, has been added. The damping parameter a is given by

$$
\begin{aligned}
a &= \frac{\gamma}{4\pi} \frac{1}{\Delta v_D} = 7.96 \times 10^{-2} \, \frac{\gamma}{\Delta v_D} \\[2mm]
&= \frac{\gamma}{4\pi} \frac{\lambda^2}{c} \frac{1}{\Delta \lambda_D} = 2.65 \times 10^{-20} \, \frac{\gamma\lambda^2}{\Delta \lambda_D}.
\end{aligned} \tag{11-69}
$$

A similar expression can be written for the hydrogen lines by using eq. 11-63 for α in eq. 11-65 (but with electron broadening included) and setting $A = 1.0$.

In the event that LTE is applicable, N/N_E can be calculated from the excitation and ionization eqs. 1-16 and 1-17.

OTHER BROADENING MECHANISMS

In this chapter we have considered line broadening that occurs on the microscopic scale defined by the condition $\tau_\nu \ll 1$ across the features under discussion. As a result of these conditions the broadening was incorporated into the absorption coefficient prior to solution of the transfer equation. There are other line broadening mechanisms that fall into this category, such as Zeeman splitting, hyperfine structure, and microturbulence.

On the other end of the geometry are the macroscopic broadening agents where the scale of the phenomenon in optical depth is $\gg 1$. The common examples of this case are macroturbulence and rotation. In such a case the transfer equation is solved in I_ν for each "element" in the macro broadening distribution. The line profile in flux is then obtained by constructing the properly weighted average of the specific intensity across the disc of the star. The absorption coefficient is not affected by these macro phenomena.

At dimensional scales $\tau_\nu \sim 1$ neither of the above approximations can be used. Turbulence with "cell" sizes $\tau_\nu \sim 1$ are in radiative exchange with the adjacent material. The transfer equation should be solved for each line of sight, taking into account the variations in temperature, pressure, and velocity along that line of sight. This type of solution has been attempted in only a few selected models where the geometry and velocity distributions were specified as part of the model.

REFERENCES

Allen, C. W. 1963. *Astrophysical Quantities*, 2nd ed., The Athlone Press, University of London.

Aller, L. H. 1963. *Astrophysics. The Atmospheres of the Sun and Stars*, 2nd ed., Ronald, New York, p. 331.

Baranger, M. 1958. *Phys. Rev.* **111**, 481, 494, and **112**, 855.

Barnard, A. J., J. Cooper, and E. W. Smith 1974. *J. Quant. Spectrosc. Radiat. Transfer* **14**, 1025.

Baschek, B. 1964. *Proceedings First Harvard–Smithsonian Conference on Stellar Atmospheres*, Smithsonian Astrophysical Observatory, Cambridge, Mass., p. 253.

Blackwell, D. E., G. Calamai, and R. B. Willis 1972. *Mon. Not. Roy. Astron. Soc.* **160**, 121.

Blackwell, D. E., J. H. Kirby, G. Smith 1972. *Mon. Not. Roy. Astron. Soc.* **160**, 189.

Brueckner, K. A. 1971. *Astrophys. J.* **169**, 621.

Burgess, D. D. and J. E. Grindlay 1970. *Astrophys. J.* **161**, 343.

Cayrel, R. and G. Traving 1960. *Z. Astrophys.* **50**, 239.

Cooper, C. F. Jr. 1968. *Phys. Rev.* **165**, 215.

Cowley, C. R. 1970. *The Theory of Stellar Spectra*, Gordon and Breach, New York.

Cowley, C. R., G. H. Elste, and H. Allen 1968. *Astrophys. J.* **158**, 1177.

Ecker, G. 1957. *Z. Phys.* **148**, 593.

Edmonds, F. N. Jr., H. Schlüter, and D. C. Wells III 1967. *Mem. Roy. Astron. Soc.* **71**, 271.

Elste, G. 1953. *Z. Astrophys.* **33**, 39.

Finn, G. D. and D. Mugglestone 1965. *Mon. Not. Roy. Astron. Soc.* **129**, 221.

Foley, H. M. 1946. *Phys. Rev.* **69**, 616.

Fowler, J. W. 1974. *Astrophys. J.* **188**, 295.

Fuhr, J. R., W. L. Wiese, and L. J. Roszman 1973. "Bibliography on Atomic Line Shapes and Shifts," *Nat. Bur. Stand. Publ.* **366**.

Fullerton, W. and C. R. Cowley 1971. *Astrophys. J.* **165**, 643.

Griem, H. 1954. *Z. Phys.* **137**, 280.

Griem, H. R. 1962. *Astrophys. J.* **136**, 422.

Griem, H. R. 1964. *Plasma Spectroscopy*, McGraw-Hill, New York.

Griem, H. R. 1967. *Astrophys. J.* **147**, 1092.

Griem, H. R. 1968. *Astrophys. J.* **154**, 1111.

Griem, H. R. 1974. *Spectral Line Broadening by Plasmas*, Academic, New York.

Griem, H. R., A. C. Kolb, and K. Y. Shen 1959. *Phys. Rev.* **116**, 4.

Harris, D. L. III 1948. *Astrophys. J.* **108**, 112.

Hjerting, F. 1938. *Astrophys. J.* **88**, 508.

Holtsmark, P. J. 1919. *Phys. Z.* **20**, 162.

Holweger, H. 1971. *Astron. Astrophys.* **10**, 128.

Holweger, H. 1972. *Sol. Phys.* **25**, 14.

Hummer, D. G. 1965. *Mem. Roy. Astron. Soc.* **70**, 1.

Hunger, K. 1960. *Z. Astrophys.* **49**, 129.

Kallmann, H. and F. London 1929. *Z. Phys. Chem.* **B2**, 207.

Kusch, H. J. 1958. *Z. Astrophys.* **45**, 1.

Landolt-Börnstein 1950. *Zahlenwerte und Funktionen*, Vol. 1, Springer-Verlag, Berlin, p. 246.

Lenz, W. 1924. *Z. Phys.* **25**, 299.

Lewis, E. L., L. F. McNamara, and H. H. Michels 1971. *Phys. Rev. A* **3**, 1939.

Lewis, E. L., L. F. McNamara, and H. H. Michels 1972. *Sol. Phys.* **23**, 287.

Lindholm, E. 1945. *Ark. Mat. Astron. Fys.* **32A**, No. 17.

London, F. 1930a. *Z. Phys.* **63**, 245.

London, F. 1930b. *Z. Phys. Chem.*, **B11**, 222.

Lorentz, H. A. 1906. *Proc. Amst. Acad.* **8**, 591.

Menzel, D. H. 1961. *Mathematical Physics*, Dover, New York, p. 338.

Mihalas, D. 1970. *Stellar Atmospheres*, W. H. Freeman, San Francisco.

Mihalas, D., A. J. Barnard, J. Cooper, and E. W. Smith 1974. *Astrophys. J.* **190**, 315.

Mozer, B. and M. Baranger 1960. *Phys. Rev.* **118**, 626.

Mugglestone, D. and B. J. O'Mara 1966. *Mon. Not. Roy. Astron. Soc.* **132**, 87.

Olson, E. C. 1968. *Astrophys. J.* **153**, 187.

Peterson, D. M. and H. L. Shipman 1973. *Astrophys. J.* **180**, 635.

Petit Bois, G., 1961. *Tables of Indefinite Integrals*, Dover, New York, p. 148.

Roueff, E., and H. Van Regemorter 1969. *Astron. Astrophys.* **1**, 69.

Roueff, E., and H. Van Regemorter 1971. *Astron. Astrophys.* **12**, 317.

Stienon, F. M. 1964. *Proceedings First Harvard–Smithsonian Conference on Stellar Atmospheres* Smithsonian Astrophysical Observatory Special Report No. 167, Cambridge, Mass., p. 317.

Stone, J. M. 1963. *Radiation and Optics*, McGraw-Hill, New York, p. 73.

Strom, S. E. and D. M. Peterson 1968. *Astrophys. J.* **152**, 859.

Struve, O. 1929. *Astrophys. J.* **69**, 173.

Traving, G. 1968. *Plasma Diagnostics*, W. Lochte-Holtgreven, Ed., North Holland, Amsterdam, p. 66.

Underhill, A. B. 1951. *Publ. Dom. Astrophys. Obs. Victoria* **8**, 385.

Underhill, A. B. and J. H. Waddell 1959. *Nat. Bur. Stand. Circ.* **603**.

Unsöld, A. 1955. *Physik der Sternatmosphären*, 2nd ed., Springer-Verlag, pp. 326 and 331.

Van Regemorter, H. 1965. *Annu. Rev. Astron. Astrophys.* **3**, 71.

Van Regemorter, H. 1973. *Reports on Astronomy 15A. Transaction of the International Astronomical Union*, C. de Jager, Ed., Reidel, Dordrecht, Holland, p. 155.

Vidal, C. R., J. Cooper, and E. W. Smith 1971. *J. Quant. Spectrosc. Radiat. Transfer* **11**, 263.

Vidal, C. R., J. Cooper, and E. W. Smith 1973. *Astrophys. J. Suppl.* **25**, 37.

Weisskopf, V. 1932. *Z. Phys.* **75**, 287.

CHAPTER TWELVE

THE MEASUREMENT OF SPECTRAL LINES

The two most important things we can measure for a line are its profile and equivalent width. If we denote by F_c and F_v the continuum and line fluxes, respectively, as they exist in the true spectrum before any instrumental distortion, then the line profile is

$$R_v = \frac{F_c - F_v}{F_c}.$$

(12-1)

The equivalent width, W, is a measure of the total absorption in a line and is defined as

$$W = \int_0^\infty (F_c - F_v) \, dv/F_c$$

(12-2)

where W is the width of a perfectly black rectangular absorption line having the same total absorption as the real line. The width defined in eq. 12-2 has frequency or wavelength units depending on the choice of units for the F_c outside the integral sign. It is customary to measure W in wavelength units— angstroms or milliangstroms.

Compared to continuum photometry, studies of spectral lines require very high resolution. The integral absorption or equivalent width can be measured in some cases when the resolution is as low as $\lambda/\Delta\lambda = 20{,}000$ (~ 0.25 Å), but for line profile work, and even for some equivalent width measurements, it is necessary to have resolution in the 50,000–100,000 range (about 0.1–0.05 Å). High resolution has been attained with diffraction gratings, echelles, and interferometers. As explained in the introduction, in this book we emphasize standard grating spectrometers. Interferometers for line work have been discussed by Connes (1970) and Vaughan (1967).

It is necessary to have ultrastability in alignment of optical and mechanical components. Thermal and gravitational flexure must be reduced to a low level. Permanently mounting high resolution equipment at the coudé focus of a telescope has proved to be the best way to meet these requirements.

THE COUDÉ GRATING SPECTROMETER

High dispersion grating spectrometers are much too large to hang on the telescope itself. The spectrograph is large because of eq. 3-21, namely,

$$\Delta\lambda = \frac{-\cos\alpha}{f_{coll}}\, W'\frac{d}{n}. \tag{12-3}$$

We want $\Delta\lambda$ small by our scientific requirements but the slit width W' large to allow as much of the stellar seeing disc through as possible. For a fixed ratio of ruling spacing to order (d/n) we have left only f_{coll} as a free parameter.[*] The collimator focal length is then made as large as possible with the important constraint that the size of the optical beam not exceed the size of the grating. For example, the largest grating now ruled has rulings 350 mm long. Working at a coudé focal ratio of $f/30$, the collimator would have a focal length of 10.5 m. The grating is mounted so that it can be rotated for selection of wavelength regions, or to direct the beam toward any of several cameras.

Photographs of coudé rooms are shown in Fig. 12-1. Some coudé rooms are built around the undeflected optical beam as it comes along the polar axis. Others have an extra optical component to bring the beam into a horizontal (or sometimes a vertical) orientation. Horizontal coudé rooms are much less expensive to build than those built on the latitude slant. The light loss at the extra reflection can be reduced by using a high reflection coating. Many coudé spectrograph design details are discussed by Dunham (1956).

An entrance slit assembly is shown in Fig. 12-2. The slit jaws are moveable to allow adjustment of the slit width. The plane in which the jaws lie is tilted so it is not perpendicular to the optical axis. In this way the portion of the stellar seeing disc that does not enter the slit is reflected off axis and can be used for guiding.

---

[*] Not until we reach the really high resolution does the diffraction spread of the grating become significant compared to the slit widths.

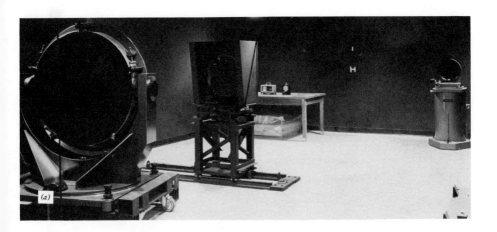

Fig. 12-1 (*a*) The coudé room for the 122 cm telescope at the University of Western Ontario. The entrance slit and grating are out of the picture toward the left. The mirror in the center is the collimator. The other mirrors are cameras. (*b*) The coudé room for the 152 cm AURA telescope at Cerro Tololo. The entrance slit (not visible) and grating are at the left. The rectangular beam turret is used to mount the cameras and to rotate them into the optical beam for use. (Courtesy of Kitt Peak National Observatory.)

Fig. 12-2   The entrance slit on the coudé spectrograph shown in Fig. 12-1a.

For photographic work the spectrogram is widened by moving the stellar seeing disc along the slit. Many such trailings are needed to obtain a uniform exposure. In some observatories a rocking plate (Dunham 1956, O'Dell 1969) is used behind the entrance slit to do the widening and then the star image is held fixed on one portion of the slit. For photoelectric scanning it is also necessary to hold the image at one position along the slit. It is convenient to use a mask to limit the effective length of the slit.

Image slicers (see Fig. 12-3) should be used whenever possible. These ingenious devices (Richardson 1966, Pierce 1965, Bowen, 1938) use multiple reflection to recycle some of the light reflected by the slit jaws, putting it in through the entrance slit. The efficiency is increased by a factor of 2–5. The height of the spectrum produced in the camera focal plane is much larger than that produced with a simple entrance slit.

In the image plane we have a choice of detectors. Photographic detectors give wide wavelength coverage with low quantum efficiency and nonlinear response. Photoelectric detectors have higher quantum efficiency but often

Fig. 12-3 A Richardson image slicer such as this one can replace the conventional entrance slit to gain efficiency by getting more of the seeing disc into the spectrograph. The length of this slicer is 12 cm.

a very small field, making the data collection efficiency much less than that of the photographic approach. The higher photometric precision obtained photoelectrically must be weighed against the inefficiency.

PHOTOGRAPHIC CAMERAS

Cameras of the Schmidt design are used to obtain as wide a wavelength field as possible. The corrector plate is placed close to the grating. Ideally the grating should be at the center of curvature of the spherical camera mirror so that all the monochromatic beams diffracted from the grating follow radii of curvature to the reflecting surface. The plate holder is midway between the center of curvature and the mirror. The focal surface is a sphere concentric to the mirror's sphere. The photographic plate is bent by the plate holder to conform to this surface, although usually the bending is done only along the wavelength coordinate. Several exposures can be put on each plate by translating the plateholder normal to the dispersion between exposures.

The aperture of a camera depends on the field to be recorded and the collimated beam size. In an optimum design the camera is large enough to record the width of the grating blaze (see the discussion of blazing in Chapter 3). It is not uncommon for the highest dispersion photographic cameras to have apertures that are a substantial fraction of the telescope aperture.

The focal length of the camera is set to give the desired linear dispersion when used with a specified grating (according to eq. 3-20). Several major coudé installations have incorporated a large number of cameras giving steps in dispersion of about a factor of, for example, 2, 4, 8 and 16 Å/mm.

The observer can decide what slit width to use with a given camera–emulsion combination using the following simple approach. In front of the entrance slit an emission line source (perhaps the comparison spectrum for low resolution work or a ^{198}Hg bulb for high resolution work) and diffusing screen are mounted so that the slit is uniformly illuminated. Several exposures of the same duration are made but with a range in entrance slit setting. After the plate has been developed, it can be placed in a microdensitometer (or looked at under a microscope) and the density of the images measured. For those exposures where the projected slit width is unresolved there is a steady increase in density with increasing slit width—flux is being measured. But when resolution occurs the density becomes constant—specific intensity is now being measured. The maximum resolution consistent with maximum efficiency is obtained by using a slit at the boundary between these domains. An example is shown in Fig. 12-4. Of course, a larger slit can be used and more light let in, but then it would be better to use a lower dispersion camera

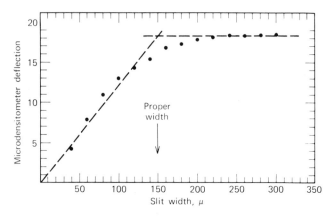

Fig. 12-4 Resolution method of setting the spectrograph entrance slit is shown. For slit widths less than ∼100 μ, flux is being recorded (slit unresolved). For slit widths greater than ∼200 μ, specific intensity is being recorded (slit fully resolved).

and shorten the exposure time. A smaller slit wastes photons and gains nothing since the resolution is then set by the emulsion grain.

Let us consider the situation where one grating is used with several cameras. Each time we switch to a longer focal length camera, the linear dispersion increases (number of Angstroms per millimeter decreases) and to keep the projected entrance slit commensurate with the plate grain it must be decreased in proportion to the dispersion. The exposure time increases approximately as the square of the dispersion—light at the entrance slit is cut and the spectrum is more spread out across the emulsion.* The resolution is wedded to the linear dispersion in this way. At times astronomers have lapsed into the habit of referring to high dispersion and high resolution as synonomous. The high resolution implied by high dispersion may not materialize if, for instance, the entrance slit is improperly set or the camera is out of focus.

For quantitative work the plate is placed in a microdensitometer that produces a graph or digital tabulation of the recorded signal as a function of plate position or wavelength. The characteristic curve for the emulsion must be carefully accounted for (see Chapter 4). Widening the spectrogram to 1 mm or more is desirable but, of course, more telescope time is needed for wider spectrograms. The full width of the spectrum should be measured on the microdensitometer.

Photographic cameras have been described by Bowen (1962), Abt (1967), and Shulte (1966).

PHOTOELECTRIC LINE MEASUREMENT

The rustic approach to photoelectric spectrophotometry consists of measuring in a single spectral band isolated by an exit slit. The light passing through the exit slit is detected by a photomultiplier. Line profiles can be scanned by moving the exit slit and photomultiplier with a precision screw or by rotating the grating in a manner akin to continuum scanners (Chapter 10).

Image motion due to seeing and changes in atmospheric transparency cause large modulation in the photon arrival rate. It is mandatory to use a compensating system. One method is to scan the spectral region of interest over and over with a period that is short compared to the light fluctuations. A better method is to add an additional (reference) channel and ratio the outputs from the profile scanning channel and this reference channel. The

---

* The greater plate scale in a direction perpendicular to the dispersion does not increase the exposure time significantly because a widened spectrum is needed anyway. The length of slit over which the seeing disc is trailed in order to widen the spectrum is reduced in proportion to the increase in dispersion.

reference channel should not differ in position by more than a few angstroms from the profile channel because the image motion is wavelength dependent. The reference channel must remain fixed in wavelength, of course, during the scan. For these reasons it is easier to use a ratioing mode when the grating is held fixed and the exit slit and detector are moved to do the scanning. When the grating is rotated, ratioing can still be accomplished by sampling the beam with a beamsplitter behind the entrance slit. An interference filter can then be used to select light from the same spectral region as measured by the profile channel.

Prismatic rather than reflecting optics must be used in manipulating the reference beam. The reason? Prismatic components do not introduce nor do they analyze polarized light. Coudé light is strongly polarized because of the oblique reflections on the coudé mirrors. Furthermore, the polarization is a function of declination and hour angle. During the observation period the polarization changes and any analyzer in the system that is not identical for the profile and reference channels introduces a secular drift in the ratio of the channel outputs. The grating itself acts as an analyzer and that is why sampling the spectrum with the reference channel next to the profile channel is superior to taking a reference beam immediately after the entrance slit.

The image motion on the entrance slit can also introduce "knife edge" noise if the telescope focus is improper as shown in Fig. 12-5. The illumination of the coudé optics and the photocathode depends on the position, relative to the entrance slit, of the seeing disc and is a strong function of time. When this is coupled with the nonuniform response across the cathode area, noise is generated. Careful focusing eliminates the effect. A similar effect occurs if the edge of the observatory dome slit is allowed to occult part of the telescope aperture. The result in this case is not enhanced noise but a systematic drift in the output.

The camera mirror can be chosen to have any convenient focal length. The linear dispersion is not connected to the resolution as it was in photographic work because the exit slit width (detector size) can be chosen by the observer. In normal operation the projected entrance slit width should be matched to the exit slit for the same reason as in photographic work— it is the most efficient combination for a given resolution.

Rapid scan systems have been described by Boyce et al. (1973) and Brault et al. (1971). Ratioing systems have been used by Gray (1971), Gray and Evans (1973), and Wilson (1968). Photometric precision of 1% is readily attainable with these systems, but with only a single spectral window being measured the efficiency is low. Furthermore, the scanning time needed to reach a fixed level of precision varies as the cube of the resolution. One factor comes from the entrance slit, a second from the exit slit, and the third because the data sampling spacing must decrease as the resolution increases. The

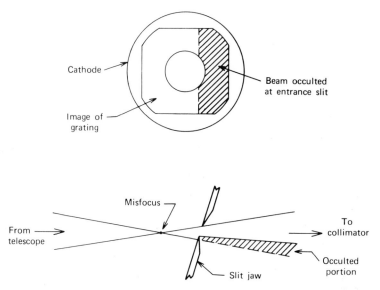

Fig. 12-5 The source of knife edge noise is illustrated. (*Bottom*) An error in telescope focus causes part of the beam to be cut off. The collimator and grating are then improperly illuminated. (*Top*) A Fabry lens images the grating on the photocathode so the photocathode is improperly illuminated. The occulted portion fluctuates as seeing moves the stellar image at the entrance slit. (The hole in the center of the beam is the shadow of the secondary mirror assembly.)

gain in photometric precision is very important in most studies of stellar photospheres and we can often justifiably accept the low efficiency. Nevertheless, the next step is to use a multichannel photoelectric detector, which becomes analogous to the photographic plate. The most promising multichannel photoelectric detectors are the image tube systems, the television cameras, and the detector rays mentioned at the end of Chapter 4. Increase in efficiency is proportional to the number of channels, provided each channel performs as well as a photomultiplier. Gains of several thousand may be obtained this way.

THE INSTRUMENTAL PROFILE

Distortion and blurring of the star's spectrum can be caused by inadequate spectral purity or detector resolution, diffraction, and aberrations. In all cases the structure in the spectrum is degraded in contrast. Measurements of the instrumental profile tell us how much degradation occurs and to some extent allows us to reconstruct the original spectrum prior to blurring. It is appropriate then to turn our thoughts toward the instrumental profile for it is clearly of paramount importance in the measurement of spectral lines.

Imagine we have a light source giving an infinitely narrow spectral emission line. Such a line can be denoted by $\delta(\lambda - \lambda_0)$. The spectrum we measure with our instrument for the δ function source is the instrumental profile (also called the delta function response). We shall denote this instrumental profile by $I(\lambda)$. It is normalized to unit area when used in the computations below. Figure 12-6 shows a measured instrumental profile with a half width of 86 mÅ (normalized to unit peak height).

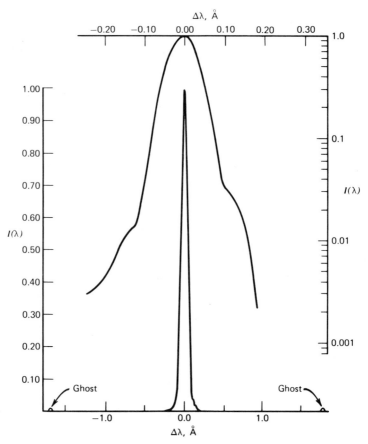

Fig. 12-6 An instrumental profile measured on the McMath solar telescope, Kitt Peak National Observatory, using a He–Ne laser $\lambda 6328$. The bottom plot is on a linear scale. The top plot is logarithmic in $I(\lambda)$ and has an expanded $\Delta\lambda$ scale. The half width is 86 mÅ which is typical of high resolution stellar observations. (Solar observations may well use resolution an order of magnitude higher than this.)

A general spectrum, $F(\lambda)$, can be viewed as a synthesis of δ functions with height modulated by $F(\lambda)$ (see Fig. 2-7). Each of these δ functions gives an $I(\lambda)$ in the output spectrum, but each $I(\lambda)$ is shifted to the wavelength of the δ function. The data actually recorded are then the sum of these shifted $I(\lambda)$'s. So we write*

$$D(\lambda) = \int_{-\infty}^{\infty} I(\lambda - \lambda_0)F(\lambda_0)\, d\lambda_0$$

or

$$D(\lambda) = I(\lambda) * F(\lambda) \tag{12-4}$$

for the measured spectrum. In other words the data is the convolution of the instrumental profile with the true flux spectrum.

We denote the Fourier transforms of $D(\lambda)$, $I(\lambda)$, and $F(\lambda)$ by $d(\sigma)$, $i(\sigma)$, and $f(\sigma)$. The Fourier frequency, σ, is measured in cycles per angstrom. The transform of eq. 12-4 is then

$$d(\sigma) = i(\sigma)f(\sigma). \tag{12-5}$$

The function $i(\sigma)$ acts as a filter on $f(\sigma)$ and in the cases of interest in this discussion $i(\sigma)$ always diminishes toward higher σ. Figure 12-7 shows $i(\sigma)$ for the instrumental profile in Fig. 12-6 along with $f(\sigma)$ for an absorption line of 180 mÅ equivalent width. Notice how the higher frequencies of $f(\sigma)$ are diminished in amplitude by the filtering process of $i(\sigma)$.

The optimum situation is when $i(\sigma)$ stays at unity up to σ sufficiently high to embrace all structure of interest in $F(\lambda)$. This utopian situation is almost never encountered. There is usually more useful information to be obtained from higher frequency measurements. Reconstruction of the function $F(\lambda)$ can be done for some of the higher frequencies by the process to which we now turn our attention.

THE RECONSTRUCTION PROCESS

If it were not for noise in the data, the reconstruction of $F(\lambda)$ would follow trivially from eq. 12-5

$$f(\sigma) = \frac{d(\sigma)}{i(\sigma)} \tag{12-6}$$

and taking the inverse transform. The points in $f(\sigma)$ where $i(\sigma) = 0$ must be omitted from the calculation since we have the indeterminate ratio of zero divided by zero. Still more restrictive is the resolution theorem (Chapter 2)

---

* We choose the zero on the wavelength scale to be at the center of the spectral region of interest.

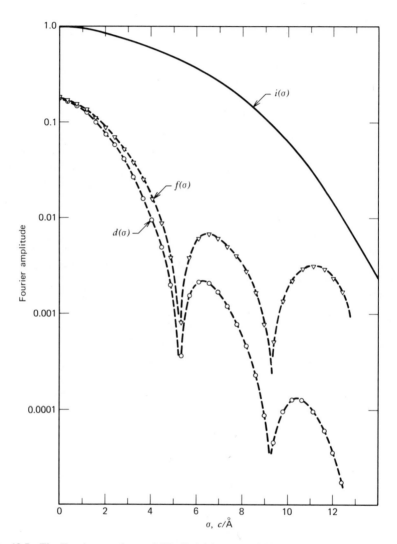

Fig. 12-7 The Fourier transform of $I(\lambda)$ diminishes toward higher σ and acts as a filter by reducing the Fourier amplitudes of $f(\sigma)$ to $d(\sigma)$. Since $i(\sigma)$ has a value of unity at $\sigma = 0$, $f(\sigma)$ and $d(\sigma)$ show the same value there. This value is the 180 mÅ equivalent width of the line.

seen in action here as $i(\sigma)$ reduces the high frequency components of $F(\lambda)$ to the noise level. Furthermore, even in principle the reconstruction process can only be done for $\sigma < \sigma_N$ to avoid aliases (see Fig. 2-12). Nevertheless, with high quality data the reconstruction can be extremely valuable.

The calculation of the transform can be simplified by using the fact that all the functions in the λ domain are real. We write $f(\sigma)$ as follows (compare eq. 2-3):

$$f(\sigma) = \int_{-\infty}^{\infty} F(\lambda)e^{2\pi i \sigma \lambda}\, d\lambda$$

$$= \int_{-\infty}^{\infty} F(\lambda) \cos 2\pi\sigma\lambda\, d\lambda + i \int_{-\infty}^{\infty} F(\lambda) \sin 2\pi\sigma\lambda\, d\lambda. \qquad (12\text{-}7)$$

The phase of $f(\sigma)$ is then the ratio of the second and first integrals. The amplitude, which will concern us most, is the length of the vector whose components are the real and imaginary terms of the above equation. The case where $F(\lambda)$ is symmetric is occasionally encountered, but generally we must expect $f(\sigma)$ to be complex. The division indicated in eq. 12-6 must then be treated appropriately. If we choose rectangular coordinates and again use R and I subscripts to denote real and imaginary components, then we find

$$f(\sigma) = f_R(\sigma) + if_I(\sigma) \qquad (12\text{-}8)$$

with

$$f_R(\sigma) = \frac{d_R(\sigma)i_R(\sigma) + d_I(\sigma)i_I(\sigma)}{i_R^2(\sigma) + i_I^2(\sigma)}$$

$$f_I(\sigma) = \frac{d_I(\sigma)i_R(\sigma) - d_R(\sigma)i_I(\sigma)}{i_R^2(\sigma) + i_I^2(\sigma)}.$$

$F(\lambda)$ follows from the inverse transform of this $f(\sigma)$.

NOISE AND ITS COMPLICATIONS

The noise in the data must be relatively small if any restoration process is to work properly. Figure 12-8 shows the Fourier transform for a 5 Å portion of the observed spectrum of Procyon. The more or less constant noise level can be seen extending to high frequencies. The spectrum signal amplitudes become less toward higher σ, resulting in a smaller signal to noise ratio. The reconstruction outlined above amounts to mathematically increasing the Fourier amplitudes to undo what the instrumental filtering has done. But the noise as well as the signal amplitude is increased by this process.

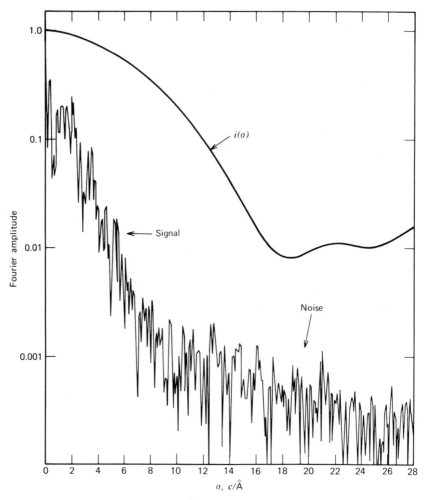

Fig. 12-8 The Fourier transform of a 5 Å portion of spectrum near $\lambda5191$ in Procyon. Eight lines are included in the region. The signal amplitudes decrease to the noise level by 10–15 $c/\text{Å}$. The transform of the instrumental profile is also shown.

This can be written explicitly if we let $n(\sigma)$ be the noise in the flux transform. Then the transform like the one in Fig. 12-8 can be written*

$$d_1(\sigma) = d(\sigma) + n(\sigma) \qquad (12\text{-}9)$$

* Generally we assume the noise in measuring $I(\lambda)$ is negligible compared to the noise in $D(\lambda)$. But if this proves not to be the case, $n(\sigma)$ in eq. 12-9 and Fig. 12-8 can be thought of as including this additional noise.

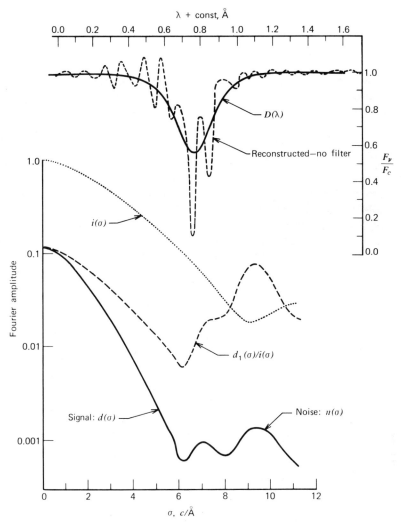

Fig. 12-9 A single line profile, λ5195 in Procyon, as observed, is shown on top as the solid line. The dashed curve shows the reconstructed profile without noise filtering. On the bottom we see the transform domain. The noise becomes greatly enhanced for $\sigma \gtrsim 6 \; c/\text{Å}$.

and the reconstruction is then

$$f_1(\sigma) = \frac{d(\sigma)}{i(\sigma)} + \frac{n(\sigma)}{i(\sigma)}.$$

The first term gives $f(\sigma)$ back again, but the second increases as $i(\sigma)$ becomes small toward high σ.

Eventually at high enough frequencies the noise becomes amplified to an intolerable level. Figure 12-9 shows this problem. Here the transform of a single line profile ($\lambda 5195$ in Procyon) has been divided by $i(\sigma)$. Beyond $\sigma \sim$ 6 cycles/Å the noise level becomes very large. The profile given by the transform of $d(\sigma)/i(\sigma)$ is virtually meaningless (Fig. 12-9, top).

In such a situation it is necessary to apply a filter, $\phi(\sigma)$, to remove the high frequency noise. We assume that the stellar spectrum and the noise have distinctly different transforms, which is usually true in practice. The stellar transform shows a concentration toward lower σ, while the noise shows an approximately constant value as in Figs. 12-8 and 12-9.

FOURIER NOISE FILTERS

The optimum filter minimizes the square of the errors between the true transform and the restored–filtered transform. We define the filtered reconstructed data transform to be

$$f_2(\sigma) \equiv \frac{d_1(\sigma)\phi(\sigma)}{i(\sigma)} \tag{12-10}$$

where $d_1(\sigma)$ is defined by eq. 12-9 and $\phi(\sigma)$ is the filter. The difference between $f(\sigma)$ and $f_2(\sigma)$ is the error in the reconstruction and filtering process,

$$\varepsilon = f(\sigma) - f_2(\sigma)$$

$$= f(\sigma) - \left[\frac{d(\sigma)\phi(\sigma)}{i(\sigma)} + \frac{n(\sigma)\phi(\sigma)}{i(\sigma)} \right].$$

And since $d(\sigma)/i(\sigma) = f(\sigma)$ this is the same as

$$\varepsilon = f(\sigma)[1 - \phi(\sigma)] - \frac{n(\sigma)\phi(\sigma)}{i(\sigma)}. \tag{12-11}$$

The first term shows how the filter introduces an error by not giving exact reconstruction to $f(\sigma)$. The second term shows that the enhanced noise, $n(\sigma)/i(\sigma)$, can be reduced by a properly chosen filter. More particularly, one requires $\int_{-\infty}^{\infty} \varepsilon^2 \, d\sigma$ to be a minimum. The result (Brault and White 1971,

Middleton 1960, Champeney 1973) is

$$\phi(\sigma) = \frac{[f(\sigma)i(\sigma)]^2}{[f(\sigma)i(\sigma)]^2 + [n(\sigma)]^2}$$

$$= \frac{d^2(\sigma)}{d^2(\sigma) + n^2(\sigma)} = \frac{1}{1 + [n(\sigma)/d(\sigma)]^2} \, . \tag{12-12}$$

It is important to notice that $\phi(\sigma)$ depends on the noise and data signal strengths and not on $f(\sigma)$ or $i(\sigma)$.

The first step then is to take the transform of the data (which might look like the examples in Fig. 12-8 or 12-9). We can adequately describe $n(\sigma)$ with a constant,

$$n(\sigma) = B.$$

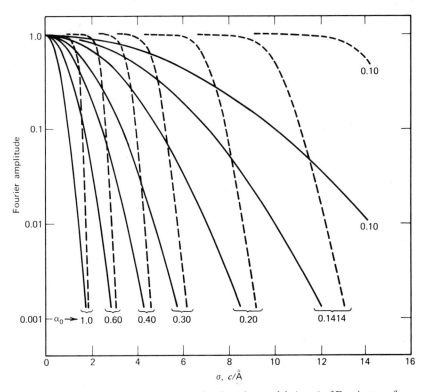

Fig. 12-10 Fourier filters (------) are shown for Gaussian models (———) of Fourier transforms. When the ratio of B/A in eq. 12-13 is 0.01, α_0 has the values as shown.

A convenient approximation for many $d(\sigma)$'s is the Gaussian, say

$$d(\sigma) \sim Ae^{-(\pi\beta\sigma)^2} \equiv A10^{-\alpha_0^2\sigma^2}.$$

Then

$$\phi(\sigma) = \frac{1}{1 + (B/A)^2 10^{2\alpha_0^2\sigma^2}} \qquad (12\text{-}13)$$

makes a reasonable filter. The ratio B/A is the noise level divided by the zero ordinate of $d(\sigma)$. The fitting parameter, α_0, is chosen according to the Gaussian that most closely matches $d(\sigma)$. Gaussian models and filters are shown in Fig. 12-10. Filters other than eq. 12-13 can be constructed as needed.

The optimum filter in eq. 12-12 was derived by minimizing the overall error. The actual function used for the filter can differ a reasonable amount from the optimum case without serious consequences. But the choice of α_0 in eq. 12-13 (or some other cutoff parameter in alternative filters) is important. Reconstructed line profiles can be very sensitive to the choice of α_0. Poor quality data, large B/A, enhance the problem.

It is sometimes useful to filter observations without removing the instrumental profile. The discussion of the optimum filter for this case follows from the above discussion by setting $i(\sigma) = 1$. Other types of filters can be used, for example, to enhance the contrast of a spectrum. Most such applications are not oriented toward quantitative analysis but toward detection and recognition and we shall not pursue the discussion of these types of filters.

THE DISCRETE FOURIER TRANSFORM

The true spectrum from the star is a continuous function, but our data consists of a set of numbers giving the flux at various wavelengths. The true spectrum runs over an infinite extent in wavelength but we measure in some wavelength window. In addition, the true spectrum is convolved with the instrumental profile as we have seen. Figure 12-11 illustrates these processes and their analog in the transform domain. We can write for the data.

$$D(\lambda) = W_1(\lambda)W_2(\lambda)I(\lambda) * F(\lambda) \qquad (12\text{-}14)$$

and for the transform

$$d(\sigma) = w_1(\sigma) * w_2(\sigma) * i(\sigma)f(\sigma) \qquad (12\text{-}15)$$

where $W_1(\lambda)$ is a box and $W_2(\lambda) = \text{III}(\lambda)$, while $w_1(\sigma) = \text{sinc } \sigma$ and $w_2(\sigma) = \text{III}(\sigma)$ (eqs. 12-14 and 12-15 should be compared to eqs. 2-16 and 2-30 of Chapter 2).

Consider the data sampling window $W_2(\lambda)$. The delta functions are spaced $\Delta\lambda$ apart, which means the repetitions of $w_1(\sigma) * i(\sigma)f(\sigma)$ are spaced $\Delta\sigma =$

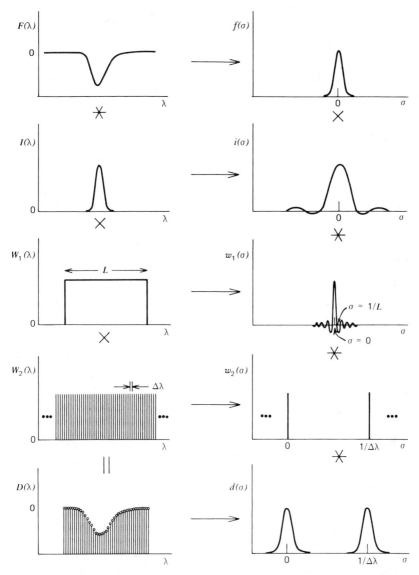

Fig. 12-11 The flux profile we seek is on the upper left. It is blurred by convolution with the instrumental profile and then sampled through the windows W_1 and W_2. In the σ domain, the transform of the flux profile on the upper right is filtered by $i(\sigma)$, blurred by $w_1(\sigma)$, and replicated by $w_2(\sigma)$ to give the data transform at the lower right. The negative signs on the amplitudes of $D(\lambda)$ and $F(\lambda)$ have been ignored.

$1/\Delta\lambda$ cycles/Å apart. Aliasing is avoided if the highest frequencies contained in $w_1(\sigma) * i(\sigma) f(\sigma)$ are less than $\sigma_N = 0.5/\Delta\lambda$, which is another way of saying that the data spacing we should use is dictated by what we are measuring and the equipment we are using. In particular, let us suppose the characteristic width of $I(\lambda)$ is h Å. Then for $\sigma \gtrsim 1/h$ the transform $i(\sigma)$ will have small amplitudes and, assuming $w_1(\sigma)$ contributes negligible width, the width of $w_1(\sigma) * i(\sigma) f(\sigma)$ will be $\lesssim 1/h$. So to avoid aliasing, we need $\sigma_N > 1/h$ or $\Delta\lambda < h/2$. This is an upper bound on reasonable data sampling spacing. Actual spacings a few times smaller may be desirable.

Should $f(\sigma)$ turn out to be comprised of Fourier components that are all of frequencies much lower than $1/h$, then $\Delta\lambda$ could be increased accordingly. In practice it would be wise in such a case to decrease the resolution of the spectrograph by opening the slit or slits in order to shorten the observing time.

The window $W_1(\lambda)$, as we have seen in Fig. 12-11, is the width of our spectrum scan. It transforms into the sinc σ function, which has strong sidelobes. Even though $W_1(\lambda)$ is very much wider than $\Delta\lambda$, these strong sidelobes could cause a broadening of $w_1(\sigma) * i(\sigma) f(\sigma)$ and a subsequent aliasing problem. $W_1(\lambda)$ can be artificially widened by adding more continuum to the real scan and the sidelobes can be apodized by rounding the corners of $W_1(\lambda)$. The cosine bell is a convenient apodizing function. When applied to a data string of length L it can be written

$$A(\lambda) = 0.5\left(1 - \frac{\cos 10\pi\lambda}{L}\right) \qquad \text{for} \quad 0 \le \lambda \le \frac{L}{10}$$

$$= 1.0 \qquad \text{for} \quad \frac{L}{10} < \lambda < \frac{9L}{10} \qquad (12\text{-}16)$$

$$= 0.5\left(1 - \frac{\cos 10\pi(L - \lambda)}{L}\right) \qquad \text{for} \quad \frac{9L}{10} \le \lambda \le L.$$

Some such function should be multiplied into the data string.

If several angstroms or more of the spectrum are involved, the mean value of the data should be subtracted before computing the transform. This avoids a large spike in the transform at $\sigma = 0$. But when a single line and adjacent continuum comprise $F(\lambda)$, then we would normally subtract the value of the continuum, that is, set the continuum level at zero.* Then the shape of $f(\sigma)$ describes the Fourier composition of the line. At the end of the Fourier manipulations, the mean or continuum is again added on.

---

* The negative sign that results for the data points in the profile causes no harm. The same Fourier frequencies describe the function whether it is positive or negative. The Fourier amplitudes also reverse in sign or the phase shifts by π. Notice that in this case the apodization function is applied to zero, making it an unnecessary calculation.

OTHER TECHNIQUES

Several older techniques have been used for instrumental profile corrections. Among these are Voigt function approximations (Elste 1953), iterative schemes (Burger and van Cittert 1933 and de Jager and Neven 1966), and graphical calculations (Bracewell 1955). All these are either not rigorous or have difficulty with noise and convergence. With modern computers the full Fourier transform restoration can be accomplished easily and is the only method recommended.

Rough estimates of the effect of the instrumental profile can often be made by assuming both $I(\lambda)$ and $F(\lambda)$ are Gaussian or dispersion profiles: In the first case the half width of $D(\lambda)$ is $\beta = (\beta_I^2 + \beta_F^2)^{1/2}$, while in the second case $\beta = \beta_I + \beta_F$ [where β_I and β_F are the half widths of $I(\lambda)$ and $F(\lambda)$] in accord with the examples of Chapter 2.

THE INSTRUMENTAL PROFILE OF AN INTERFEROMETER

An ideal Fourier spectrometer measures the Fourier components of the spectrum out to the maximum path difference, δ_m, up to which point the amplitudes are essentially unfiltered and beyond which they are zeroed. The Fourier transform of the instrumental profile is then a box of width δ_m. The instrumental profile itself is sinc $(\pi\delta_m/\lambda)$ as already alluded to in Chapter 3. Reconstruction concepts do not apply in this case because the Fourier amplitudes already have their full value.

The finite size of the diaphragm in front of the detector (Fig. 3-20), for example, introduces some deviation from the sinc form of instrumental profile. The amplitudes at δ_m may show considerable reduction from their true values, especially if the interferometer is designed for extended sources.

The strong sidelobes of the sinc function make a poor instrumental profile for some applications, and apodization filtering is then introduced to taper the corners of the δ_m box smoothly to zero, but at the expense of a broader instrumental profile.

Noise removal can be accomplished in the usual way whenever the signal drops below the noise level.

δ FUNCTION SPECTRA

There are several sources suitable for generating very narrow spectral lines. These narrow lines are so small compared to the instrumental profile that they can be considered to be δ functions. One source is the isotopic mercury bulb, a second the stabilized laser, a third telluric absorption lines.

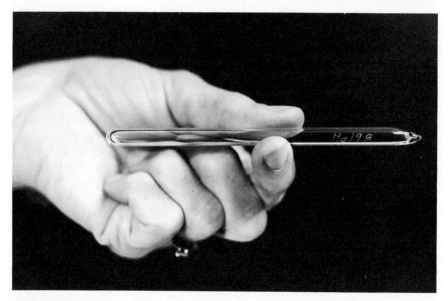

Fig. 12-12   An isotopic mercury bulb used for measuring instrumental profiles is excited by microwave radiation.

A suitable mercury bulb is shown in Fig. 12-12. It is simply a sealed glass tube in which is deposited a small amount of [198]Hg. The mercury is excited to emission by means of a microwave generator. Heating it or using electrical discharge to obtain emission broadens the lines by thermal Doppler shifts. In fact cooling of the bulb is still necessary for the highest resolution work. We can expect a characteristic line width by twice $\Delta\lambda_D$, (eq. 11-50)

$$\frac{\Delta\lambda}{\lambda} = \frac{2}{c}\sqrt{\frac{2kT}{m}} = 8.6 \times 10^{-7}\sqrt{\frac{T}{\mu}} \qquad (12\text{-}17)$$

where m is the mass in grams and μ the atomic weight in atomic mass units of the mercury atom. In the visible region $\Delta\lambda \sim 5$ mÅ at a room temperature of 300°K.

If the instrumental profile is narrower than 50 mÅ, a Gaussian correction can be applied for the intrinsic width of the mercury line, otherwise none is likely to be needed. Table 12-1 lists the strong lines in the [198]Hg visible spectrum. $\lambda5461$ is used most frequently and is referred to as the mercury green line. The strength of the other lines are compared to that of $\lambda5461$.

The laser gives a narrow emission line by virtue of the phase coherence of the stimulated emission. The wavelength spread of an untuned laser

Table 12-1 ^{198}Hg Spectrum and Relative Line Strength*

λ, Å	Strength	λ, Å	Strength
3125.67	0.72	4046.57	2.14
3131.55	0.52	4077.84	0.17
3131.84	0.63	4339.22	0.01
3341.48	0.12	4347.50	0.03
3650.16	1.51	4358.34	2.04
3654.84	0.56	4916.07	0.004
3662.88	0.09	5460.73	1.00 (reference)
3663.28	0.39	5769.60	0.18
3704.17	0.002	5789.67	0.004
3906.37	0.006	5790.66	0.16

Source. Adapted from the *American Institute of Physics Handbook*, D. E. Gray, Ed. Copyright 1963 by McGraw-Hill Book Company. Used with permission of McGraw-Hill Book Company.

encompasses several modes of oscillation of wavelengths given by

$$\lambda_j = \frac{2L}{n} \qquad (12\text{-}18)$$

where L is the length of the laser resonant cavity and n is the integer number of half wavelengths in the spacing L. In a typical 1 m laser $n = 3 \times 10^6$ and, because of the thermal breadth of the lasing line along with other broadening mechanisms, as many as a dozen or more resonant modes may exist simultaneously, that is from eq. 12-18

$$d\lambda_j = -2L\frac{dn}{n^2} \sim 2 \times 10^{-3} \, dn \text{ Å}.$$

Thus if $d\lambda \sim 6328$ Å $\times 8.6 \times 10^{-7}$ $(T/\mu)^{1/2} \sim 0.054$ Å for neon gas at 2000°K according to eq. 12-17, then $dn \sim 27$ is indicated. To use such a source as a δ function, it is necessary to select a single mode of operation. This is accomplished by inserting an etalon, slightly tilted to the optical axis, into the beam. The mode of oscillation can be selected by changing the tilt of the etalon, the temperatue of the etalon, or by adjusting the length of the laser cavity. Needless to say, these parameters must be carefully controlled after mode selection to ensure no mode changing occurs during measurement of the instrumental profile.

The highly coherent and collimated beam of the laser poses a problem. In order to reduce diffraction at the entrance slit and to fill the collimator with the proper f ratio beam, it is necessary to use diffusing screens and baffles. Griffin (1968) has discussed this problem.

The telluric absorption lines of molecular oxygen are strong lines with dispersion profiles but have half widths of 13 mÅ according to the measurements of Panofsky (1943). Coupled with the sun, moon, or other bright object as a background source, these lines are suitable δ functions for most work. Panofsky measured $\lambda\lambda 6287$, 6290, 6293, 6296, 6299, 6307, and 6313 and gives profiles should allowance need to be made for the finite width of these lines. Griffin (1969 a, 1969b, 1969c) has used the 6276, 6867, and 7593 Å lines for instrumental profile measurement.

MEASUREMENT OF THE INSTRUMENTAL PROFILE

Photographically the instrumental profile can be measured by recording the δ function spectra with the same emulsion type and so forth used for the stellar spectrogram. It is also important to use the microdensitometer in exactly the same way for the stellar and instrumental profile measurements. Several exposures at different densities are necessary to bridge the range in light level between the main peak and the wings of $I(\lambda)$.

Photoelectrically the δ function spectrum can be scanned without complication. Measurements to 10^{-4} or 10^{-5} of the main peak are readily attainable. In a ratioing system where the reference channel is taken in the spectral plane, a cube beam-splitter can be introduced before the exit slit to send a fraction of the emission line to the reference channel while scanning in the usual way with the profile channel. A small change in focus may have to be made to compensate for the optical thickness of the beam–splitter.

In general we can expect the instrumental profile to change with wavelength. Wavelength dependence arises through the diffraction contribution of the broadening. It is also possible that transmission optics may introduce λ dependent terms. The wavelength dependence is measurable if several different emission lines are measured and compared. Table 12-1 shows Hg has several suitable lines. Adding the red measurements of the He–Ne laser and the telluric lines pretty well covers the visible window. Interpolation to the appropriate wavelength can then be made.

In some cases the grating must be orientated differently for the measurement of the narrow emission line and the stellar line. Then eq. 12-3 shows that the projected slit width changes and hence $I(\lambda)$ changes. Such differences in α should be kept to a minimum so that little or no correction needs to be made. The projected slit width, measured in millimeters in the focal plane, is independent of the order according to eq. 3-18. In situations where $I(\lambda)$ is dominated by the slit width, we may wish to measure the δ function source and the stellar spectrum at nearly the same grating orientation even if they are not measured in the same grating order. The instrumental profile measurements are converted to a λ scale using the dispersion of the order in which the stellar data are taken. The ghosts will scale with the square of the order (eq. 3-17) and, if they are significant, they should be adjusted for any order differences.

Focus or other changes in spectrograph adjustment should be minimized between data collection and instrumental profile measurement. The wider $I(\lambda)$ is compared to the stellar lines, the more precisely it must be known. It may prove necessary to measure $I(\lambda)$ more than once a night should the spectrograph show any unstable behavior.

SCATTERED LIGHT

Some of the light entering the spectrograph may appear in the spectrum well away from the "proper" position of focus for photons of their wavelength. These scattered photons may have been deflected by optical imperfections in the grating and the mirrors, a piece of dust in the air, and so on, or they may be in the low amplitude tail of a diffraction pattern from the entrance slit, grating, or other aperture. The presence of scattered light in a spectrograph results in a systematic error in the line strength and shape measurements and in general causes a loss of contrast as the scattered light fills in the absorption lines.

The instrumental profile removal would properly account for scattered light if $I(\lambda)$ could be measured to a wavelength interval from line center that is comparable to the spectral band pass of the system. In a wide band system this is impossible, and yet the cumulative effect of a weak wing coupled with the continuous spectrum of the star can amount to a significant amount of light. In a system that can detect radiation from 3000 to 7000 Å, for example, an average $I(\lambda)$ wing of only 2.5×10^{-6} gives a 1% contribution. This cumulative radiation from beyond the wavelength interval where $I(\lambda)$ can be measured is called scattered light.

Suppose we have some absorption line whose profile we wish to measure. We choose $\lambda = 0$ at the center of the line. The measurements of the true spectrum, $F(\lambda)$, are

$$D(\lambda) = F(\lambda) * I(\lambda)$$

$$= \int_{-\infty}^{\infty} I(\lambda - \lambda_0)F(\lambda_0)\,d\lambda_0. \tag{12-19}$$

The flux entering eq. 12-19 is not the flux as it comes from the star, but rather the flux as detected in the spectrometer. That is, it includes the spectral response of the spectrograph, the transmission losses in the terrestrial atmosphere, reflection losses in the telescope, filter transmission, and so on. Suppose we denote the spectral band where $F(\lambda)$ is nonzero by* $\pm \lambda_b$. Further, we let the span over which we measure the instrumental profile be $\pm \lambda_1$. Then

$$
D(\lambda) = \int_{-\lambda_b}^{-\lambda_1} I(\lambda - \lambda_0)F(\lambda_0)\, d\lambda + \int_{-\lambda_1}^{\lambda_1} I(\lambda - \lambda_0)F(\lambda_0)\, d\lambda_0
$$

$$
+ \int_{\lambda_1}^{\lambda_b} I(\lambda - \lambda_0)F(\lambda_0)\, d\lambda_0
$$

$$
\equiv S(\lambda, \lambda_1, \lambda_b) + \int_{-\lambda_1}^{\lambda_1} I(\lambda - \lambda_0)F(\lambda_0)\, d\lambda_0 \qquad (12\text{-}20)
$$

where $S(\lambda, \lambda_1, \lambda_b)$ is the scattered light since it is comprised of the contribution from outside the measured region $(-\lambda_1, \lambda_1)$.

We still require

$$
\int_{-\infty}^{\infty} I(\lambda)\, d\lambda = 1
$$

so the true zero ordinate value of $i(\sigma)$ is unity. But if we measure $I(\lambda)$ only between $-\lambda_1$ and λ_1 we obtain an area that is less than unity. Suppose we write

$$
\int_{-\infty}^{\infty} I(\lambda)\, d\lambda = \int_{-\infty}^{-\lambda_b} I(\lambda)\, d\lambda + \int_{-\lambda_b}^{-\lambda_1} I(\lambda)\, d\lambda + \int_{-\lambda_1}^{\lambda_1} I(\lambda)\, d\lambda
$$

$$
+ \int_{\lambda_1}^{\lambda_b} I(\lambda)\, d\lambda + \int_{\lambda_b}^{\infty} I(\lambda)\, d\lambda
$$

$$
\doteq \int_{-\lambda_b}^{-\lambda_1} I(\lambda)\, d\lambda + \int_{-\lambda_1}^{\lambda_1} I(\lambda)\, d\lambda + \int_{\lambda_1}^{\lambda_b} I(\lambda)\, d\lambda. \qquad (12\text{-}21)
$$

This last step holds in many practical cases.† The first and last integrals

---

* If the spectral feature being studied is not near the center of the band pass, $\pm \lambda_b$ can be replaced with separate symbols and the following equations trivially modified accordingly.
† Typically $I(\lambda)$ varies as $1/\lambda^2$ in the wings. Direct integration shows that

$$
\int_{\lambda_b}^{\infty} I(\lambda)\, d\lambda \Big/ \int_{\lambda_1}^{\lambda_b} I(\lambda)\, d\lambda = (\lambda_b/\lambda_1 - 1)^{-1}.
$$

So if λ_1 is a few angstroms and λ_b is a few tens of angstroms, for example, the ratio is $\sim 10^{-1}$, which is sufficient for our purposes.

remaining in the expression are related to the scattering light defined in eq. 12-20,

$$S(\lambda, \lambda_1, \lambda_b) = \int_{-\lambda_b}^{-\lambda_1} I(\lambda - \lambda_0)F(\lambda_0)\, d\lambda_0 + \int_{\lambda_1}^{\lambda_b} I(\lambda - \lambda_0)F(\lambda_0)\, d\lambda_0$$

or

$$\frac{S(0, \lambda_1, \lambda_b)}{\bar{F}} = \int_{-\lambda_b}^{-\lambda_1} I(-\lambda_0)\, d\lambda_0 + \int_{\lambda_1}^{\lambda_b} I(-\lambda_0)\, d\lambda_0$$

$$= \int_{\lambda_1}^{\lambda_b} I(\lambda)\, d\lambda + \int_{-\lambda_b}^{-\lambda_1} I(\lambda)\, d\lambda$$

where $\bar{F}$ is, strictly speaking, the average flux in the intervals $(-\lambda_b, -\lambda_1)$ and (λ_1, λ_b), but which we assume to be characteristic of the whole spectral

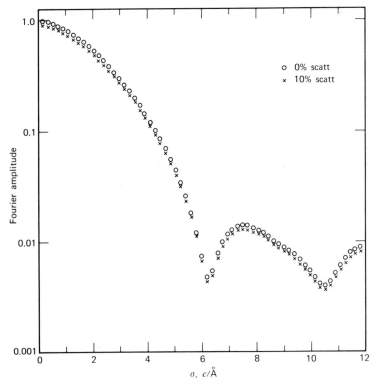

Fig. 12-13 All the Fourier amplitudes are reduced by the fractional scattered light resulting in increased filtering of the spectrum. (Adapted from Gray 1974. Courtesy of the *Publications* of the Astronomical Society of the Pacific.)

region in the vicinity of the line. We then can rewrite eq. 12-21 as

$$1 = \int_{-\infty}^{\infty} I(\lambda)\, d\lambda \doteq \frac{S(0, \lambda_1, \lambda_b)}{\bar{F}} + \int_{-\lambda_1}^{\lambda_1} I(\lambda)\, d\lambda. \qquad (12\text{-}22)$$

This equation expresses the important and interesting result that the measured instrumental profile should be normalized to unity less the fractional scattered light—the scattered light being measured at the wavelength of the feature of interest.

At this point let us consider the instrumental profile and scattered light in the Fourier transform plane. Figure 12-13 shows $i(\sigma)$ having normalizations with and without the scattered light term. Because $i(\sigma)$ is multiplied into the transform of $F(\lambda)$, we see that the effect of the scattered light is to reduce the Fourier amplitudes of the spectrum by increased filtering. Notice that $i(0)$, which scales the total absorption produced by the lines, is decreased by the fractional scattered light along with the other components. But before we can use this important fact, we must be able to measure the fractional scattered light.

MEASUREMENT OF SCATTERED LIGHT

No easy means of measuring the scattered light $S(\lambda, \lambda_1, \lambda_b)$ in spectrographs has been devised. But because the scattered light is usually small (i.e., a few per cent at most) it need not be known with high precision. It is convenient to think of the scattered light being comprised of two components, (1) the general scattered light caused by dust on mirror and in the air, bubbles in lenses, and other such randomly oriented imperfections and (2) the so-called linearly scattered light that goes in the direction of dispersion and arises from the slit and grating.

The general scattered light is measurable by looking above and below the continuous spectrum formed by an incandescent light bulb (cf. Gray and Evans 1973) or by covering a portion of the entrance slit and measuring the light in the obscured portion of the spectrum (Edmonds 1965, Conway 1952, and Shane 1933). By definition this component of scattering is uniform over the output field, so scattering measured near the spectrum is the same as scattering in the position of the spectrum. Some care must be exercised to make certain that optical aberrations, such as astigmatism, and any diffraction perpendicular to the dispersion are not mistaken for general scattered light.

A determination that includes the linear scattering cannot be made so trivially because the spectrum itself is in the way. Waddell (1956) has used an absorption cell technique with some success. An absorption cell containing sodium was placed in front of the entrance slit and the residual flux in the core of the highly saturated D_1 line was measured as a function of the line widths. This residual flux is scattered light since the emission from sodium at the low temperatures of the absorption cell is negligible. The D_1 line widths were controlled by altering the sodium vapor pressure in the cell. Waddell then measured the relations between the residual flux, grating angle, and half width of the D_1 lines. The half width is associated with λ_1 so that the measured relations were taken to be $S(\lambda, \lambda_1, \lambda_b)$.

A similar estimate of $S(\lambda, \lambda_1, \lambda_b)$ can be made by using a rejection band interference filter at the entrance slit to introduce artificial absorption lines into a continuous spectrum. Since the rejection bands are known to be opaque to better than one part in 10^3 typically, any light above that level must be due to scattering. An integration of the filter transmission allows one to calculate $\bar{F}$ and thus the per cent scattered light. In this case $2\lambda_1$ is the width of the rejection band. This scheme does not give the λ_1 dependence but is much easier to use than the absorption cell method. To be consistent then, the measurements of $I(\lambda)$ must extend over a width equal to the width of the rejection band. A similar approach has been used by Teske (1967) and Goldberg et al. (1959) in which a Fabry–Perot interferometer replaces the rejection band filter.

It is clear from eq. 12-20 that the scattered light can be reduced by bringing the limits λ_1 and λ_b as close together as possible. Measurement of $I(\lambda)$ to low flux levels is needed to increase λ_1. Photoelectric instruments can measure $I(\lambda)$ to $\sim 10^{-4}$ to 10^{-5} of the main peak, whereas photographic measurements give difficulties at fractional levels as large as 10^{-2}. The band pass of photographic observations is often hundreds of angstroms wide while the band pass for photoelectric spectrophotometry frequently extends over just a few tens of angstroms. Scattered light corrections are, as a result, often larger for photographic observations.

Reduction of the band pass of the instrument can be highly advantageous. Use of a predisperser, which is a low dispersion spectrograph placed in front of the main instrument, is one way. A similar scheme is to place a rectangular band pass interference filter in the beam. The double pass system (Evans and Waddell 1962, Brault et al. 1971) used in solar spectrographs is the ultimate in this regard. Here the dispersed light is sent through the spectrograph a second time so the instrument acts as its own predisperser. The band pass can be made as narrow as desired and, specifically, λ_b can be reduced to λ_1. Scattered light is reduced to zero for such a case. Unfortunately the large number of reflections and the second pass over the grating make the total photon throughput too low for stellar work (see Griffin and Griffin 1969).

Reduction of the scattered light by decreasing λ_b to λ_1 is most effective when λ_1 is large. There is the danger that decreasing λ_b may cause the far wing area, neglected in eq. 12-21, to become significant. The normalization condition equivalent to eq. 12-22 is then

$$1 = \int_{-\infty}^{-\lambda_1} I(\lambda)\, d\lambda + \int_{\lambda_1}^{\infty} I(\lambda)\, d\lambda + \int_{-\lambda_1}^{\lambda_1} I(\lambda)\, d\lambda \qquad (12\text{-}23)$$

where now we have $\lambda_b = \lambda_1$. In general the first and second integrals cannot be measured, but we see that they can be reduced to a small fraction of the measured third integral by making λ_1 large enough. If the instrumental profile could be represented, for example, by a dispersion function, then a $\lambda_1 > 35$ half widths reduces the lost wing area to less than 1%. By analogy with the effect of scattered light as expressed in Fig. 12-13, the equivalent widths as well as the other Fourier frequencies are then distorted by less than 1% due to this effect.

CORRECTIONS FOR SCATTERED LIGHT

It should be clear from the discussion of Fig. 12-13 that corrections for scattered light must be made prior to removal of the instrumental profile.

Perhaps the simplest method of correction is to alter the normalization of the measured $I(\lambda)$ according to eq. 12–22 or 12–23. Then the scattered light correction is automatically included in the removal of $I(\lambda)$ from the data (Gray 1974). When wide wavelength spans of data are to be reconstructed, the renormalization of $I(\lambda)$ must be redone as often as the scattered light shows a significant wavelength variation. This implies that a reduction of a wide spectral band must be made in portions, using a different $S(0, \lambda_1, \lambda_b)/\bar{F}$ for each. This is not an additional burden because $I(\lambda)$ is itself likely to be wavelength dependent, forcing us to reduce the spectrum in pieces anyway.

A slightly more laborious approach to scattered light correction is to correct the observed data spectrum directly (e.g., Teske 1967). Let us denote the fractional scattered light by

$$s \equiv \frac{S(\lambda, \lambda_1, \lambda_b)}{\bar{F}}. \qquad (12\text{-}24)$$

We can reasonably take $\bar{F}$ to be the average measured spectrum $\bar{D}$. Then by calling an observed data point D_{obs} and the value that would exist without scattering D_{true}, we can write

$$D_{obs} = D_{true} - sD_{true} + s\bar{D}$$

$$= (1 - s)D_{true} + s\bar{D}$$

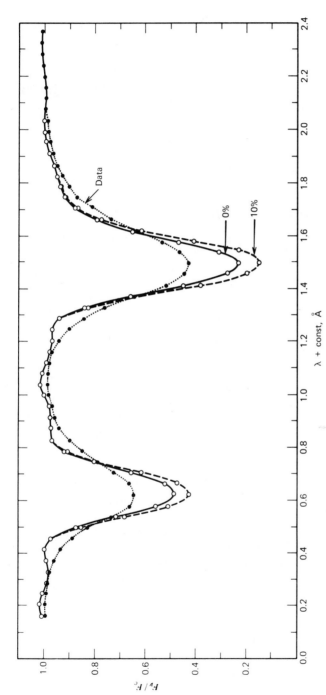

Fig. 12-14 Incorporating a correction for scattered light deepens the profile. An *arbitrary* 10% correction has been applied to the data for Fe I λ6065 in Arcturus to show the effect. The instrumental profile has been removed. The transforms of the zero and 10% scattering curves are shown in Fig. 12-13. (Adapted from Gray 1974. Courtesy of the *Publications* of the Astronomical Society of the Pacific.)

or

$$D_{\text{true}} = \frac{D_{\text{obs}} - s\bar{D}}{1 - s}$$

where sD_{true} is the light scattered out of the profile and $s\bar{D}$ is the light scattered into the profile. The continuum is also altered in a similar way,

$$D_{\text{true}}^c = \frac{D_{\text{obs}}^c - s\bar{D}}{1 - s}.$$

The true profile is given by $D_{\text{true}}/D_{\text{true}}^c$,

$$\frac{D_{\text{true}}}{D_{\text{true}}^c} = \frac{D_{\text{obs}} - s\bar{D}}{D_{\text{obs}}^c - s\bar{D}} = \frac{D_{\text{obs}}/D_{\text{obs}}^c - s\bar{D}/D_{\text{obs}}^c}{1 - s\bar{D}/D_{\text{obs}}^c}. \tag{12-25}$$

Each of the ratios $D_{\text{obs}}/D_{\text{obs}}^c$ are points on the observed profile. In an uncrowded spectral region where $\bar{D} \doteq D_{\text{obs}}^c$ we have

$$\frac{D_{\text{true}}}{D_{\text{true}}^c} = \frac{D_{\text{obs}}/D_{\text{obs}}^c - s}{1 - s}. \tag{12-26}$$

Each of the observed data points is then processed through eq. 12-25 or 12-26 after which the instrumental profile can be removed.

Figure 12-14 shows the effect of scattered light on the spectrum. The correction becomes smaller for points farther from line center. The scattered light "fills in" the line, reducing the contrast—an effect we already noticed with regard to Fig. 12-13. In eq. 12-26, the correction amounts to raising the zero signal level by s and re-normalizing to unit continuum. Let us notice again, as we did when discussing Fig. 12-13, that the equivalent width of a line is changed by s.

LINE MEASUREMENTS WITH LOW RESOLUTION

We expect to be able to measure the wings of broad lines and equivalent widths of narrower lines using only low resolution. The bulk of stellar observations are comprised of low resolution measurements because they require less telescope time or because faint objects are measured.

In eq. 12-2 the equivalent width is defined as

$$W = \int_0^\infty (F_c - F_v)\, dv/F_c.$$

The definition implies that the spectrum contains only one line and the integration is carried over the full spectrum to make certain the wings of the line are fully accounted for. The definition has limited connection to the observed spectrum because we do not measure F_v, but $D_v = I(\lambda) * F_v$.

The measured profile is

$$R'_v = \frac{D_c - D_v}{D_c} = \frac{I(\lambda) * (F_c - F_v)}{D_c}. \tag{12-27}$$

The measured equivalent width is

$$W' = \int_{-\Delta}^{\Delta} \frac{D_c - D_v}{D_c}\, dv = \int_{-\Delta}^{\Delta} \frac{I(\lambda) * (F_c - F_v)}{D_c}\, dv \tag{12-28}$$

where 2Δ is the spectral range over which the profile can be traced.

Now it may seem straightforward to relate R'_v to R_v by removing the instrumental profile according to the previous discussion. And certainly we would expect W' and W to be equal because $I(\lambda)$ has unit area, provided the constraints of eq. 12-23 are accommodated. One difficulty is that in practice D_c and F_c may not be the same. A significant error often results from choosing D_c amid the noise. A second contribution to error occurs in crowded regions because D_c is depressed below F_c. Consider Figs. 12-15 and 12-16. The former shows a "nice" case, the line is not blended and the continuum is well defined. Then, as long as Δ in eq. 12-27 is made sufficiently large, $W' = W$ and the reconstruction of R' can be considered. The more realistic case is that of Fig. 12-16—the "not so nice" case. There are only a

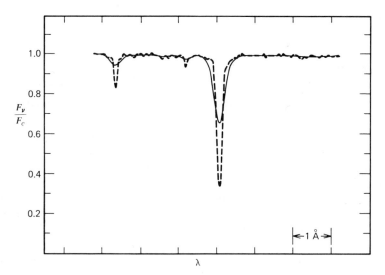

Fig. 12-15 In a portion of a spectrum like this one, the continuum is well defined independent of the resolution: (——) 0.25 Å resolution; (------) 0.03 Å resolution.

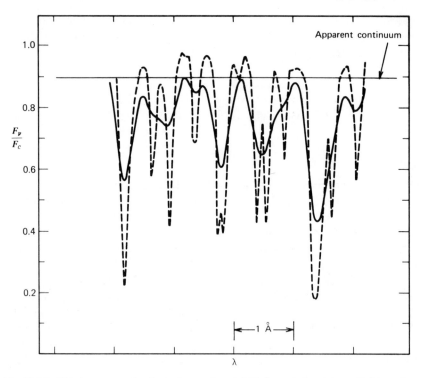

Fig. 12-16 This is an example of a bad case: (———) 0.25 Å resolution (about 17 Å/mm for a spectrogram having 15 μ resolution); (------) 0.03 Å resolution.

few lines that are unblended and the continuum is badly cut up. D_c is not at the level of F_c. $W' \neq W$ and reconstruction of R' will be systematically in error.

The proper procedure for the not so nice case is to reconstruct *the whole spectral region* so the true continuum can be measured. If the resolution being used is inadequate for this, or if the noise level is too high, the observations are unsuitable for precise equivalent width measurements.

The crowding of spectral lines is less in the red region of the spectrum where there are fewer lines. The most severe problems occur in K and M stars. A totally unblended line is a rarity in stellar spectra.

The broad lines, particularly the hydrogen lines, pose a different problem when it comes to choosing F_c. The wings of these lines extend very far and reach the continuum only asymptotically. In an early A star, for example, at 70 Å from the line center the profile is still 0.5% below the continuum

(Peterson 1969). Identifying the far wing at 70 Å with the continuum produces an error of more than 140 Å $\times$ 0.5% = 0.7 Å or about a 5% error in equivalent width. If the continuum cannot be judged to any better than $\pm 1\%$ because of noise, an additional uncertainty of 10% occurs. Continuum placement is obviously very important for broad lines. The profile data suffer less from continuum uncertainty than W does.

The situation can be improved by adopting a reference flux level in the line wing, at say $\Delta\lambda \sim 30$–50 Å (chosen at the convenience of the investigator). Both the measurements and the theoretical profiles are normalized to the same point as if it were the true continuum. Analysis can then be made independent of the true continuum. Noise associated with measurement of the reference point still remains of course.

A comparison of sets of equivalent widths measured at different observatories may show systematic differences of 10% or more. Usually values of W obtained from low dispersion spectrograms are larger than those from high dispersion spectrograms (Cayrel de Strobel 1967, Wright 1958, 1966, Gathier and Houtgast 1952). Even in solar measurements the errors are substantial. An interesting comparison has been made by Helfer et al. (1963).

Photoelectrically measured equivalent widths have not yet been obtained in sufficient numbers to see if they fare any better. Photoelectric measurements of hydrogen line profiles, however, show excellent agreement among independent investigations (Gray and Evans 1973).

CHOICE OF FOURIER DOMAINS FOR ANALYSIS

So far we have considered the σ domain with regard to removal of the instrumental profile and the filtering of noise. These operations can be performed on a wide spectral region or on a single line profile.

Now we consider specifically a single line profile and its transform. The choice of filter to remove the high σ noise can be a sensitive one in some cases. The result is that we do not know how much to trust the reconstructed profile. On the other hand, if we compute the transform of the theoretical line and analyze in the σ domain, we need not choose a filter at all and the data points can be trusted according to their error bars, which increase with increasing σ. An example is shown in Fig. 12-17. This type of analysis is very useful, especially in Chapter 18.

Another good alternative to removing the instrumental profile from the data is to convolve it into the theoretical profiles. Filtering the theoretical profiles with the instrumental profiles can be done in either Fourier domain. The enhancement of noise by the removal of $I(\lambda)$ is again avoided.

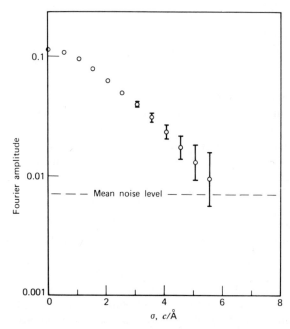

Fig. 12-17 The Fourier transform of an absorption line is shown with error bars according to the signal to noise ratio. The instrumental profile has been removed. In fitting the data with theoretical calculations, much more weight is given to the low frequency points in view of the errors.

SPECTROPHOTOMETRIC STANDARD STARS

Standard stars have been proposed to facilitate intercomparison of data and for calibration work. The standards form a fundamental link between the most detailed physical theory and basic stellar properties. The stars chosen must be bright so that the highest resolution can be employed with high precision for the measurement of line profiles and equivalent widths. Stars suitable for basic standards should also have parallaxes large enough to measure, no interstellar reddening, and whenever possible a mass determination. Each star's energy distribution, effective temperature, radius, surface gravity, chemical composition, turbulence, and rotation are then measured in the most fundamental manner possible.

Table 12-2 is a list of standard stars. Most of the entries fulfill most of the prerequisites for standard stars. None of them is a known unresolved spectroscopic binary or variable. Only α Boo is listed as peculiar. Several of them have well determined masses. Some appeared on earlier lists by

Table 12-2   Spectrophotometric Standard Stars

Star	HR	HD	Sp	B − V	V	Radial velocity, km/sec	v sin i, km/sec	Parallax
η Aur	1641	32630	B3 V	−0.18	3.17	+7	139	0″013
γ Ori	1790	35468	B2 III	−0.23	1.64	+18	64	0.026
α Lyr	7001	172167	A0 V	0.00	0.03	−14	17	0.124
γ Gem	2421	47105	A0 IV	0.00	1.93	−13	37	0.031
θ²Tau	1412	28319	A7 III	+0.18	3.41	+40	80	0.025
γ Vir A	4825	110379	F0 V	+0.36	3.65	−20	27	0.099
α CMi A	2943	61421	F5 IV–V	+0.42	0.37	−3	6	0.285
ζ Leo	4031	89025	F0 III	+0.31	3.43	−15	82	0.009
η Cas A	219	4614	G0 V	+0.57	3.45	+9	<6	0.170
Sun	—	—	G2 V	+0.63	—	—	2	—
μ Her A	6623	161797	G5 IV	+0.75	3.42	−16	20	0.124
ε Vir	4932	113226	G9 III	—	2.81	−14	<17	0.036
70 Oph A	6752	165341	K0 V	+0.86	4.21	−7	16	0.195
σ Dra	7462	185144	K0 V	+0.80	4.69	+27	<17	0.176
γ Cep	8974	222404	K1 IV	+1.03	3.21	−42	<17	0.065
β Gem	2990	62509	K0 III	+1.00	1.14	+4	<17	0.093
α Boo	5340	124897	K2 IIIp	+1.23	−0.05	−5	<17	0.092

Source.   Gliese (1969), Hoffleit (1964), and Uesugi and Fukuda (1970).

Wright (1955, 1958) and Cayrel de Strobel (1967). Extensive photographic measurements have been published by Wright et al. (1964) and Cayrel and Cayrel (1963) for a few of them. Griffin's (1968) *Atlas of the Spectrum of Arcturus* is a valuable work for a standard star.

The sun is a valuable standard star but is troublesome to compare to other stars because of its extended image and, to some extent, because of its very brightness. Any of the usual optical systems form an image of the sun at the entrance slit of the spectrograph. The light from only a small portion of the solar image goes through the entrance slit. This is good for the solar astronomer who wants high spacial resolution, but it is poor for the stellar astronomer who wants to see the whole of the sun at once like he does for other stars. In short, we must measure the solar *flux* spectrum not a specific intensity spectrum. At various times people have resorted to the moon, asteroids, and sky light to measure the solar flux spectrum. There is evidence that these sources do not faithfully represent the solar flux spectrum (e.g. Hunten 1970). Direct measurement of the sun can be made using the flux heliostat as described by Gray (1972).

REFERENCES

Abt, H. 1967. *Facilities of the Kitt Peak National Observatory*, Tucson, Arizona.

Bowen, I. S. 1938. *Astrophys. J.* **88**, 113.

Bowen, I. S. 1962. *Astronomical Techniques*, W. A. Hiltner, Ed., University of Chicago Press, Chicago, p. 34.

Boyce, P. B., N. M. White, R. Albrecht, and A. Slettebak 1973. *Publ. Astron. Soc. Pac.* **85**, 91.

Bracewell, R. N. 1955. *J. Opt. Soc. Am.* **45**, 873.

Brault, J. W., C. D. Slaughter, A. K. Pierce, and R. S. Aykens 1971. *Sol. Phys.* **18**, 366.

Brault, J. W. and O. R. White 1971. *Astron. Astrophys.* **13**, 169.

Burger, H. C. and P. H. van Cittert 1933. *Z. Phys.* **81**, 428.

Cayrel, R., and G. Cayrel 1963. *Astrophys. J.* **137**, 431.

Cayrel de Strobel, G. 1967. *Transactions of the International Astronomical Union Reports on Astronomy*, L. Perek, Ed., Reidel, Dordrecht, Holland, p. 639.

Champeney, D. C. 1973. *Fourier Transforms and Their Physical Applications*, Academic, New York, p. 121.

Connes, P. 1970. *Annu. Rev. Astron. Astrophys.* **8**, 209.

Conway, M. T. 1952. *Mon. Not. Roy. Astron. Soc.* **112**, 55.

de Jager, C. and L. Neven 1966. *Bull. Astron. Inst. Neth.* **18**, 306.

Dunham, T. Jr. 1956. *Vistas in Astronomy*, Vol. 2, A. Beer, Ed., Pergamon, London, p. 1223.

Edmonds, F. N. Jr. 1965. *Astrophys. J.* **142**, 278.

Elste, G. 1953. *Z. Astrophys.* **33**, 39.

Evans, J. W. and J. Waddell 1962. *Appl. Opt.* **1**, 111.

Gathier, P. and J. Houtgast 1952. *Int. Astron. Union Trans.* **8**, 568.

Gliese, W. 1969. *Catalogue of Nearby Stars Edition 1969*, Veröffentlichungen des Astronomischen Rechen-Instituts, Heidelberg.

Goldberg, L., O. C. Mohler, and E. A. Müller 1959. *Astrophys. J.* **129**, 119.

Gray, D. E. 1963. *American Institute of Physics Handbook*, McGraw-Hill, New York, pp. 7–122.

Gray, D. F. 1971. *Bull. Am. Astron. Soc.* **3**, 387.

Gray, D. F. 1972. *Publ. Astron. Soc. Pac.* **84**, 721.

Gray, D. F. 1974. *Publ. Astron. Soc. Pac.* **86**, 526.

Gray, D. F. and Evans, J. C. 1973. *Astrophys. J.* **182**, 147.

Griffin, R. F. 1968. *A Photometric Atlas of the Spectrum of Arcturus*, Cambridge Philosophical Society.

Griffin, R. F. 1969a. *Mon. Not. Roy. Astron. Soc.* **143**, 319.

Griffin, R. F. 1969b. *Mon. Not. Roy. Astron. Soc.* **143**, 349.

Griffin, R. F. 1969c. *Mon. Not. Roy. Astron. Soc.* **143**, 361.

Griffin, R. and R. Griffin 1969. *Mon. Not. Roy. Astron. Soc.* **143**, 341.

Helfer, H. L., G. Wallerstein, and J. L. Greenstein 1963. *Astrophys. J.* **138**, 97.

Hoffleit, D. 1964. *Catalogue of Bright Stars*, 3rd ed., Yale University Observatory, New Haven, Conn.

Hunten, D. M. 1970. *Astrophys. J.* **159**, 1107.

Middleton, D. 1960. *Introduction to Statistical Communication Theory*, McGraw-Hill, New York.

O'Dell, C. R. 1969. *Publ. Astron. Soc. Pac.* **81**, 854.

Panofsky, H. A. A. 1943. *Astrophys. J.* **97**, 180.

Peterson, D. M. 1969. Smithsonian Astrophysical Observatory Special Report No. 293. *The Balmer Lines in Early Type Stars.*

Pierce, A. K. 1965. *Publ. Astron. Soc. Pac.* **77**, 216.

Richardson, E. H. 1966. *Publ. Astron. Soc. Pac.* **78**, 436.

Shane, C. D. 1933. *Lick Obs. Bull.* **16**, 76.

Shulte, D. H. 1966. *Appl. Opt.* **5**, 457.

Teske, R. G. 1967. *Publ. Astron. Soc. Pac.* **79**, 110.

Uesugi, A. and I. Fukuda 1970. *A Catalogue of Rotational Velocities of the Stars.* Contributions of the Institute of Astrophysics and Kwasan Observatory, University of Kyoto, No. 189.

Vaughan, A. H. Jr. 1967. *Annu. Rev. Astron. Astrophys.* **5**, 139.

Waddell, J. 1956. *An Empirical Determination of the Turbulence Field in the Solar Photosphere Based on a Study of Weak Fraunhofer Lines*, thesis, University of Michigan.

Wilson, O. C. 1968. *Astrophys. J.* **153**, 221.

Wright, K. O. 1955. *Int. Astron. Union Trans.* **9**, 527.

Wright, K. O. 1958. *Int. Astron. Union Trans.* **10**, 558.

Wright, K. O. 1966. *Abundance Determinations in Stellar Spectra IAU Symposium No. 26*, H. Hubenet, Ed., Academic, London, p. 15.

Wright, K. O., E. K. Lee, T. V. Jacobson, and J. L. Greenstein 1964. *Publ. Dom. Astrophys. Obs. Victoria* **12**, 173.

CHAPTER THIRTEEN

THE BEHAVIOR
OF SPECTRAL LINES

We know that the absorption lines that appear in stellar spectra show differences in shape and strength according to the physical conditions in the star's atmosphere. Some of the toughest and most fascinating problems arise in the study of the interaction of the line absorption with the temperature, pressure, radiation field, and so on of the gas. We are not yet able to quantify and calculate the full interaction of these variables with the current level of understanding. On the other hand, the panoramic view of spectral line behavior can be understood in relatively simple terms which we shall presently describe.

One of the goals of study in stellar atmospheres is to understand the varied profiles and line strengths that arise. Another is to use our knowledge of line behavior to interpret the fundamental properties of many stars. Under this second category one would place the measurement of stellar effective temperatures, atmospheric pressure or surface gravity, radii, and chemical composition. Our attention is directed to the first of these goals in this chapter while the others have been postponed to the following chapters.

THE LINE TRANSFER EQUATION

The transfer equation derived in Chapter 7 holds for the case including both line and continuum radiation provided the variables are suitably defined. We let l_v be the line absorption coefficient, j_v^l the line emission coefficient, and κ_v and j_v^c the corresponding continuum parameters. We should recall that scattering and real absorption are included in these coefficients. Then by defining

$$d\tau_v = (l_v + \kappa_v)\rho\, dx \qquad (13\text{-}1)$$

and the source function

$$S_v = \frac{j_v^l + j_v^c}{l_v + \kappa_v}, \qquad (13\text{-}2)$$

we can write

$$\frac{dI_v}{d\tau_v} = -I_v + S_v \qquad (13\text{-}3)$$

as the transfer equation along the coordinate line x. The source functions for the line and continuum may be defined separately,

$$S_l \equiv \frac{j_v^l}{l_v}$$

$$S_c \equiv \frac{j_v^c}{\kappa_v},$$

in which case the total source function can also be written

$$S_v = \frac{(l_v/\kappa_v)S_l + S_c}{1 + l_v/\kappa_v} \quad \text{or} \quad \frac{S_l + (\kappa_v/l_v)S_c}{1 + \kappa_v/l_v}. \qquad (13\text{-}4)$$

The integral forms of the transfer equation, eqs. 7-4 and 7-11 still hold. In particular, the specific intensity at the surface is (eq. 7-12),

$$I_v(0) = \int_0^\infty S_v e^{-\tau_v \sec\theta} \sec\theta\, d\tau_v. \qquad (13\text{-}5)$$

When the granulation cell structure or other inhomogeneities are to be included, the flux can only be obtained from the proper weighting of I_v computed for many lines of sight. This means using eq. 13-5 with the spacial and depth model under study. In the simpler case where inhomogeneities are ignored, the surface flux is given by (eq. 7-16),

$$\mathscr{F}_v(0) = 2\pi \int_0^\infty S_v(\tau_v)E_2(\tau_v)\, d\tau_v. \qquad (13\text{-}6)$$

But the central problem is still as it was in Chapter 7: how to obtain $S_v(\tau_v)$ so that the integration can actually be done.

THE LINE SOURCE FUNCTION

Let us explore the intimate relation between the source function and the measurable (surface) flux, $\mathcal{F}_\nu(0)$, with an approach similar to that used for the continuum in the discussion of eq. 9-3. To avoid a purely numerical problem, we adopt the source function with the form of the Eddington approximation (eq. 7-41),

$$S_\nu(\tau_\nu) = \frac{3}{4\pi} \mathcal{F}_\nu(0)\left[\tau_\nu + \frac{2}{3}\right]. \tag{13-7}$$

In particular, when $\tau_\nu = (4\pi - 2)/3 \sim 3.5$, let us call it τ_1, we see that the source function and the surface flux are equal,

$$S_\nu(\tau_1) = \mathcal{F}_\nu(0). \tag{13-8}$$

Across a line profile l_ν changes, being larger toward the center of the line. The condition $\tau_\nu = \tau_1$ is true higher up in the atmosphere for ν near line center and holds for progressively deeper layers for frequencies farther into the wing. Under the assumption that S_ν is a slowly varying function of ν so it can be considered constant over the width of the line, there is a mapping according to eq. 13-8 between S_ν as a function of τ_ν, and $\mathcal{F}_\nu$ as a function of ν. This is shown in Fig. 13-1. Because the source function decreases outward an absorption line is formed.

Microturbulence alters the depth at which $\tau_\nu = \tau_1$. Our rather poor understanding of turbulence (see Chapter 18) hampers the quantitative use of the concept expressed in eq. 13-8. Macroturbulence and rotation should be removed from $\mathcal{F}_\nu$ prior to any interpretation of the $\mathcal{F}_\nu$-S_ν mapping.

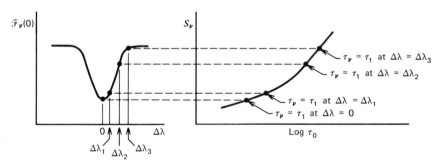

Fig. 13-1 The mapping between the source function and the line profile. As with limb darkening, it is the drop in the source function with height that results in the characteristic behavior. In this case an absorption line is formed.

More detailed relations between S_v and $\mathcal{F}_v(0)$ are considered later in the chapter for some special cases. In particular, we see that the flux at any point in the line profile comes from a considerable range in depth and cannot be completely characterized by one depth such as τ_1.

We now derive a very general expression for the line source function S_v by generalizing the thoughts that led to eq. 6-8. We denote the frequency dependence of the spontaneously emitted radiation by $\psi(v)$. Then the probability that an atom will emit a photon in the time dt, in the solid angle $d\omega$, and in the frequency interval dv is $A_{ul} \, dt \, d\omega \, \psi(v) \, dv$. We have given, via $\psi(v)$, a frequency dependence or profile to A_{ul}. The profile $\psi(v)$ is normalized to unit area. The emission coefficient is then

$$j_v^l \rho = N_u A_{ul} \psi(v) hv$$

which is the frequency dependent form of eq. 5-19.

The frequency dependence of the absorption and stimulated emission processes we call $\phi(v)$. Then the absorption coefficient is

$$l_v \rho = N_l B_{lu} \phi(v) hv - N_u B_{ul} \phi(v) hv,$$

an expression similar to eq. 5-20.

Ratioing these expressions gives

$$S_l = \frac{j_v^l}{l_v} = \frac{N_u A_{ul} \psi(v)}{N_l B_{lu} \phi(v) - N_u B_{ul} \phi(v)}.$$

Then by using the relations between the transition probabilities, eqs. 6-7, $B_{ul} g_u = B_{lu} g_l$ and $A_{ul} = 2hv^3 B_{ul}/c^2$, we recast the source function to read

$$S_l = \frac{2hv^3}{c^2} \frac{1}{(N_l/N_u)(g_u/g_l) - 1} \frac{\psi(v)}{\phi(v)}. \tag{13-9}$$

The relation shows how the line source function depends on the ratio of populations in the levels involved in the transition and also the dependence on the ratio of the emission to absorption profiles. The complexity of the situation is now apparent. In order to calculate I_v (eq. 13-5) we need S_l, but S_l depends in part on I_v because radiation affects the excitation of the levels N_l and N_u. Collisions also affect the excitation. In the special case where collisions completely dominate the excitation, the cyclic coupling of I_v and S_l is broken and we can expect N_l/N_u to be given by the usual excitation equation

$$\frac{N_u}{N_l} = \frac{g_u}{g_l} e^{-hv/kT}.$$

Furthermore, if we have an equilibrium situation, then for each emitted photon there must be an identical photon that is absorbed—in other words, $\psi(v) = \phi(v)$. In this way we see that for the special case where equilibrium holds, the source function is given by

$$S_l = \frac{2hv^3}{c^2} \frac{1}{e^{hv/kT} - 1} = B_v(T) \tag{13-10}$$

that is, the line formed in LTE. As a first and very useful approximation, LTE has been used extensively in model atmosphere calculation. With it, eqs. 13-5 and 13-6 can be evaluated if T is known as a function of τ_v. We recall that the temperature, in the LTE approximation, is the same for all physical processes—thermal velocity distributions, ionization equilibrium, excitation of all atomic populations, and source function according to eq. 13-10. What a tremendous simplification over the real problem where the population of every atomic level in each atomic species has to be calculated separately and where each affects the other through the radiation field! No wonder the LTE approximation has been and still is used in many analyses.

Usually the general treatment, where S_l is given by eq. 13-9, is referred to as non-LTE. The term means "not restricted to" LTE, but it does include LTE as a special case.

THE LEVEL POPULATIONS

The ratios of populations are needed in the source function, eq. 13-9. The general approach to calculating the population of any atomic level is to obtain the sum of all the incoming and outgoing rates. In the steady state we require the population of each level to be constant in time. Specifically we can combine the rates of transition, for all physical processes, from the level i to the level j, denoted by P_{ij}, with the level populations N_i and N_j by writing

$$\frac{dN_j}{dt} = 0 = \sum_{i=1}^{M} (N_i P_{ij} - N_j P_{ji}) = \sum_{i=1}^{M} N_i P_{ij} - N_j \sum_{i=1}^{M} P_{ji} \tag{13-11}$$

for each level j. The term $N_i P_{ij}$ represents the number of atoms arriving into the level j from the levels i, while the term $N_j P_{ji}$ represents the atoms lost from the level j. The summation must be done over all possible combinations including the continuum (of course the $i = j$ case does not make physical sense and is omitted from the summation).

The transition rates P_{ij} and P_{ji} include radiative rates and collisional rates. In terms of the Einstein coefficients* we can write

$$P_{ij} = 4\pi A_{ij} + 4\pi B_{ij} \int_0^\infty \bar{I}_\nu \phi(\nu) \, d\nu + C_{ij} \qquad \text{for} \quad i \neq j \qquad (13\text{-}12)$$

but where $A_{ij} = 0$ for $i < j$. The first term is the spontaneous transition from upper levels. The second term is the rate of radiatively induced transitions for $i > j$ but is the rate of absorption for $i < j$. The collisionally induced transitions represented by C_{ij} are dependent on the density of the particles causing the collision, their velocity distribution, and atomic constants such as the collision cross sections and the excitation potentials. Similarly for P_{ji} we can write

$$P_{ji} = 4\pi A_{ji} + 4\pi B_{ji} \int_0^\infty \bar{I}_\nu \phi(\nu) \, d\nu + C_{ji} \qquad \text{for} \quad i < j$$

and

$$P_{ji} = 4\pi B_{ji} \int_0^\infty \bar{I}_\nu \phi(\nu) \, d\nu + C_{ji} \qquad \text{for} \quad i > j. \qquad (13\text{-}13)$$

The system of $M - 1$ equations, (eqs. 13-11), is to be solved simultaneously for the $M - 1$ values of N_j. Since the equations are equal to zero, however, there are only $M - 2$ independent relations so we can solve only for population ratios. This is sufficient for S_l.

It is easy to see how iterative solutions are frequently used when eqs. 13-11 are coupled with the transfer equation. To calculate S_l, the population ratios are needed; to calculate the population ratios, $\bar{I}_\nu$ is needed; and to calculate $\bar{I}_\nu$, S_l is needed.

OTHER FORMALISMS FOR S_ν

The two physical processes entering the absorption and emission coefficients are scattering and real absorption. If no frequency shift occurs during the scattering, the scattering is said to be coherent. The source function for pure coherent scattering is the mean intensity, $\bar{I}_\nu$. Coherent scattering is unlikely though because the atoms and ions are in motion and since the direction of the scattered photon is arbitrary, the Doppler shift is different for the initial and scattered photons and thus destroys the coherence. If all the coherence is destroyed, the scattering is called noncoherent and the

---

* We have defined the Einstein probabilities per unit solid angle, but P_{ij} includes all transitions regardless of direction. We therefore multiply the Einstein coefficients by 4π.

corresponding source function is $\int_0^\infty \bar{I}_\nu \phi(\nu) \, d\nu$. The partially coherent case is investigated by Avery and House (1968) for example. Generally both scattering and absorption appear in the source function.

It has been shown by Milne (1930), Jefferies and Thomas (1958, 1959, 1960), and Jefferies (1968) how, for a hypothetical atom consisting of two levels, the source function can be written*

$$S_\nu = \frac{\int_0^\infty \bar{I}_\nu \phi(\nu) \, d\nu + \varepsilon(\tau) B_\nu(T)}{1 + \varepsilon(\tau)} \tag{13-14}$$

in which $\varepsilon(\tau)$ is related to the ratio of collisional deexcitation to spontaneous deexcitation. When $\varepsilon(\tau)$ is small, the scattering dominates and there are large deviations from LTE. In sufficiently deep layers of the photosphere where $\tau_\nu \gg 1$ at all frequencies, the source function must become $B_\nu(T)$, and according to eq. 7-26b we can write $S_\nu = \bar{I}_\nu = B_\nu(T)$.

In the higher layers $\varepsilon(\tau)$ becomes smaller as the collisional deexcitations become less probable with decreasing density. Depending on the actual size of $\varepsilon(\tau)$, there is some depth above which the scattering term in eq. 13-14 dominates S_ν. The higher in the photosphere, the closer to the "boundary" where photons are being lost from the star. The loss of photons necessarily means $\bar{I}_\nu$ decreases with height and hence S_ν must decrease with height. As we have seen (Fig. 13-1), this decrease in S_ν leads to an absorption line. In short, the natural result of the photosphere's outer boundary is an absorption line.

A strong line, as is discussed shortly, is formed over an extensive depth. The wings are formed in the deep layers, the core in higher layers. On the basis of the above discussion we might expect the wing to be formed in LTE, the shape of which is controlled by the outward decrease in $B_\nu(T)$. We might expect the core to be formed in non-LTE with its shape controlled by the outward decrease in $\bar{I}_\nu$. This is found to be the case for the Na D lines in the sun, for example. The boundary between the two is found to occur at $\Delta\lambda \sim 0.2$ Å and $\tau_{5000} \sim 0.01$ (Johnson 1965, Curtis, 1965, Mugglestone 1965, Curtis and Jefferies 1967).

Atomic models including more than two levels lead to a somewhat more complicated source function and several interesting physical interactions. The reader is referred to Athay's (1972) monograph. Computations have been done on several multilevel atomic models. Athay and Canfield (1969) have studied the solar Mg b and Na D lines in this way. Auer and Mihalas (1972) calculate hydrogen and helium spectra for O stars. A detailed study of a 15 level iron atom model has been done by Athay and Lites (1972) with application to the sun (see also Lites 1973).

---

* This form should also be compared to eq. 5-18.

COMPUTATION OF A LINE PROFILE IN LTE

We rely heavily on the approximation of local thermodynamic equilibrium in the remainder of the chapter. Scattering is ignored and pure absorption is assumed as the mechanism of interaction between the gas and the radiation. Many interesting results can be obtained in this way, making it a very useful device for an introductory treatment. Sometimes the approximation can be very good and at other times very poor. How can we tell which? The only rigorous answer is to first solve the general case of line formation for the line of interest and compare the results with the LTE restricted solution. But without so much effort, we can learn from the non-LTE calculations done to date and from the general considerations above, that LTE is an acceptable approximation when the ratio of collisionally induced transitions to radiatively induced transitions is large. In cases where the ratio is not large, local equilibrium conditions can still be approximated if the loss of radiation is small compared to the radiation field in the volume. We must expect LTE to be a poor approximation in the outer photospheric layers where the radiation can escape freely into space. The cores of strong lines are formed in these high layers and an LTE approach is totally inapplicable for them. In O stars the radiation field is so large, it must dominate over the collisional interactions by a large factor, and in supergiants where the density is very low, collisions are few in number. For these cases one can expect a much better interpretation of the data with an analysis that is not restricted to LTE.

We proceed under the assumption that the temperature distribution is known as a function of τ_0, a reference optical depth scale. [Methods for obtaining $T(\tau_0)$ were discussed in Chapter 9]. The surface flux is then computed using

$$\mathcal{F}_\nu = 2\pi \int_0^\infty B_\nu(T)E_2(\tau_\nu)\,d\tau_\nu = 2\pi \int_0^\infty B_\nu(T)E_2(\tau_\nu)\frac{d\tau_\nu}{d\tau_0}\,d\tau_0$$

$$= 2\pi \int_{-\infty}^\infty B_\nu(T)E_2(\tau_\nu)\frac{l_\nu + \kappa_\nu}{\kappa_0}\tau_0\frac{d\log\tau_0}{\log e}. \tag{13-15}$$

To compute τ_ν one uses

$$\tau_\nu(\tau_0) = \int_{-\infty}^{\log\tau_0}\frac{l_\nu + \kappa_\nu}{\kappa_0}\tau_0\frac{d\log\tau_0}{\log e} \tag{13-16}$$

with the line absorption coefficient, l_ν, given by eq. 11-68 and the continuous absorption coefficient, κ_ν, given by eqs. 8-16 and 8-17. Notice that it is the ratio of line to continuous absorption coefficients that enters eqs. 13-15 and 13-16 and not l_ν alone.

In the situation where the portion of the model spectrum being computed includes several lines, it is only necessary to add the absorption coefficients for all the lines. Then l_ν in eqs. 13-15 and 13-16 is replaced by the sum of line absorption coefficients, with each individual l_ν evaluated at the proper $\Delta\lambda$, namely, the wavelength difference from the center of the line to the position of $\mathscr{F}_\nu$. A computation of this kind is termed a spectrum synthesis.

In the case of a weak line, we expect the flux profile to mimic the shape of l_ν (or α). This result is borne out by detailed model calculations, but we can see it conceptually in the following analytical approximation. We use the relation in eq. 13-8 to write the profile

$$\frac{\mathscr{F}_c - \mathscr{F}_\nu}{\mathscr{F}_c} \doteq \frac{S_\nu(\tau_c = \tau_1) - S_\nu(\tau_\nu = \tau_1)}{S_\nu(\tau_c = \tau_1)} \tag{13-17}$$

where τ_c and τ_ν are optical depths without and with l_ν, respectively, and where we make no distinction between the continuum source function and the line source function [both of which are $B_\nu(T)$ in LTE]. We take eq. 13-16 and break it into two parts, thereby defining a line optical depth, τ_l, and the continuum optical depth τ_c,

$$\tau_\nu = \int_0^{\tau_0} \frac{l_\nu}{\kappa_0}\, dt_0 + \int_0^{\tau_0} \frac{\kappa_\nu}{\kappa_0}\, dt_0$$

$$\equiv \tau_l + \tau_c.$$

We ignore any change in l_ν/κ_0 and κ_ν/κ_0 with depth and write

$$\tau_l \doteq \frac{l_\nu}{\kappa_0}\, \tau_0$$

$$\tau_c \doteq \frac{\kappa_\nu}{\kappa_0}\, \tau_0. \tag{13-18}$$

Now for eq. 13-17 we need $S_\nu(\tau_\nu = \tau_1) = S_\nu(\tau_l + \tau_c = \tau_1) = S_\nu(\tau_c = \tau_1 - \tau_l)$. Since we are dealing with the weak line approximation, $\tau_l \ll \tau_c$, we can evaluate S_ν at $\tau_c = \tau_1 - \tau_l$ by a Taylor expansion from the point $\tau_c = \tau_1$,

$$S_\nu(\tau_\nu = \tau_1) \doteq S_\nu(\tau_c = \tau_1) + \left(\frac{dS_\nu}{d\tau_c}\right)_{\tau_1} (-\tau_l) + \cdots.$$

Equation 13-17 then becomes

$$\frac{\mathscr{F}_c - \mathscr{F}_\nu}{\mathscr{F}_c} \doteq \frac{\tau_l}{S_\nu(\tau_c = \tau_1)} \left(\frac{dS_\nu}{d\tau_c}\right)_{\tau_1} = \tau_l \left(\frac{d\ln S_\nu}{d\tau_c}\right)_{\tau_1}.$$

Now using the relations 13-18, we get

$$\frac{\mathscr{F}_c - \mathscr{F}_v}{\mathscr{F}_c} \doteq \left(\frac{d \ln S_v}{d\tau_c}\right)_{\tau_1} \frac{l_v}{\kappa_0} \tau_0 = \left(\frac{d \ln S_v}{d\tau_c}\right)_{\tau_1} \frac{l_v}{\kappa_v} \tau_c$$

or

$$\frac{\mathscr{F}_c - \mathscr{F}_v}{\mathscr{F}_c} \doteq \tau_1 \left(\frac{d \ln S_v}{d\tau_c}\right)_{\tau_1} \frac{l_v}{\kappa_v} \tag{13-19}$$

which indeed shows that the profile of a weak line mimics the shape of l_v since the v dependence of κ_v across the line is negligible. Expression 13-19 also shows that the weak line strength is directly proportional to the ratio l_v/κ_v. Further, we saw in Chapter 9 how $T(\tau_0)$ is approximately the same for a wide range of T_{eff} and g. We can then treat $\tau_1 (d \ln S/d\tau_c)_{\tau_1}$ as a constant with changing temperature and pressure. This is an important result and is useful in several contexts.

CONTRIBUTION FUNCTIONS AND THE DEPTH OF FORMATION OF SPECTRAL LINES

The integrand of eq. 13-15 goes to zero for small and large τ_0 so that it is possible to say something about the depth of formation for the flux at each $\Delta\lambda$ across the profile. One way to view the matter is shown in Fig. 13-2. Here data is given for a 0 V excitation Fe I line in a solar type model. The profile itself is shown on the left. The number of absorbers in the line, N_l, is seen to be a slowly varying function of depth with a maximum near log $\tau_{5000} = -2.2$. The contribution functions are shown for the points indicated on the profile. These contribution functions are identical to those used to discuss the depth of formation of the continuum in Chapter 9. The curves show that the continuum flux ($\Delta\lambda = 1$ Å) is formed at $\tau_{5000} \sim 1$. For the points closer to line center, the flux comes from higher layers. The rapid shift to higher layers starts when the residual flux drops to $\lesssim 50\%$. Compare this with Fig. 13-3 which shows the same material for a line of 5 V excitation. The density of line absorbers is now a very steep function of depth and the flux emission shows a steady progression to higher layers as $\Delta\lambda$ goes toward zero. These examples are elaborations of eq. 13-8 for the LTE case.

Another example is given for a set of weak Fe II lines in Fig. 13-4. The two lines have the same equivalent width. It is clear that the residual flux for the 5 V line originates in higher layers compared to the 0 V line.

On the other hand, we can equally well look at the *absorption* caused by the line by comparing the contribution function for the continuum with any of the contribution functions in the line to see how much flux has been removed at

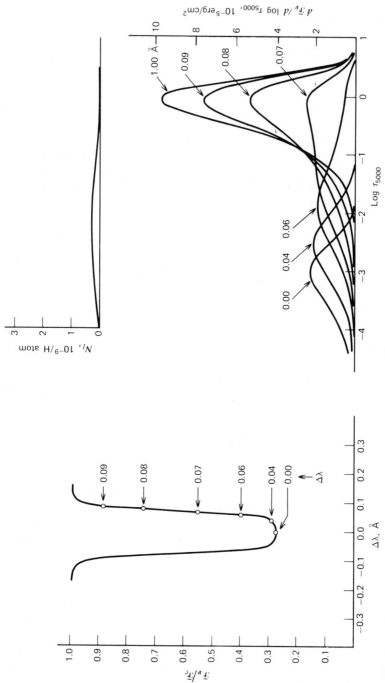

Fig. 13-2 The residual flux contribution functions are shown for Fe I $\lambda6065$ having $\chi = 0$ in a model having $S_0 = 1.02$ and $g = 10^4$. The line profile is shown at the left. The number of line absorbers and the integrands of eq. 13-15 are on the right. The points on the profile correspond to the contribution functions as labeled.

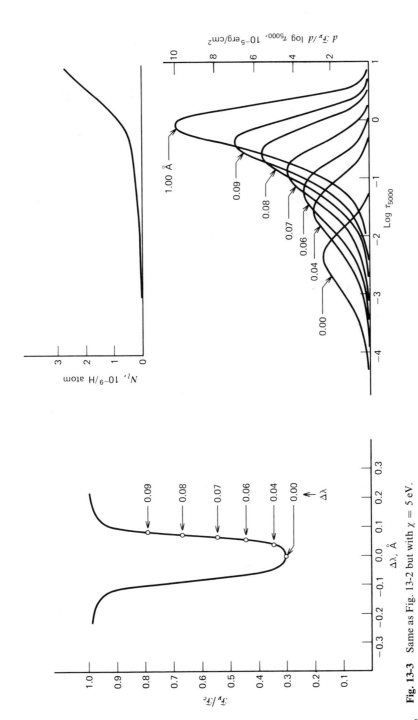

Fig. 13-3 Same as Fig. 13-2 but with $\chi = 5$ eV.

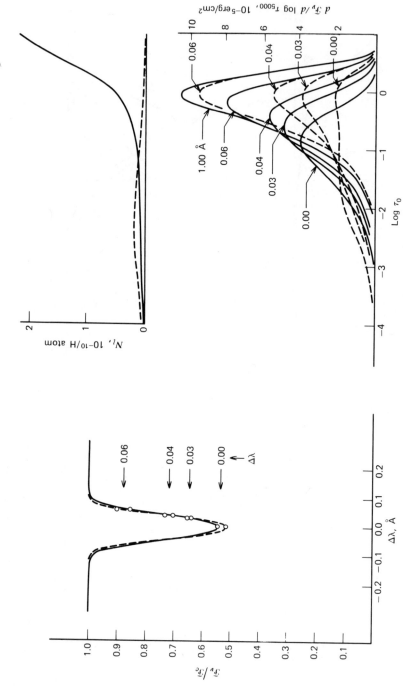

Fig. 13-4 Same as Figs. 13-2 and 13-3 but for a weak Fe II line: (——) $\chi = 5$ eV; (-----) $\chi = 0$ eV.

310

each depth. Specifically, should we write, using continuum and total optical depths denoted by τ_c and τ_v,

$$\frac{(\mathscr{F}_c - \mathscr{F}_v)}{\mathscr{F}_c} = \frac{2\pi}{\mathscr{F}_c} \int_0^\infty \left[S_v(\tau_c) E_2(\tau_c) \frac{\kappa_v}{\kappa_0} - S_v(\tau_v) E_2(\tau_v) \frac{l_v + \kappa_v}{\kappa_0} \right] \tau_0 \frac{d \log \tau_0}{\log e},$$

we see that the contribution to the absorption at any depth is the contribution function in the continuum minus the contribution function in the line. Figure 13-5 shows this absorption contribution function for the same lines as in Fig. 13-4. The peak absorption for the $\chi = 0$ line occurs at $\log \tau_{5000} \doteq 0.00$ to -0.05, that is, nearly at the same depth independent of $\Lambda\lambda$. The absorption functions for the $\chi = 5$ line behaves the same but has a maximum at $\log \tau_{5000} \doteq 0.00-+0.15$, that is, somewhat deeper than the $\chi = 0.00$ line. We are justified in saying that the 5 V line is formed deeper than the 0 V line, although the difference is very small.

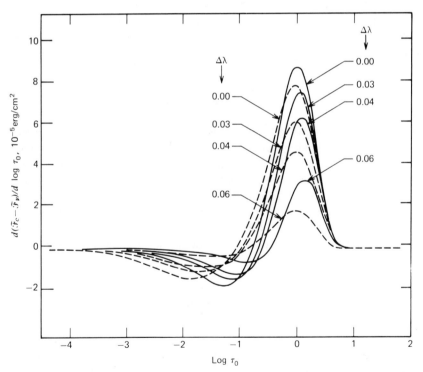

Fig. 13-5 The contribution functions for absorption or line depth are shown. These functions correspond to those of Fig. 13-4. They are for the same two Fe II lines having $\chi = 5$ eV (——) and $\chi = 0$ eV (-----).

The negative values of the absorption contribution functions show that the flux emitted in the line is greater than the flux emitted in the continuum at these depths. As $\Delta\lambda$ becomes large, the absorption contribution function becomes zero at all depths.

We can summarize these results as follows. In the cores of the stronger lines we see only the upper layers in the atmosphere and generally the stronger the absorption, the higher in the atmosphere does the surface flux originate (visualized in Figs. 13-2 and 13-3). The line absorption takes place in layers below those where the residual fluxes originate. High excitation and ionization are favored by the increase in temperature with depth, with the result that such lines do their absorbing preferentially in deeper layers (visualized in Fig. 13-5).

The gradient formulation of the flux computation (eq. 9-10) can also be used here, but the integrand is not an increment of flux and should not be used to define depths of formation of spectral lines. The formulation does show that the minimum residual flux in the line computed from an LTE model is $\pi B_\nu(T_0)$ since the integral is never negative.

Absorption lines are formed over a large range in the atmosphere as we have just seen. It is probably more or less meaningless to attempt to specify a single depth of formation to an equivalent width or to the line as a whole.

THE BEHAVIOR OF THE LINE STRENGTH

The strength of the spectral line depends in part upon the width of the absorption coefficient α, which in turn is a function of the thermal and microturbulence velocities. The strength also is dependent on the number of absorbers, which, under the LTE approximation, is calculated from the excitation and ionization eqs. 1-16 and 1-17 and so depends on the temperature, the electron pressure, and atomic constants. In strong lines the wings may depend on the electron and gas pressures through the damping constant. Now let us consider in turn the more important variables and their effect on the absorption lines.

THE TEMPERATURE DEPENDENCE

Temperature is the variable that most strongly controls the line strength. This sensitivity arises from the exponential and power dependence with T in the excitation–ionization processes. Figure 13-6 shows the behavior of typical weak metal lines with effective temperature. Most lines go through some sort of maximum in strength. Usually the increase with temperature is

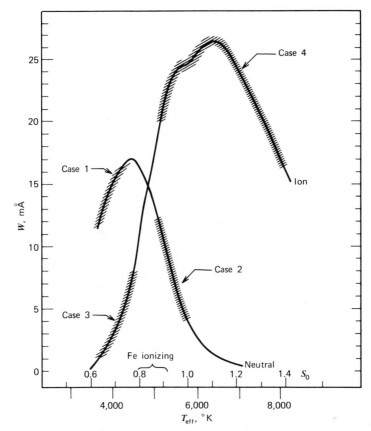

Fig. 13-6 Typical metal line behavior is shown. Ionic lines grow at the expense of the neutral lines with increasing temperature. The cases for eqs. 13-21 are shown.

due to increased excitation. The decrease beyond the maximum results from the increase in continuous opacity of the negative hydrogen ion, which in turn comes from the increase in electron pressure with temperature. In other situations (see case 2 below) the decrease results from ionization of the absorbing species.

When we deal with a strong line, the atomic absorption coefficient is proportional to γ, which is a function of temperature (eqs. 11-33 and 11-35). This dependence must then be considered in addition to those of the weak line. In Fig. 13-7 the observed behavior of the Na D lines as a function of temperature is shown.

Hydrogen lines have an atomic absorption coefficient that is also temperature sensitive through the Stark effect (eq. 11-63). Figure 13-8a shows the

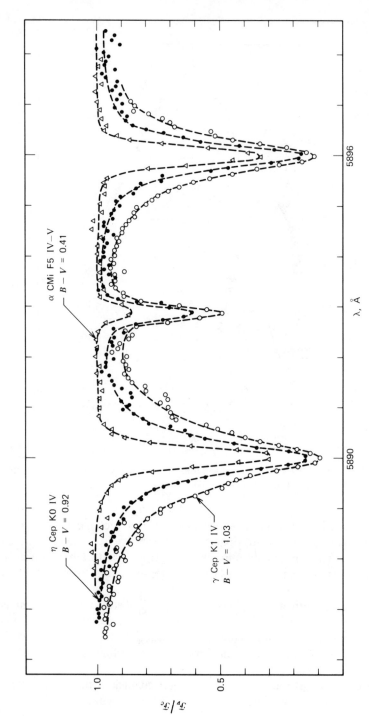

Fig. 13-7 The sodium D lines weaken with increasing temperature. Data shown are measured with a photoelectric line scanner having a resolution of 0.27 Å. The dashed lines are simply drawn in by hand to fit the data.

change in Hγ profiles with temperature. Figure 13-8b shows the equivalent width behavior. The relatively high excitation of the Balmer series (10.2 eV) causes the excitation growth to continue to quite high temperatures so that the maximum occurs at $T_{\text{eff}} \sim 9000°$K. At this temperature most of the other elements are multiply ionized and the lines they give are in the ultraviolet.

It is through an expansion of this discussion that the spectral classification sequence is understood. But we shall pursue the temperature dependence only one step further by asking how we can predict whether a given line will grow or weaken with a change in temperature and with what sensitivity. The ratio of l_v to κ_v must then be considered (eq. 13-19). We shall restrict the discussion to the visible region in cooler stars so that κ_v can be taken to be due to the negative hydrogen ion bound–free absorption. A slight modification of eq. 8-12 gives the κ_v temperature dependence as

$$\kappa_v = \text{const } T^{-5/2}\, e^{0.75/kT}\, P_e.$$

We now consider the following four cases: (1) neutral line with the element mostly neutral, (2) neutral line with the element mostly ionized, (3) ionic line with the element mostly neutral, and (4) ionic line with the element mostly ionized.

In cases 1, the number of absorbers in the level l is given by

$$N_l = \text{const } N_0 e^{-\chi/kT}$$
$$\doteq \text{const } e^{-\chi/kT}.$$

The number of neutrals, N_0, is approximately constant with temperature until ionization occurs because the number of ions, N_i, is small compared to N_0. The ratio of line to continuous absorption is then

$$R = \text{const } \frac{l_v}{\kappa_v} = \text{const } \frac{T^{5/2}}{P_e} e^{-(\chi+0.75)/kT}. \tag{13-20}$$

The atomic absorption coefficient, α, is lumped in the constant and its small dependence on temperature through Δv_D is ignored. The electron pressure dependence on temperature can be written

$$P_e = \text{const } e^{\Omega T}$$

as we saw in Chapter 9 (eq. 9-18). Equation 13-20 becomes

$$\ln R = \text{const} + \frac{5}{2}\ln T - \frac{\chi + 0.75}{kT} - \Omega T$$

and

$$\frac{1}{R}\frac{dR}{dT} = \frac{2.5}{T} + \frac{\chi + 0.75}{kT^2} - \Omega \qquad \text{(case 1, neutral line, } N_0 \gg N_i) \tag{13-21a}$$

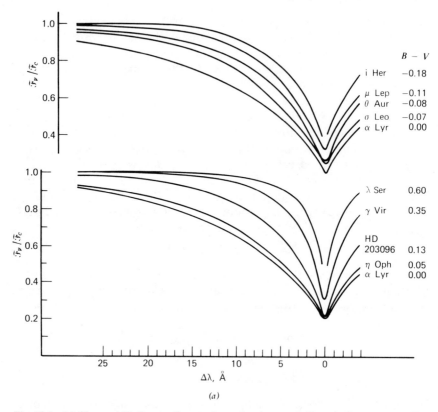

Fig. 13-8 (*a*) Observed Hγ line profiles are shown for the stars indicated: On top are profiles for stars hotter than where Hγ reaches its maximum strength. Below are profiles for stars on the cooler side of the Hγ maximum. (Data from Gray and Evans 1973.)

is the per cent change in line strength with temperature. The temperature used in such an equation should be characteristic of the line forming region—usually about 15% below T_{eff}.

Derivation of the other cases are left to the reader. The results are

$$\frac{1}{R}\frac{dR}{dT} = \frac{\chi + 0.75 - I}{kT^2} \qquad \text{(case 2, neutral line, } N_0 \ll N_i \text{)} \qquad (13\text{-}21b)$$

$$\frac{1}{R}\frac{dR}{dT} = \frac{5}{T} + \frac{(\chi + 0.75 + I)}{kT^2} - 2\Omega \qquad \text{(case 3, ionic line } N_0 \gg N_i \text{)} \qquad (13\text{-}21c)$$

$$\frac{1}{R}\frac{dR}{dT} = \frac{2.5}{T} + \frac{(\chi + 0.75)}{kT^2} - \Omega \qquad \text{(case 4, ionic line } N_0 \ll N_i \text{)}. \qquad (13\text{-}21d)$$

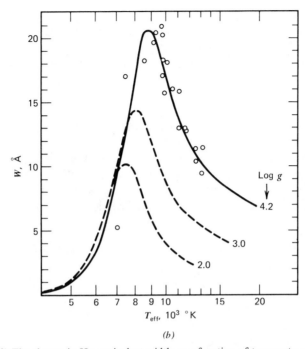

Fig. 13-8 (*b*) The change in Hγ equivalent width as a function of temperature is shown for three surface gravities (lines) according to the calculations of Carbon and Gingerich (1969). The circles represent observed equivalent widths based on the data of Gray and Evans (1973). The effective temperatures were assigned to the stars using their $B - V$ color index and the temperature scale of Morton and Adams (see Chapter 15).

The weak temperature dependence of the partition functions has been ignored in obtaining the solutions for cases 2 and 3. Notice that cases 1 and 4 behave identically. The direction and strength of change depends upon the temperature of the gas in the line forming layers and the excitation potential of the line. We see explicitly in cases 1, 3, and 4 that the electron pressure term, Ω, is against the positive terms. In the solar case most elements are ionized, making cases 2 and 4 applicable. With an excitation temperature of 5000°K, the ionic line (case 4) change is $R^{-1} \, dR/dT \sim -30\%/10^{3}\,°K$ if $\chi = 0$ eV, but $R^{-1} \, dR/dT \sim +200\%/10^{3}\,°K$ change for $\chi = 5$ eV. [A value of $\Omega = 0.00117$ has been used in accordance with eq. 9-18 for dwarfs.] Neutral lines (case 2) almost always decrease in strength since $\chi + 0.75$ is usually less than I. In the case of Fe I one finds $R^{-1} \, dR/dT \sim -300\%/10^{3}\,°K$ for $\chi = 0$ and $R^{-1} \, dR/dt \sim -100\%/10^{3}\,°K$ for $\chi = 5$.

The temperature sensitivity of *strong* lines can be investigated by including the T dependence of γ according to eqs. 11-33 and 11-35, or in the case of hydrogen, eq. 11-63. But this task shall be left to the reader. In the case of saturated lines not dominated by strong wings it is probably best to proceed numerically rather than analytically.

Additional material describing temperature effects may be found in Chapter 15.

THE PRESSURE DEPENDENCE

Pressure effects are visible in three ways. The first is due to a change in the ratio of line absorbers (ionization equilibrium) to the continuous opacity, the second arises in the pressure sensitivity of γ for strong lines, and the third is the pressure dependence of the Stark broadening in hydrogen. The gas

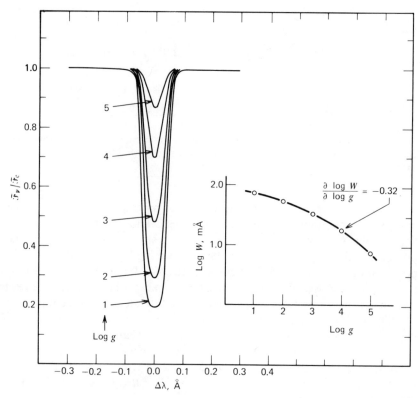

Fig. 13-9 The Fe II $\lambda 4508$ theoretical behavior is shown. Surface gravities as labeled. Models use $S_0 = 1.00$ and helium abundance of 0.10. The insert shows the equivalent width variation with g.

pressure and electron pressure are related approximately by eq. 9-13, $P_g \doteq P_e^2/\text{const}$. The pressure in turn is controlled by the surface gravity and given approximately, according to eqs. 9-11 and 9-12, as $P_g = \text{const } g^{2/3}$ and $P_e = \text{const } g^{1/3}$. Therefore the pressure dependences that we now consider can be translated into gravity dependences, for the F, G, and K stars using these relations.

The first type of pressure effect is illustrated in Fig. 13-9. The Fe II $\lambda4508$ line is used in this example to show how an ionic line in a solar-type star increases in strength with decreasing gravity. The enhancement with lower pressure is due in this case to a decrease in κ_ν as we explain below. The second pressure effect is illustrated with the broad wings of the Mg I b lines in Fig. 13-10. If the b lines were weak, they would show no pressure dependence in stars like these. But in Fig. 13-10 we see that they are strong and their absorption coefficient is then proportional to γ. On the main sequence, γ is dominated by the pressure dependent broadening, but with decreasing surface gravity this interaction decreases, eventually leaving only the natural

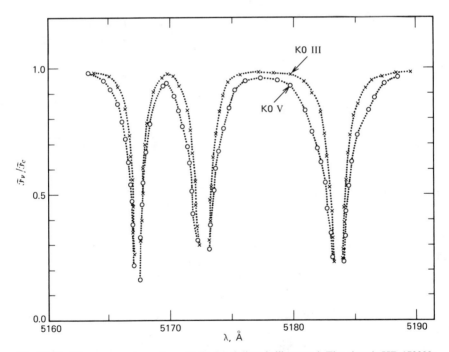

Fig. 13-10 The pressure dependence in the Mg b lines is illustrated. The giant is HD 173398; the dwarf HD 185144 = σ Dra. (Adapted from Cayrel de Strobel 1969 with permission of MIT press.)

damping which is much smaller. The third pressure dependence is shown in Fig. 13-11. The hydrogen line pressure sensitivity becomes significant for T_{eff} greater than $\sim 7500°$K. The strength of the lines in Figs. 13-10 and 13-11 increases with increasing pressure, that is, opposite the weaker lines of Fig. 13-9.

It is important to realize that the range in surface gravity shown in Figs. 13-9, 13-10, and 13-11 is several hundred per cent. A similar change in profile has been obtained for a few tens of per cent change in temperature. Pressure effects in stellar spectra are simply much weaker than temperature effects.

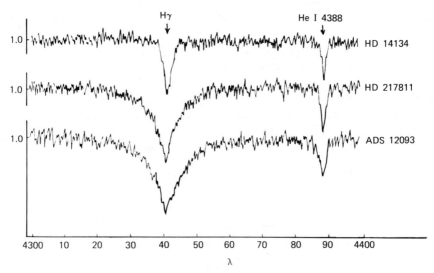

Fig. 13-11 The Hγ pressure variation as seen in B3 stars: (upper) supergiant; (lower) dwarf. Densitometer tracings adapted from Petrie (1953). Resolution is about 2 Å.

The pressure sensitivity can be estimated by again considering the ratio of line to continuous absorption coefficients. As in the previous section, let us consider first several cases in cooler stars for metal lines that do not show strong wings. We can state the following rules:

1. Lines formed by any ion or atom where most of the element is in the next higher ionization stage are insensitive to pressure changes.

2. Lines formed by any ion or atom where most of the element is in that same ionization stage are pressure sensitive. Lower pressure causes greater line strength.

3. Lines formed by any ion or atom where most of the element is in the next lower ionization stage are very pressure sensitive. Again lower pressure enhances the lines.

These rules can be derived as follows. We assume the ionization stage is not altered by the change in pressure so that the atom does not switch cases. We write the ionization equation in the form

$$\frac{N_{r+1}}{N_r} = \frac{\Phi(T)}{P_e}$$

where $\Phi(T)$ includes all the terms not dependent on pressure. Then for rule 1 the line is formed by an element in the rth ionization stage and N_{r+1} is approximately equal to the total number of the element. Hence

$$N_r \doteq \text{const } P_e$$

and

$$l_v \doteq \text{const } P_e,$$

making the ratio of l_v to κ_v independent of P_e when the negative hydrogen ion dominates κ_v since then κ_v is also proportional to P_e.

For case 2 we reason that for a line formed by an element in the rth ionization stage, $N_r \doteq N_{\text{total}} = \text{const}$, making

$$\frac{l_v}{\kappa_v} = \text{const}/P_e \doteq \text{const } g^{-1/3}. \tag{13-22}$$

Notice that the ionic lines in this case are enhanced not because l_v changes but because the H^- continuum opacity becomes less as P_e decreases. The "photosphere" deepens in geometrical depth and so includes more line absorbers. From eq. 13-22, $\partial \log (l_v/\kappa_v)/\partial \log g = -0.33$ in agreement with the change shown in Fig. 13-9. The proof of case 3 is similar. The more commonly encountered cases are 1 and 2, which apply in solar-type stars for neutrals and first ions, respectively. Fe I lines, for example, are insensitive to pressure while lines of Fe II are pressure indicators; iron is mostly ionized at solar temperatures. The infrared lines of OI($\lambda\lambda7774$, 8446) are pressure sensitive; oxygen is neutral at solar temperatures.*

With increasing effective temperature, neutral hydrogen takes over from the negative hydrogen ion as the major continuous absorber. Since it is not proportional to P_e, the pressure dependences change accordingly. These cases can be worked out by the reader.

---

* But these comments should not be misconstrued to mean that the detailed behavior of these lines is completely explained and understood. (see Osmer 1972 and Johnson et al. 1974).

The same approximate analysis can be applied to the wings of strong lines by noting that l_ν is proportional to γ (eq. 11-13) and that for van der Waals broadening

$$\gamma_6 = \varphi_6(T)P_g,$$

while for the quadratic Stark broadening

$$\gamma_4 = \varphi_4(T)P_e.$$

The functions $\varphi_6(T)$ and $\varphi_4(T)$ are simply shorthand for the pressure independent terms of eqs. 11-35 and 11-33. So we find, for example, that the wings of a strong line caused by a neutral atom where most of the element is ionized behave according to

$$\frac{l_\nu}{\kappa_\nu} \doteq \text{const } \gamma$$

$$= \text{const } [\varphi_6(T)P_g + \varphi_4(T)P_e + \gamma_{\text{nat}}]$$

$$\doteq \text{const } g^{2/3} + \text{const } g^{1/3} + \text{const.} \qquad (13\text{-}23)$$

Depending on the relative sizes of the damping constants, there may be no gravity dependence (γ_{nat} dominant) or a maximum $g^{2/3}$ dependence (van der Waals dominant). For the Na D lines in solar type stars, this last case holds.

But for ionic lines of the same ionized element we find

$$\frac{l_\nu}{\kappa_\nu} \doteq \text{const } \frac{\gamma}{P_e}$$

$$= \text{const } \left[\varphi_6(T)\frac{P_g}{P_e} + \varphi_4(T) + \frac{\gamma_{\text{nat}}}{P_e} \right]$$

$$\doteq \text{const } g^{1/3} + \text{const} + \text{const } g^{-1/3}. \qquad (13\text{-}24)$$

Again several gravity dependences are possible. The Ca II H and K lines have been found to show an inverse gravity dependence for stars above the main sequence (Lutz et al. 1973). From eq. 13-24 we deduce that γ_{nat} is the dominant term. Formal calculation of the three damping constants bears this out. These examples can be elaborated on to suit the situation.

Hydrogen Balmer lines are pressure sensitive when they are strong. The line absorption coefficient in the wing is proportional to P_e according to eq. 11-43 (we can neglect the weak P_e dependence in R). When the neutral hydrogen absorption dominates κ_ν, then κ_ν, like l_ν, will be proportional to the number of neutral hydrogen atoms and

$$\frac{l_\nu}{\kappa_\nu} \doteq \text{const } P_e.$$

In hotter stars electron scattering becomes an important part of κ_ν and hydrogen is ionized. Since the number of hydrogen ions is equal to the number of electrons, the ionization equation applied to this case gives the number of neutral hydrogen atoms proportional to P_e, making

$$\frac{l_\nu}{\kappa_\nu} \doteq \text{const } P_e^3/\text{const } P_e$$

$$= \text{const } P_e^2.$$

To convert these approximate pressure dependences analytically to surface gravity dependences is not a very feasible task because as we saw in discussing eqs. 9-11 and 9-12 the relation between pressure and surface gravity changes rapidly with temperature in the hotter stars. Some of the pressure sensitive cases are considered in more numerical detail in Chapter 16.

As we saw in Fig. 13-8, the pressure sensitivity of the hydrogen lines becomes nil at cooler temperatures. One can show by using these same types of arguments that this is a result of κ_ν being dominated by the negative hydrogen absorption.

THE ABUNDANCE DEPENDENCE

The line strength can be expected to increase with an increase in the chemical abundance of the absorber. The change in line profile and equivalent width is not a simple one but depends on the optical depth in the line. In Fig. 13-12 we see the theoretical behavior of the profile and equivalent width for Fe I λ 6065 deduced from a model photosphere. A is the total number abundance of iron expressed as a fraction of hydrogen.

There are three phases of line growth. For weak lines the Doppler core dominates. The width is set by $\Delta\lambda_D$ and the depth grows in proportion to the abundance, A. The equivalent width varies directly as A does. The second phase is entered as the depth reaches the maximum value and the line saturates. In the LTE formulation the minimum residual flux is fixed by the boundary temperature T_0, at $\pi B_\nu(T_0)$ and therefore is dependent on the temperature distribution. The equivalent width grows asymptotically toward a constant value. The third stage of the behavior starts as the optical depth in the line wings becomes significant compared to κ_ν. The abundance at which this starts is very much dependent on the size of γ for the line being studied. If γ were constant for a line, then the equivalent width would grow in proportion to $A^{1/2}$. But γ is a function of depth since it depends on the temperature and pressure. As the line strength changes, the average γ value for the line changes. The $A^{1/2}$ relation holds approximately.

Each and every spectral line shows some such behavior, describing its growth with abundance. A graph specifying the change in equivalent width, as in Fig. 13-12, is called a *curve of growth*. The details depend on such things as the continuous absorption at the position of the line, the atomic constants for the line (including hyperfine structure and Zeeman splitting), the surface gravity, the temperature distribution, and the type and strength

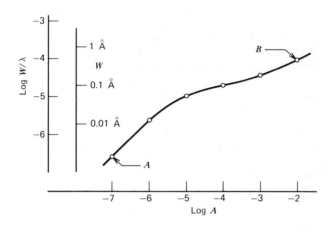

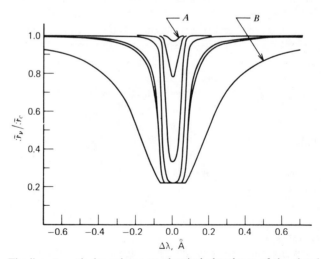

Fig. 13-12 The line strength dependence on chemical abundance of the absorbing species is shown. At the bottom the increasing profile strength is illustrated for decade steps in abundance. At the top is the curve of growth for this same line. The dots on the curve of growth correspond to the profiles below. (Computations based on a model with $S_0 = 0.87$ and $g = 10^4$.)

of the damping. But in every case a curve of growth exists. We do not restrict the meaning of curve of growth to any particular type of model photosphere. All curves of growth look qualitatively similar to the one shown in Fig. 13-12. It is of great practical significance that the shape is approximately the same quantitatively for a large number of lines.

Analytically we can say something about the change in line strength with abundance by having recourse to extremely simple model photospheres. For example, if we assume an absorption cell model where the lines are formed in a cool nonemitting gas placed above the source of continuous emission, then the surface flux in the line is

$$\mathscr{F}_v = \mathscr{F}_c e^{-\tau_v} \tag{13-25}$$

where τ_v is the optical depth in the cool gas and $\mathscr{F}_c$ is the flux in the continuum. If we take the thickness of the cool gas to be L, then the optical depth can be calculated from

$$\tau_v = \int_0^L l_v \rho \, dx = \int_0^L N\alpha \, dx$$

$$= A \int_0^L \left(\frac{N}{N_E}\right) N_H \alpha \, dx \tag{13-26}$$

where a development similar to eq. 11-66 has been used. The ratio N/N_E is the number of line absorbers per cubic centimeter divided by the total number per cubic centimeter of the element E. We see that τ_v is proportional to the abundance A. According to eq. 13-25 the flux in the line then varies exponentially with A.

For a weak line $\tau_v \ll 1$ and eq. 13-25 can be written

$$\mathscr{F}_v = \mathscr{F}_c(1 - \tau_v) \quad \text{or} \quad \frac{\mathscr{F}_c - \mathscr{F}_v}{\mathscr{F}_c} \doteq \tau_v$$

and therefore the line depth is proportional to τ_v or A. The equivalent width of the weak line is

$$W = \int_0^\infty \frac{(\mathscr{F}_c - \mathscr{F}_v) \, dv}{\mathscr{F}_c} = \int_0^\infty \tau_v \, dv$$

$$= A \int_0^\infty \int_0^L \left(\frac{N}{N_E}\right) N_H \alpha \, dx \, dv$$

$$= \frac{\pi e^2}{mc} Af \int_0^L \left(\frac{N}{N_E}\right) N_H \, dx \tag{13-27}$$

so that W is proportional to A.

For a strong line the wings of the line dominate the profile, so according to eq. 11-57, where the Gaussian component acts like a δ function, we can write $\alpha = (\pi e^2/mc)(\gamma/4\pi^2)f/\Delta v^2$. Then eq. 13-26 gives

$$\tau_v = \frac{\pi e^2}{mc}\frac{Af}{\Delta v^2}\int_0^L \left(\frac{N}{N_E}\right)N_H\frac{\gamma}{4\pi^2}\,dx$$

$$= \frac{Af\,\bar{\gamma}h}{\Delta v^2}.$$

The average damping constant is denoted by $\bar{\gamma}$ and the function h denotes the remaining constants and the integral. The line depth varies as (from eq. 13-25)

$$\frac{\mathscr{F}_c - \mathscr{F}_v}{\mathscr{F}_c} = 1 - e^{-\tau_v}$$

and the equivalent width of the strong line is

$$W = \int_0^\infty (1 - e^{-\tau_v})\,dv = \int_0^\infty (1 - e^{-Af\bar{\gamma}h/\Delta v^2})\,dv.$$

But since dv and $d\Delta v$ are the same, we can substitute

$$u^2 = \frac{\Delta v^2}{Af\,\bar{\gamma}h}$$

with the result

$$W = (Af\,\bar{\gamma}h)^{1/2}\int_0^\infty (1 - e^{-1/u^2})\,du. \tag{13-28}$$

Thus W for the strong line is proportional to the square root of the abundance.

An analytical approach has limited usefulness and we use the full model calculations from now on, especially in Chapter 14.

We have treated the above behavior of the line on chemical abundances as if only A for the line were variable. If, instead, we wish to study the case where the whole chemical composition of the model is changing along with A, then the behavior may be different than that described above. In the first place, the ionization equilibrium may change, making the relation between N and A a function of A. In the second place, the continuous absorption may change with the change in electron donors. And in the third place, the collisional damping may be different because of the changes in P_e and P_g.

The direction of the change can be worked out by considering the relations among P_e, P_g, and the metal abundances as done in Chapter 9 and then applying the pressure dependences given in the previous section of this chapter. But we shall not do so explicitly. Instead, Fig. 13-13 shows the result of a model calculation for the curves of growth of Fe II. The heavy curves delineate the expected behavior when all the electron donors change abundance in proportion to $A(Fe)$. W increases at a slower rate than when the model abundances are held constant.

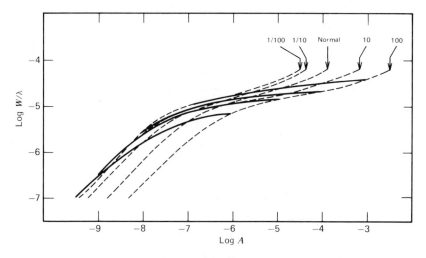

Fig. 13-13 The equivalent width behavior of pressure sensitive lines, such as this Fe II line, depends upon the behavior of the chemical composition of the model. Curves of growth are shown for five different model abundances (-----) where unity implies the normal solar composition, 10 means the metals are increased 10 times relative to hydrogen, and so on. The heavy curves show the equivalent width change with abundance when the whole model composition changes in step with the chemical being studied.

The behavior of strong lines is particularly graphic. Three cases for the Na D_2 line are shown in Fig. 13-14: (a) change with Na abundance for a constant model, (b) change for a constant Na abundance but different model metal abundances, and (c) Na abundance and other metal abundances changed together and in proportion. These examples serve to illustrate the point—the measurement of A for one element is coupled to what the remainder of the elements are doing.

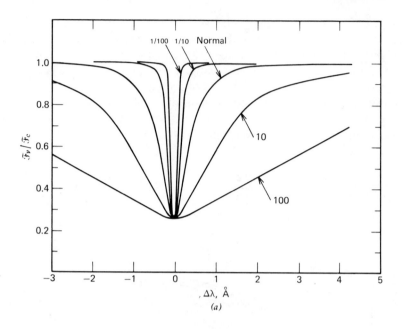

(a)

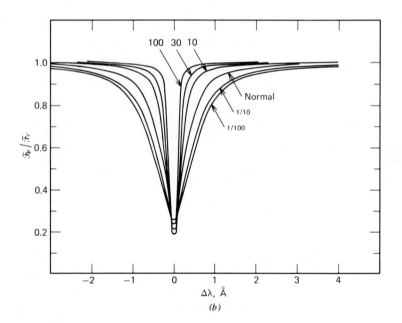

(b)

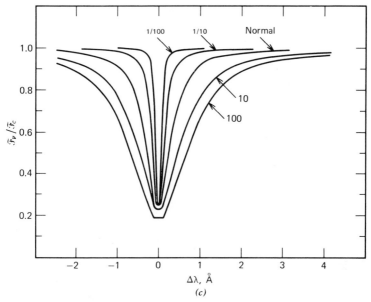

Fig. 13-14 The abundance behavior for the strong Na D$_2$ line is shown for the three cases: (*a*) model constant, Na abundance changed as labeled, (*b*) Na abundance held constant, model abundances changed as labeled, and (*c*) Na abundance and model abundances changed in step as labeled.

A COMPARISON OF THEORY AND OBSERVATION

Before we move to the next chapter where we begin to deduce fundamental properties of stars from their spectra, it is prudent to consider a few examples of how well the models reproduce the spectral line observations. In Fig. 13-15 we see photoelectric observations of Hγ in θ Leo compared to a theoretical profile computed under the assumption of LTE. The fit to the wing is good but the observed line core is deeper than the theoretical core. This illustrates a commonly seen failing in the LTE theory and is to be expected according to the discussion earlier in the chapter. The non-LTE calculation improves the situation markedly as shown by Auer and Mihalas (1970) and Mihalas (1972a). Higher resolution photoelectric observations are still needed to see what discrepancies remain between the non-LTE calculated and the observed line cores. Evidence so far (such as the examples cited in Mihalas 1972a) indicates that additional line broadening occurs that is not accounted for by the non-LTE models. Peterson and Strom (1969) have measured the equivalent widths of Hγ and Hα in a few B stars and find marginal evidence that

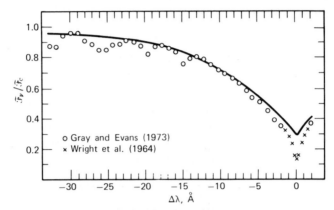

Fig. 13-15   The observed H_γ profile in θ Leo is matched with the LTE profile calculated by Carbon and Gingerich (1969). The depressions in the observed line wing are weaker lines broadened by the instrumental profile. The theoretical profile fits the wings adequately. In the core, however, the observed line is much deeper than the model profile.

departures from LTE affect the equivalent widths for luminosity class I and II stars.

More generally, a high degree of internal consistency has been attained by Schild et al. (1971) in their analysis of Vega and Sirius using LTE models to fit the Paschen continuum, the Balmer jump, and the Balmer lines (exclusive of the core). A detailed fitting of Sirius by Fowler (1974) is successful in a similar way. (The line blanketing has been treated differently in the two analyses of Sirius and this leads to different effective temperatures being deduced for the star.)

The story for other strong lines is similar. The Na D lines (Fig. 13-16) can be matched well by the LTE models except within the core. An analysis of the D line cores in the sun has been done by Curtis and Jefferies (1967) and, along with the Mg b lines, by Athay and Canfield (1969).

The non-LTE calculations usually show deeper cores than the LTE calculations, but they still do not fit the observed core profiles nor do they reproduce the observed behavior across the solar disc. The discrepancy has often been accounted for by invoking turbulence (nonthermal motions). The solar D line observations of Slaughter and Wilson (1972) show the profile cores to be variable with time and with position on the disc—the situation is indeed more complex than our models. In other stars we lose this temporal and spacial resolution forcing us to the unhappy situation of dealing with averages of nonlinear quantities.

We do not escape all complications by looking at weaker lines. An empirical curve of growth, the construction of which is discussed in the next

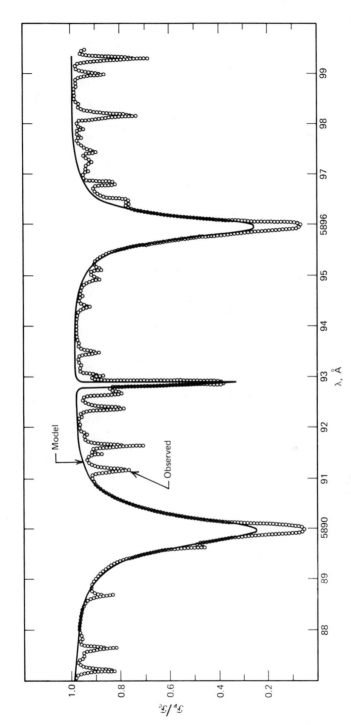

Fig. 13-16 An LTE model with $S_0 = 1.00$, $g = 2.74 \times 10^4$ gives Na D line profiles in good agreement with the observed solar spectrum. Obvious failure to fit occurs in the line cores. Almost all the weaker lines are due to telluric water vapor. (Courtesy of Kitt Peak National Observatory.)

331

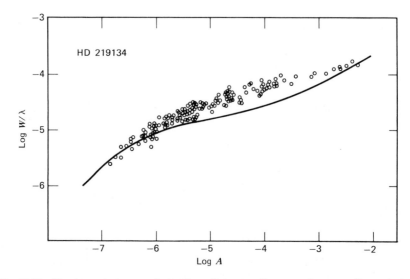

Fig. 13-17 The theoretical curve of growth predicts saturation occurring at smaller equivalent widths than is actually observed. Microturbulence is usually invoked to explain the discrepancy. (Data from Cayrel de Strobel 1966.)

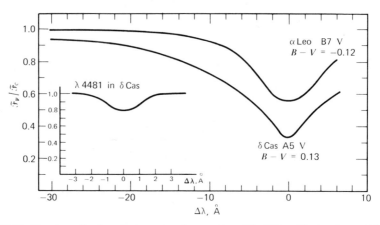

Fig. 13-18 Very large line broadening occurs in some stars. The Hβ profiles in α Leo and δ Cas are shown as examples. The core region is shallow and bowl shaped (compare with Fig. 13-8a). The insert shows the Mg II λ4481 line in δ Cas. It too shows a shallow broad profile.

chapter, is shown by the points in Fig. 13-17. Matched to it is a theoretical curve of growth. The discrepancy is obvious. The saturated lines are stronger than we expect from our model. Again turbulence can be introduced into the model and it has the effect of modifying the theoretical curve of growth in the right direction. The implications are discussed in Chapter 18. The non-LTE effects in weaker lines may also be significant in some cases (see Mihalas 1972b and Smith 1974, for example).

There are some stars in which the spectral lines are broadened by a large amount—weak lines and strong lines alike. Examples are given in Fig. 13-18. The effect is generally interpreted as being due to the rotation of the star or to turbulence.

Our calculations have brought us a long way in understanding the general features of stellar spectra. But it should be clear that application of our photospheric models must be done with care.

REFERENCES

Athay, R. G. 1972. *Radiation Transport in Spectral Lines*, D. Reidel, Dordrecht, Holland.

Athay, R. G. and R. C. Canfield 1969. *Astrophys. J.* **156**, 695.

Athay, R. G. and B. W. Lites 1972. *Astrophys. J.* **176**, 809.

Auer, L. H. and D. Mihalas 1970. *Astrophys. J.* **160**, 233.

Auer, L. H. and D. Mihalas 1972. *Astrophys. J. Suppl.* **24**, 193.

Avery, L. W. and L. L. House 1968. *Astrophys. J.* **152**, 493.

Carbon, D. F. and O. Gingerich 1969. *Theory and Observation of Normal Stellar Atmospheres*, O. Gingerich, Ed., MIT Press, Cambridge, Mass., p. 377.

Cayrel de Strobel, G. 1966. *Ann. Astrophys.* **29**, 413.

Cayrel de Strobel, G. 1969. *Theory and Observation of Normal Stellar Atmospheres*, O. Gingerich, Ed., MIT Press, Cambridge, Mass., p. 35.

Curtis, G. W. 1965. *The Formation of Spectrum Lines. Proceedings of the Second Harvard–Smithsonian Conference on Stellar Atmospheres*, Smithsonian Astrophysical Observatory, Cambridge, Mass., Special Report No. 174. p. 301.

Curtis, G. W. and J. T. Jefferies 1967. *Astrophys. J.* **150**, 1061.

Fowler, J. W. 1974. *Astrophys. J.* **188**, 295.

Gray, D. F. and J. C. Evans 1973. *Astrophys. J.* **182**, 147.

Jefferies, J. T. 1968. *Spectral Line Formation*, Blaisdell, Waltham, Mass.

Jefferies, J. T. and R. N. Thomas 1958. *Astrophys. J.* **127**, 667.

Jefferies, J. T. and R. N. Thomas 1959. *Astrophys. J.* **129**, 401.

Jefferies, J. T. and R. N. Thomas 1960. *Astrophys. J.* **131**, 695.

Johnson, H. R. 1965. *The Formation of Spectrum Lines. Proceedings of the Second Harvard–Smithsonian Conference on Stellar Atmospheres*, Smithsonian Astrophysical Observatory, Cambridge, Mass., Special Report No. 174. p. 333.

Johnson, H. R., R. W. Milkey, and L. W. Ramsey 1974. *Astrophys. J.* **187**, 147.

Lites, B. W. 1973. *Sol. Phys.* **32**, 283.

Lutz, T. E., I. Furenlid, and J. H. Lutz 1973. *Astrophys. J.* **184**, 787.

Mihalas, D. 1972a. *Astrophys. J.* **176**, 139.

Mihalas, D. 1972b. *Astrophys. J.* **177**, 115.

Milne, E. A. 1930. *Handb. Astrophys.* **3**, 159. (Reprinted in *Selected Papers on the Transfer of Radiation*, D. H. Menzel, Ed. 1966, Dover, New York, pp. 173–178.)

Mugglestone, D. 1965. *The Formation of Spectrum Lines. Proceedings of the Second Harvard–Smithsonian Conference on Stellar Atmospheres*, Smithsonian Astrophysical Observatory, Cambridge, Mass., Special Report No. 174. p. 347.

Osmer, P. S. 1972. *Astrophys. J. Suppl.* **24**, 255.

Peterson, D. M. and S. E. Strom 1969. *Astrophys. J.* **157**, 1341.

Petrie, R. M. 1953. *Publ. Dom. Astrophys. Obs. Victoria* **9**, 251.

Schild, R., D. M. Peterson, and J. B. Oke 1971. *Astrophys. J.* **166**, 95.

Slaughter, C. D. and A. M. Wilson 1972. *Sol. Phys.* **24**, 43.

Smith, M. A. 1974. *Astrophys. J.* **190**, 481.

Wright, K. O., E. K. Lee, T. V. Jacobson, and J. L. Greenstein 1964. *Publ. Dom. Astrophys. Obs. Victoria* **12**, 173.

CHAPTER FOURTEEN

CHEMICAL ANALYSIS

When we look at a stellar spectrum, especially for a cooler star, it is immediately clear that a large number of chemicals are present in the star. One of the tasks of the stellar photosphere analyst is to disentangle the effects of temperature, pressure, turbulence, and so on from the effects of chemical composition. If one does not press the interpretation of the results beyond the precision involved in the analysis, we can say the methods now in use give us believable results. The broad results can be simply summarized. Hydrogen is overwhelmingly the most abundant element, comprising $\sim 90\%$ of the atoms in a normal stellar atmosphere. Helium comes next with $\sim 10\%$ of the atoms. The remaining elements comprise a sprinkling—like salt in a bowl of broth—adding savor and interest. Most of the spectral lines are caused by these low abundance elements.

Besides our natural curiosity to know what stars are made of, we are motivated to perform chemical composition studies on other fronts. Almost any study in stellar atmospheres presupposes some chemical composition. Our model photosphere construction depends upon it. Some evidence may be obtained concerning the type of nuclear reactions and other physical processes occurring during a star's lifetime. The evolution of our galaxy may be traceable in part through the differences in chemical composition among stars. We expect stars formed at different times and in different places to retain information about the composition of the material from which they formed.

Unfortunately we have proceeded only part way toward these goals as yet, and the impatience of human nature being what it is, too often details are read into the results that are simply beyond the precision of the measurements. It is unlikely, when one looks critically at the uncertainties in the measurements and the theory used to interpret them, that even the best answers can be trusted to better than a factor of 2 or 3.

By now the reader is well aware of the compromise situation between spectral resolution and the faintness of the star to be recorded. As we saw in Chapter 12 most of the measurements are made on spectrograms where the resolution is too low to obtain any information about the line profiles and only the integrated absorption is recorded. The analysis in this case must proceed by means of the curve of growth.

There are a few bright stars for which more detailed measurements can be made. Then one can attempt to fit the line profiles or even a whole spectral region with calculated spectra. The synthesis of a portion of the spectrum simply involves a calculation where all the lines in the region of interest are included. This approach is particularly useful in situations where the lines are crowded together and overlap. Blended lines are a major source of error in the analysis of cooler stars.

Several summaries and reviews of chemical analysis have been compiled in recent years, including a book by Aller (1961), the reviews by Cayrel and Cayrel de Strobel (1966) and Traving (1966) and studies on magnetic stars by Sargent (1964), on lithium and beryllium by Wallerstein and Conti (1969), and on late type stars by Vardya (1970). The solar chemical composition has been rediscussed by Pagel (1973) and Withbroe (1971). A more general review of chemical composition studies is presented by Taylor (1967).

WHAT CAN WE DETERMINE?

We start with the absorption caused by the spectral line and work our way to the abundance of the element. The next few pages of the chapter detail how this is done. Never do we obtain the abundance independent of the physical variables, particularly temperature, nor of the atomic constants, particularly the oscillator strengths. As we have seen in eqs. 13-27 and 13-28, it is the product Af that enters.

Uncertainties in f values have caused many limitations in practical analyses. Laboratory measurements of the most careful kind are needed to obtain dependable f values. Many techniques have been used, including shock tubes, arcs of various sorts, atomic collisions, and resonance excitation. We do not review this topic here. Relevant material can be found in Hubenet (1960), Aller (1963), Cocke et al. (1973), Corliss and Bozman (1962), Huber and Parkinson (1972), Unsöld (1971), Corliss and Warner (1964), Garstang (1971), Weiss (1970), Crossley (1969), and Fuhr and Wiese (1971) to name just a few.

By means of a differential analysis, that is, comparison of one star to another, the oscillator strength is canceled out and the ratio of the abundances in the stars is obtained. Usually the differential approach is applied to stars of similar effective temperature and surface gravity in the hope that the same curves of growth can be applied to both stars. If errors in the analysis exist, then they are made in the same way for both stars and by approximately the same amount. Internal consistency can be much better for a differential analysis than for an absolute analysis. The most certain differential analysis is for a set of stars having identical spectra. Then one infers an identity in chemical composition as well. The chemical analysis has been performed without any physics and without any calculation!

Finally, when we have completed a chemical analysis, we must realize that this chemical composition holds only in the atmosphere of the star. The composition of the interior may be dramatically different. Even within the atmosphere, diffusion or settling processes may make the composition a function of depth if turbulence, for instance, does not continually mix the material (Michaud 1970).

THE CURVE OF GROWTH: SCALING RELATIONS

Economy of effort can be attained by using one curve of growth for many spectral lines. This can be done because many of the differences in shape that exist between curves of growth are small compared to the scatter in the data. We now describe how it is possible to scale one or more of the physical parameters. We assume that each ion is to be analyzed separately.

Let us first generalize our concept of the curve of growth as presented in the preceding chapter. Specifically, it is not the abundance alone that determines the strength of the weak line but the ratio of line to continuous absorption coefficient according to eq. 13-19. The total flux removed by a weak line is then proportional to the ratio* (compare eqs. 11-12, 11-15, and 11-16)

$$R \equiv \frac{1}{\kappa_v} \int_0^\infty l_v \rho \, d\lambda = \frac{\pi e^2}{mc} \frac{\lambda^2}{c} f \frac{N}{\kappa_v} \tag{14-1}$$

where the number of line absorbers per cm^3 is denoted by N.

Equation 14-1 can be reduced to a more useful form. We introduce the number abundance, A, the excitation eq. 1-15, and the ratio of the number

---

* We assume the equivalent width is measured in wavelength units and integrate $l_v \rho$ on a wavelength scale.

of the element in the rth stage of ionization to the total number of the element, N_r/N_E, and write

$$N = A \frac{N_r}{N_E} \frac{g_l}{u(T)} N_H e^{-\chi/kT}. \tag{14-2}$$

N_H is the number of hydrogen particles per cubic centimeter and is the same quantity used in eq. 11-66. Combining eqs. 14-1 and 14-2 yields

$$\log R = \log \left[\frac{\pi e^2}{mc^2} \frac{N_r/N_E}{u(T)} N_H \right] + \log (\lambda^2 \, Afg_l) - \log \kappa_v - \theta_{\mathrm{ex}} \chi. \tag{14-3}$$

The quantity in the brackets is constant for all the lines of any ion in any given star. The temperature from the excitation equation is contained in $\theta_{\mathrm{ex}} = 5040/T$.

It is customary to reduce the equivalent width to velocity units by dividing by the wavelength of the line. This removes the scaling by λ that occurs for Doppler dependent phenomena like thermal broadening. Since the ratio of absorption coefficients, R, is proportional to W for the weak lines, we change eq. 14-3 into

$$\log \frac{R}{\lambda} = \log \mathrm{const} + \log A + \log \lambda f g_l - \log \kappa_v - \theta_{\mathrm{ex}} \chi. \tag{14-4}$$

The generalization of the curve of growth concept is then that $\log R/\lambda$ (aside from a zero point arbitrariness) could be the abscissa to the curve of growth rather than $\log A$ alone. We shall continue to think of $\log A$ as the abscissa of the curve of growth, but now we see explicitly in eq. 14-4 that the physically meaningful coordinate, R/λ, depends on other quantities as well. In the case of the curve of growth for a single spectrum line, all the quantities except A are constant, while in other cases we explicitly wish to consider *many* lines of a given element. Sometimes eq. 14-4 is written

$$\log \left(\frac{R}{\lambda} \right) \equiv \log \mathrm{const} + \log A + \log X$$

where $\log X$ stands for the last three terms in eq. 14-4. So we can plot $\log W/\lambda$ versus $\log A$, which is the case of one line or $\log X = $ constant, and obtain the curve of growth as we have been doing, or we can plot $\log W/\lambda$ against $\log X$, which is the case for many lines, and again obtain the curve of growth for the element, since A, of course, has one unique value for the star.

The scaling rules come from eq. 14-4. We assume that $\log W/\lambda$ is given as a function of $\log A$ (for example, Table 14-2 shown later in the chapter) for some specified line. Then eq. 14-4 can be used to see what translation in the $\log A$ coordinate is appropriate for curves of growth of other lines having different values of $\log X$ to bring them all onto the initial curve of growth.

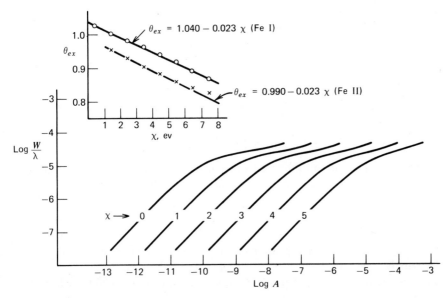

Fig. 14-1 Increasing the excitation potential χ lowers the fraction of the element that causes the absorption line. The equivalent width therefore decreases or the curve of growth moves to the right. The curves shown are for Fe I in a solar temperature model with a surface gravity of 10^4. The insert graph shows the shift between adjacent curves, $\Delta \log A$, divided by their mean χ as a function of χ. For comparison, the Fe II relation is shown.

The effect of χ on the curve of growth is shown in Fig. 14-1. These curves of growth are computed using a model photosphere that takes into account the detailed LTE relation between A and the number of line absorbers per cm^3, N. The progressive shift in the log A coordinate is a result of the lower population associated with higher χ values. The shape of the curve of growth does not change significantly until well into the damping portion. Although we know θ_{ex} is a function of depth, we see in Fig. 14-1 that one curve is the same as the next under a coordinate shift, $\Delta \log A$. According to eq. 14-4 the terms $\Delta \log A$ and $\Delta(\theta_{\text{ex}}\chi)$ are equivalent except for sign. The amount of shift, $\Delta \log A$, from the $\chi = 0$ curve can be expressed as a function of χ for the curves shown in Fig. 14-1. The slope of such a relation is θ_{ex}. The insert in Fig. 14-1 shows θ_{ex} as a function of χ. The fact that χ does not range widely for certain groups of lines has led some researchers to adopt a single θ_{ex} to describe all lines. In any case, one scaling that can be done then is to reduce all the curves of growth for an element to, for instance, the $\chi = 0$ curve of growth by subtraction of

$$\Delta \log A = \theta_{\text{ex}}(\chi)\chi. \tag{14-4a}$$

The wavelength dependence of κ_v in eq. 14-4 is often neglected. It may be justifiable to neglect the term when all the lines come from a small portion of the spectrum. For example, we can easily show using eqs. 8-12 and 8-6 or by looking at tables of absorption coefficients that between 4000 and 6000 Å

$$\frac{\partial \log \kappa_v}{\partial \lambda} \sim 0.1 \text{ cm}^2/\text{g}/10^3 \text{ Å} \qquad \text{for} \quad T \lesssim 7500°\text{K}$$

$$\sim 0.3 \text{ cm}^2/\text{g}/10^3 \text{ Å} \qquad \text{for} \quad T \gtrsim 7500°\text{K}.$$

The investigator should judge for himself whether or not this factor can be neglected. To take it into account, we scale by

$$\Delta \log A = \log \frac{\kappa_v(\lambda)}{\kappa_v(\lambda_0)} \tag{14-4b}$$

where λ and λ_0 are the wavelengths of the lines and the reference wavelength for which the standard curve of growth is computed. It is sufficient to evaluate the κ_v's at one temperature and one pressure characteristic of the line forming layers.

The scaling for the f value term is trivial but important with

$$\Delta \log A = \Delta \log \lambda f g_l$$

$$= \log \frac{\lambda f g_l}{\lambda_0} \tag{14-4c}$$

if we take $f g_l = 1$ for our standard curve of growth.

THE CURVE OF GROWTH: OTHER CHARACTERISTICS

The pressure effects on the curve of growth are shown in Fig. 14-2. The effect enters from the ionization equilibrium N_r/N_E of eq. 14-2 and in κ_v. For neutral lines in solar type stars we expect the two effects to cancel as explained in Chapter 13, so Fe II is used as an illustration in the figure. Again there is a nearly linear relation between $\Delta \log A$ and $\Delta \log g$ as seen in Fig. 14-2 and in agreement with eq. 13-22. Cayrel and Jugaku (1963) have computed $\partial \log A/\partial \log g$ for a line at 4167 Å with $\chi = 2$ eV. Table 14-1 is adapted from their calculations. As long as the element is mostly ionized, $\partial \log A/\partial \log g \doteq 0$. If the surface gravity is known from independent measurements, a separate and independent chemical analysis can be done for the neutrals and for the ions. One would expect the same A to be indicated by both if the solution is internally consistent. When the surface gravity is

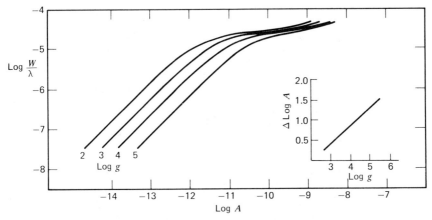

Fig. 14-2   Curves of growth for Fe II showing the gravity dependence.

unknown, g becomes a free parameter that may be determined by forcing the ion and neutral solutions to give the same abundance. This method of determining surface gravity is discussed in Chapter 16. The chemical abundance analysis in such a case is based on the neutrals alone.

Occasionally a strong line is used in an abundance analysis. Figure 14-3 shows an example of how the surface gravity alters the damping portion of the curve of growth. As g diminishes to very low values, the pressure broadening terms become small and γ approaches the minimum value given by the natural damping.

Microturbulence, which is discussed more fully in Chapter 18, is almost always incorporated into curve of growth analyses to explain the observed fact that the equivalent widths of saturated lines are greater than predicted by models based on thermal and damping broadening alone. Figure 14-4 shows how the model lines incorporating microturbulence (isotropic, Gaussian velocity distribution with dispersion ξ km/sec, see Chapter 18 for details) do not saturate as quickly. In all cases, ξ is determined as an ad hoc free parameter as part of the abundance analysis. This is done by simply

Table 14-1 $\partial \log A / \partial \log g$

T_{eff}	Ca I	Ca II	Cr I	Cr II	Fe I	Fe II
7200	0.02	−0.33	0.02	−0.33	0.01	−0.33
5040	0.00	−0.39	0.00	−0.40	−0.11	−0.45
3870	−0.06	−0.43	−0.26	−0.52	−0.35	−0.60

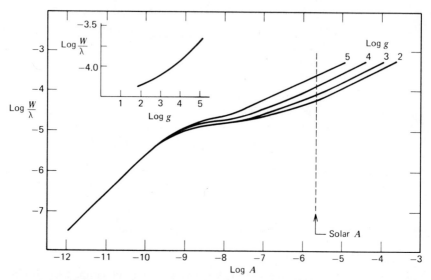

Fig. 14-3 Curves of growth for Na D$_2$ λ5890 in a model having $S_0 = 1.00$ and surface gravity as shown. The position of the solar sodium abundance is shown. The insert graph shows the equivalent width–gravity plane at the solar abundance.

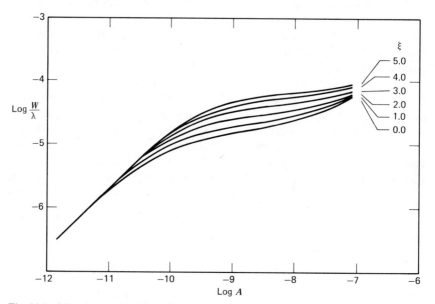

Fig. 14-4 Microturbulence delays saturation by Doppler shifting the absorption over a wider spectral band. Values of velocity dispersion, ξ, for the assumed Gaussian velocity distribution are in kilometers per second.

fitting the equivalent widths of the weakest lines (where the equivalent widths show no change with microturbulence) and adopting the theoretical curve of growth having the best fit to the data in the saturation portion. That curve's ξ value is taken as the star's microturbulence.

One change in the curve of growth that has hardly been investigated at all is the effect caused by differing $T(\tau)$'s. The temperature distribution can be expected to differ from one star to another because of differences in line blanketing, in the strength of convection, and in the mechanical energy dissipation. How important these or other effects are is not completely known. Figure 14-5 shows two hypothetical temperature distributions. They

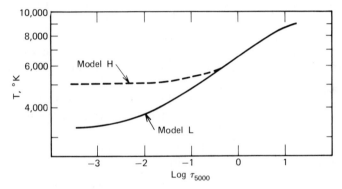

Fig. 14-5 Hypothetical temperature distributions with very different behavior in the upper photospheric layers.

are concocted to keep the energy distribution unchanged [by making $T(\tau)$ the same in the continuum forming regions] while causing strong changes in line strength. Figure 14-6 shows the curves of growth for these models. An obvious translation in the curves of growth for the same abundance is seen. It means that an error ~ 0.7 dex could conceivably occur if continuum measurements are used to choose the model. A less obvious difference between the curves is seen by comparison of the shape of the curves. The saturation portion is at a different level in the two curves. Saturation occurs ealier in the hotter model because of the higher residual flux, $\pi B_\nu(T_0)$. The difference interpreted as microturbulence amounts to 1.3 km/sec.

When the source function deviates from $B_\nu(T)$, the saturation portion of the curve of growth is again affected as shown by Athay and Skumanich (1968). The reason is similar to the reason $T(\tau)$ differences caused such an effect. The greater the deviation from the LTE case, the deeper the line may

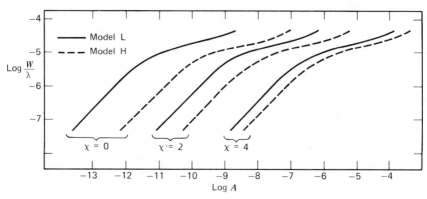

Fig. 14-6 Curves of growth for models based on $T(\tau_0)$ according to Fig. 14-5.

grow before saturation, and the higher the curve of growth will be. And again either the deducted abundance or microturbulence or both must be reduced.

We finish our general discussion of the curve of growth by tabulating an LTE model curve of growth in Table 14-2 based on the temperature distribution of Table 9-2. The atomic constants for iron are used with $\chi = 0$, $\log g_l f = 0$, and $\lambda = 4234$ Å. Depth dependant damping is included according to eqs. 11-33 and 11-35 with $\log C_4 = -15.2$ and C_6 computed

Table 14-2   A Curve of Growth for $\lambda 4243$*

			$\log W/\lambda$		
$\log A$	$\xi = 0$	$\xi = 1$	$\xi = 2$	$\xi = 3$	$\xi = 5$
−12.0	−6.68	−6.68	−6.68	−6.68	−6.68
−11.5	−6.18	−6.18	−6.18	−6.18	−6.18
−11.0	−5.70	−5.70	−5.70	−5.69	−5.69
−10.5	−5.31	−5.27	−5.26	−5.25	−5.24
−10.0	−5.05	−4.98	−4.92	−4.86	−4.81
−9.5	−4.92	−4.83	−4.72	−4.63	−4.50
−9.0	−4.77	−4.70	−4.60	−4.49	−4.34
−8.5	−4.54	−4.54	−4.48	−4.39	−4.24
−8.0	−4.31	−4.31	−4.31	−4.29	−4.17
−7.5	−4.06	−4.06	−4.06	−4.05	−4.05
−7.0	−3.84	−3.84	−3.84	−3.84	−3.84

* Fe I, $\chi = 0$, $\log g_l f = 0$, ξ as shown.

for a line of 4 V excitation. But the damping portion of this curve of growth is included more for completeness than for utility because a comparison between model computations and observations for such strong lines should be done line by line, taking into accunt the individual γ of each line. The $\xi = 2$ curve of growth from Table 14-2 is compared in Fig. 14-7 to the empirical solar curve of growth according to Cowley and Cowley (1964) and Warner (1964, $\lambda4250$ curve).

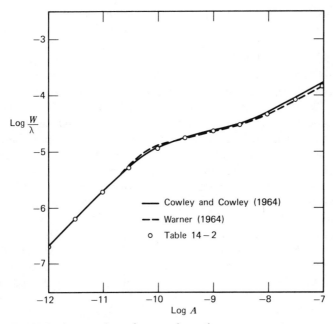

Fig. 14-7 A comparison of curves of growth.

CURVES OF GROWTH FOR ANALYTICAL MODELS:
A HISTORICAL NOTE

Early curve of growth calculations were done by Voigt (1912) and van der Held (1931) with application by Minnaert and Mulders (1931) and Minnaert and Slob (1931). These and some of the analyses that followed were made with very simple models; often the whole line forming region was characterized by a single temperature and electron pressure. Through the intervening years, some have continued to use the very simple models while others have not. Some astronomers mean a single layer model when they refer to "... the curve of growth method." We have not adopted this narrow meaning nor have we restricted the terminology to any type of model atmosphere. Instead, we use the curve of growth as a general concept.

Curves of growth based on simple models have been published by Struve and Elvey (1934), Menzel (1936), Unsöld (1955), Strömgren (1940), Pannekoek and van Albada (1946), Greenstein (1948), and Wrubel (1949, 1954).

DERIVATION OF ABUNDANCES VIA THE CURVE OF GROWTH

The simplest approach conceptually is to ignore all the scaling rules and compute a curve of growth for each measured line using a model photosphere that takes into account the relation between the number of absorbers in the line and the total abundance. That is, the model calculation explicitly takes into account the ionization equilibrium, the excitation, the damping constant depth dependence, and the numerical integration of the equation of transfer. The abundance is simply read back from the $\log W/\lambda$ versus $\log A$ theoretical curve of growth upon entering the observed value of W.

The simplicity is paid for by the large amount of calculation that must be done and by the difficulty in handling microturbulence or other desaturating mechanisms. For example, suppose an initial guess is made for ξ. Then the theoretical curves of growth are computed for all the measured W's of iron or some other element showing a large number of lines. From each line an A is obtained. Now a plot of A versus W is made and it is found that A appears to be a function of W. Under the reasonable assumption that the photosphere is chemically homogeneous, such an effect is interpreted as an erroneous choice of ξ. Then a new ξ must be chosen and the process repeated until all lines give the same abundance—a rather lengthy procedure to say the least.

The scaling rules can be used to eliminate the computation of many curves of growth. We assume that one standard curve of growth similar to the one in Table 14-2 will be computed for the effective temperature of the star to be studied. Think of the following simplified procedure.

A measured W is taken and entered into the standard curve. The abundance obtained would hold for a line with $\chi = 0, \log g_l f = 0$ at $\lambda = \lambda_0$, the standard wavelength. We call this abundance A_0. Then according to eqs. 14-4a, 14-4b, and 14-4c we obtain the real abundance by subtracting

$$\Delta \log A = \log g_l f + \log\left(\frac{\lambda}{\lambda_0}\right) - \log\left(\frac{\kappa_\nu}{\kappa_0}\right) - \theta_{\mathrm{ex}}\chi, \qquad (14\text{-}5)$$

namely,

$$\log A = \log A_0 - \Delta \log A. \qquad (14\text{-}6)$$

The parameters $g_l f$, λ, and χ are for the line whose W was used to get A_0.

Instead of doing this line by line, we plot the observed $\log W/\lambda$ against $\Delta \log A$. Each line is then reduced to the same curve of growth and we have constructed what is sometimes referred to as an observed or an empirical curve of growth. Superposition of this curve of growth with the standard curve of growth is equivalent to entering simultaneously all the observed W's into the standard curve. When the observed curve of growth is fit to the standard curve of growth, the difference in abscissa is $\log A$ according to eq. 14-6. An example of such a fit is shown in Fig. 14-8.

Should the standard curve of growth be computed for χ other than zero and for $g_l f$ other than unity, $\Delta \log A$ becomes

$$\Delta \log A = \log g_l f \lambda - \log \kappa_\nu - \theta_{ex}\chi - (\log g_l f \lambda - \log \kappa_\nu - \theta_{ex}\chi)_{std}$$
$$\equiv \log X - \log X_{std}. \tag{14-7}$$

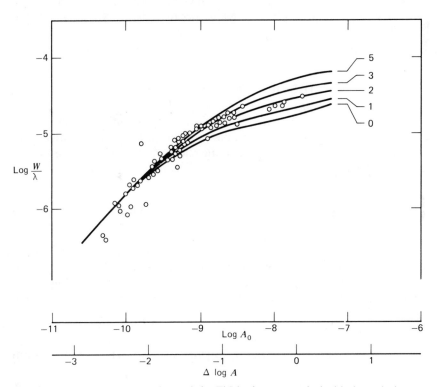

Fig. 14-8   The empirical curve of growth for Ti I in the sun matched with theoretical curves of growth. The $\xi = 2$ km/sec curve shows the best overall fit. Both the $\Delta \log A$ and the $\log A_0$ abscissas are shown. An abundance $\log A = \log A_0 - \Delta \log A = -7.72$ is indicated. (Data from Cowley and Cowley 1964.)

An important advantage of the empirical curve of growth is the simplicity in choosing the microturbulence parameter. The $\xi = 2$ curve of growth in Fig. 14-8 fits the data and is adopted as the proper theoretical curve of growth for the star. So we see that the second problem of the line by line model photosphere approach has also been avoided by use of the scaling procedure.

The values of θ_{ex} that one adopts for each line to compute $\Delta \log A$ in eq. 14-5 should be obtained from the model photosphere used to compute the curve of growth (as in Fig. 14-1). Other, more approximate schemes are in use. One is to choose a single θ_{ex} that gives the smallest scatter in the empirical curve of growth. Another is to plot the empirical curve of growth separately for groups of lines selected to have similar excitation potentials. Instead of $\Delta \log A$ as in eq. 14-5, one uses $\log g_l f + \log (\lambda/\lambda_0) - \log (\kappa_v/\kappa_0)$ $= \Delta \log A + \theta_{ex}\chi$ as abscissa. The shift into the standard curve of growth is then $\log A - \theta_{ex}\chi$ instead of eq. 14-6. Finally, a plot of these ($\log A - \theta_{ex}\chi$) values versus χ has $\log A$ as the intercept and θ_{ex} as the slope. The method is difficult to apply in practice because the measureable lines of most elements do not show a sufficiently large range in χ.

THE DIFFERENTIAL ANALYSIS

The popularity of the differential abundance analysis has been high because this method avoids the uncomfortable problem of uncertain f values. We combine eq. 14-6 for the unknown and the reference star, using a superscript r to denote the latter,

$$\log \left(\frac{A}{A^r} \right) = \log \left(\frac{A_0}{A_0^r} \right) - \Delta \log A + (\Delta \log A)^r. \qquad (14\text{-}8)$$

A glance at eq. 14-5 shows the $\Delta \log A$ difference reduces to the difference in the $\theta_{ex}\chi$ terms because we understand that eq. 14-8 is to be applied to the same lines in the two stars. In other words, $\log g_l f + \log (\lambda/\lambda_0) - \log (\kappa_v/\kappa_0)$ is the same, or very nearly so, for any line in the two stars. So we rewrite eq. 14-8 as

$$\log \left(\frac{A}{A^r} \right) = \log A_0 - [\log A_0^r - \Delta\theta_{ex}\chi] \qquad (14\text{-}9)$$

where $\Delta\theta_{ex} \equiv \theta_{ex} - \theta_{ex}^r$. Although $\Delta\theta_{ex}$ is a weak function of depth, it can be treated as a constant for any given ion and still be consistent with the approximations of the differential analysis.

The first step is to read the log A_0^r values from a standard curve of growth*
by entering the equivalent widths measured in the reference star's spectrum.
The empirical curve of growth is constructed next by plotting log W/λ for
the unknown against log $A_0^r - \Delta\theta_{ex}\chi$. Matching the empirical curve to the
standard curve gives the difference in abscissa which is the right side of eq.
14-9.

The answers, of course, are the abundance ratios A/A^r and can only be
reduced to absolute abundances when the A^r's are known. A microturbulence
parameter and average damping constant for the unknown may be estimated from the standard curve of growth in the usual way.

The standard curve of growth is introduced for one purpose. It is a way of
taking into account the curve of growth shape, which we must do because
most measured stellar lines are saturated. *If* we could compare only weak
lines between the stars, we could write

$$\log\left(\frac{A}{A^r}\right) = \log\left(\frac{W}{W^r}\right) - \Delta\theta_{ex}\chi$$

directly with no reference to a curve of growth.

The following subtleties should be noticed. The standard curve of growth
must have the proper microturbulence and average damping parameter so
it is the correct theoretical curve of growth for the reference star. This
implies the empirical curve of growth for the reference star must first be
constructed using f values. It is true that no f values are needed for those
particular lines measured in the unknown–reference star comparison.
Nevertheless, relative f values must be known for *some* set of lines observed in
the reference star and they must range sufficiently in strength to define
the curve of growth for the reference star. The f value uncertainties have not
really been eliminated but, shall we say, have been swept under the rug.

By using the same standard curve of growth for both stars, we implicitly
assume the stars have the same temperature distributions. This may be a
reasonable assumption when the stars have the same chemical composition
but becomes suspect otherwise (as Fig. 14-6 has convinced us).

THE SYNTHESIS METHOD

The most complete studies are done on high quality data by computing
simultaneously all the lines over a portion of the spectrum. The method works
best when all the observed spectral lines can be identified and have known

---

* The standard curve of growth can be one such as that in Table 14-2, but it must have ξ and
damping appropriate to the reference star. The zero point of the abscissa is of no consequence
however, since we use only differences along the abscissa.

f values. The individual abundances (and perhaps the temperature and pressure) are adjusted until a fit is obtained. Usually turbulence must also be introduced into the model. Even when lines are blended together, one simply computes the blend and then a comparison is made with the data. The LTE model synthesis technique is described by Ross and Aller (1968) with application to Cr I in the solar spectrum. Grevesse and Swings (1970) have studied the solar nickel abundance with the synthesis method. Lithium in sunspots was studied by Traub and Roesler (1971) using an interesting variation of the synthesis approach. The complex spectra of cool stars have been synthesized by Phillips (1966) and Climenhaga (1966), for example, while the abundances in the metal poor star HD 122563 were studied by Sneden (1973) using synthesized spectra. Figure 14-9 shows the $\lambda 3984$ region of the solar spectrum as fit by Ross and Aller (1968). The stronger lines are labeled. The discrepancy at 3984.25 Å is a result of not including a weak blend in the calculations. A few angstroms of the fantastically crowded spectrum of the carbon star TX Psc are shown in Fig. 14-10. The synthesized spectrum is for a model having $T_{\text{eff}} = 3200°\text{K}$ and $\log g = 0$. Most of the lines belong to the CN spectrum.

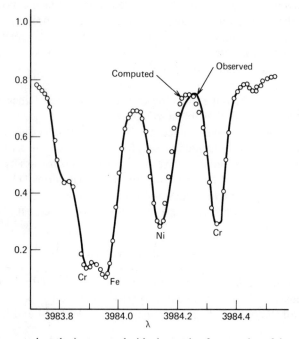

Fig. 14-9 A spectral synthesis compared with observation for a portion of the solar spectrum. (Adapted from Ross and Aller 1968 with permission from the Astrophysical Journal, Copyright 1968 by the University of Chicago.)

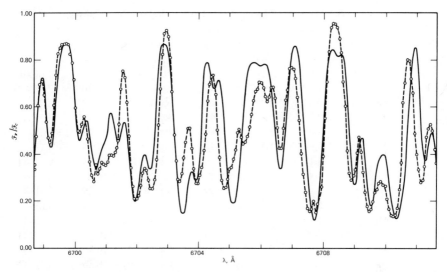

Fig. 14-10   The observed spectrum of TX Psc (-----) is compared to the synthesized spectrum (———). (Courtesy of H. R. Johnson.)

THE SOLAR CHEMICAL COMPOSITION

Many many studies have been done for the sun. They are mostly photospheric analyses similar to what we have been discussing, but some important contributions have also come from chromospheric and coronal measurements (see, for example, Athay 1968, Zirin 1968, and Pottasch 1964). One of the most comprehensive photospheric studies was done by Goldberg, Müller, and Aller (1960). Their work is commonly quoted and is frequently denoted by GMA. An excellent review of solar abundance measurements has been given by Müller (1966). More recent reviews have been given by Pagel (1973) and Withbroe (1971). A summary of the solar composition is given in Table 14-3. Typical uncertainties are ± 0.3 dex. Systematic errors can occur as stunningly illustrated by the case of iron where a downward revision of the f values led to an order of magnitude increase in the deduced abundance (Unsöld 1971, Garz e al. 1969, Baschek et al. 1970).

The uncertainties in these measurements are not so large that general patterns of behavior are obliterated. In Fig. 14-11 the general drop in abundance with atomic number can be seen. Deviations from the general trend are sometimes dramatic, such as the $\sim 10^{10}$ deficiency of lithium and beryllium. The elements around iron show enhanced abundance forming the "iron peak." The remarkable alternation caused by even number elements

Table 14-3   The Solar Chemical Composition Relative to Hydrogen

No.	Element	log A	A	No.	Element	log A	A
1	H	0.00	1.00	41	Nb	−9.70	2.0E-10
3	Li	−11.40	4.0E-12	42	Mo	−10.10	7.9E-11
4	Be	−10.94	1.1E-11	44	Ru	−10.43	3.7E-11
5	B	≤ −9.20	≤6.3E-10	45	Rh	−10.45	3.5E-11
6	C	−3.43	3.7E-4	46	Pd	−10.43	3.7E-11
7	N	−3.94	1.1E-4	47	Ag	−11.33	4.7E-12
8	O	−3.17	6.8E-4	48	Cd	−10.03	9.3E-11
9	F	−7.44	3.6E-8	49	In	−10.29	5.1E-11
10	Ne	−4.55	2.8E-5	50	Sn	−10.29	5.1E-11
11	Na	−5.76	1.7E-6	51	Sb	−11.25	5.6E-12
12	Mg	−4.46	3.5E-5	55	Cs	≤ −10.21	≤6.2E-11
13	Al	−5.60	2.5E-6	56	Ba	−10.20	6.3E-11
14	Si	−4.45	3.5E-5	57	La	−10.19	6.5E-11
15	P	−6.57	2.7E-7	58	Ce	−10.36	4.4E-11
16	S	−4.79	1.6E-5	59	Pr	−10.37	4.3E-11
17	Cl	−6.35	4.5E-7	60	Nd	−10.18	6.6E-11
18	Ar	−5.27	5.4E-6	62	Sm	−10.34	4.6E-11
19	K	−6.95	1.1E-7	63	Eu	−11.51	3.1E-12
20	Ca	−5.67	2.1E-6	64	Gd	−10.88	1.3E-11
21	Sc ·	−8.93	1.2E-9	66	Dy	−10.89	1.3E-11
22	Ti	−7.26	5.5E-8	68	Er	−11.24	5.8E-11
23	V	−7.90	1.3E-8	69	Tm	−11.57	2.7E-12
24	Cr	−6.30	5.0E-7	70	Yb	−11.19	6.5E-12
25	Mn	−6.80	1.6E-7	71	Lu	−11.16	6.9E-12
26	Fe	−4.60	2.5E-5	74	W	−9.43	3.7E-10
27	Co	−7.50	3.2E-8	76	Os	−11.25	5.6E-12
28	Ni	−5.72	1.9E-6	77	Ir	−9.79	1.6E-10
29	Cu	−7.55	2.8E-8	79	Au	−11.68	2.1E-12
30	Zn	−7.58	2.6E-8	80	Hg	≤ −9.00	≤1.0E-9
31	Ga	−9.16	6.9E-10	81	Tl	−11.80	1.6E-12
32	Ge	−8.68	2.1E-9	82	Pb	−10.13	7.4E-11
37	Rb	−9.37	4.3E-10	83	Bi	≤ −11.20	6.3E-12
38	Sr	−9.18	6.6E-10	90	Th	−11.18	6.6E-12
39	Y	−10.38	4.2E-11	92	U	−11.40	4.0E-12

Source. Withbroe (1971).

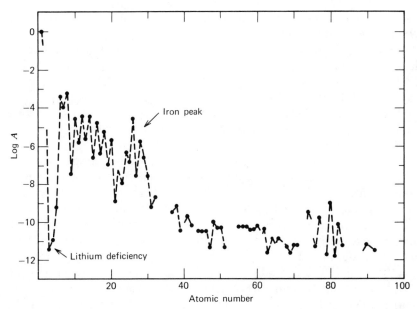

Fig. 14-11 The solar abundances are plotted as a function of atomic number from the data in Table 14-3.

being more abundant than their adjacent odd number neighbors is also clearly visible.

These solar abundances are used as the standards for most differential stellar analyses. We also use them in the construction of model photospheres.

STELLAR ABUNDANCES

Many of the population I stars near the sun show a composition that is essentially solar. The subdwarfs, high velocity stars, and globular clusters show strong population II characteristics and have been found to be metal deficient relative to the sun by factors of up to $\sim 10^2$. It has come to be accepted that one of the characteristics of population II is metal deficiency. Most of the population–abundance studies have been done with filter photometry because the population II stars are so faint (see Dixon 1965 and Peat and Pemberton 1968).

There is disagreement as to whether the elements heavier than sodium show depletion by the same factor or whether significant individual abundance variations occur (Unsöld 1970, Wallerstein 1971). The ratio of individual element abundances can be used to identify the types of nuclear reactions important in stars (see Burbidge et al. 1957, Burbidge and Burbidge 1957, and Hubenet 1966).

A seemingly endless list of references to investigations of chemical abundances could be given; but instead we cite the summaries listed at the end of the first section in this chapter and suggest the following selected papers: Cayrel and Cayrel (1963) on ε Vir, Wright (1948) on four solar-type stars, Parsons (1967 and 1970) on yellow supergiants, Strom et al. (1968) on α C Ma, Griffin (1971) on α C Mi, Cayrel de Strobel (1966) on K stars, and Pagel (1966) on population characteristics.

Helium abundances are of considerable interest but are measurable only in the hot stars. The measurements are still somewhat inconclusive with values ranging from $A = 0.03$ to 0.15. Reviews on the subject are Herstmonceux (1967), Taylor (1967), Danziger (1970), and de Jager (1971).

For some of the advances made by including departures from LTE, see Mihalas and Athay (1973) and Smith (1974). Significant revisions are made in O and B star abundances.

Several groups of chemically anomalous stars have been identified. Some of these groups are barium stars (Bidelman and Keenan 1951, Gordon 1968, Warner 1965), hydrogen deficient stars (Hill 1960, Klinglesmith et al. 1970, Aller 1954), carbon stars (Wallerstein 1973, Myerscough 1968, Bidelman 1956), technetium stars (Peery 1971), CH stars (Wallerstein 1969, Keenan 1942, 1958, Bidelman 1956), super metal rich stars (Sprinrad and Luebke 1970, Spinrad and Taylor 1969, Blanc-Vaziaga et al. 1973), peculiar A stars (Adelman 1973, Sargent et al. 1969), and metallic line stars (Van't Veer-Menneret 1963, Sargent 1964). In addition, several special elements have been studied in detail. The interesting behavior of lithium has been studied by Herbig (1964), Wallerstein and Conti (1969), and Zappala (1972). Radioactive elements may have been identified in some stars (Aller and Cowley 1970, Wolf and Morrison 1972, Cowley and Aller 1972, Davis 1971, Cowley and Hartoog 1972).

COMMENTS

The use of line profiles for abundance work has not been sufficiently stressed by workers in the field. The synthesis method is a step in the proper direction but emphasis there has been on the problem of blends in a crowded region. Careful profile analysis, on the other hand, opens the door for differentiating between such phenomena as turbulence and non-LTE increases in equivalent width, eventually leading to abundance determinations that are more soundly based.

Abundance measurements are more entwined with the model used for the analysis than most methods in common use would lead us to believe. A complete abundance analysis is an iterative procedure; the model itself depends on the chemical composition. The first guess at the composition is used to compute a model that is used to obtain the chemical composition. The new composition alters the model ionization equilibrium, the temperature distribution through the line blanketing, and the radiation field through both the continuous and line absorption. Even in solar work, differences in models can result in abundances that differ by nearly a factor of 2 (Weidemann 1955, Andrews and Mugglestone 1963, Hubenet 1960, Withbroe 1971).

Not all astrophysical abundance measurements are made by stellar atmosphere analysis. Stellar composition also affects the overall structure of the star. Stellar interior models can be used to measure the fractions of hydrogen, helium, and heavier elements (Demarque and Heasley 1971, Aizenman et al. 1968).

Nonstellar information on the cosmic distribution of elements is obtained from analyses of gaseous nebulae (Aller and Liller 1968), from interstellar absorption lines (Münch 1968), and from cosmic rays (Meyer 1969). Besides the abundance measurements in the sun, the solar system composition has been established from measurements of the earth, from planetary atmospheres, and especially from meteorite analysis (Suess 1964, Suess and Urey 1956).

REFERENCES

Adelman, S. J. 1973. *Astrophys. J.* **183**, 95.

Aizenman, M. L., P. Demarque, and R. H. Miller 1968. *Astron. J.* **73**, S161.

Aller, L. H. 1954. *Mem. Soc. Roy. Sci. Liege* **14**, 337.

Aller, L. H. 1961. *The Abundance of the Elements*, Interscience, New York.

Aller, L. H. 1963. *The Atmospheres of the Sun and Stars*, 2nd ed., Ronald, New York.

Aller, L. H. and W. Liller 1968. *Nebulae and Interstaller Matter*, B. M. Middlehurst and L. H. Aller, Eds., University of Chicago Press, Chicago, p. 483.

Aller, M. F. and C. R. Cowley 1970. *Astrophys. J. Lett.* **162**, L145.

Andrews, M. and D. Mugglestone 1963. *Mon. Not. Roy. Astron. Soc.* **125**, 347.

Athay, R. G. 1968. *Astrophys. Lett.* **1**, 71.

Athay, R. G. and A. Skumanich 1968. *Astrophys. J.* **152**, 211.

Baschek, B., T. Garz, H. Holweger, and J. Richter 1970. *Astron. Astrophys.* **4**, 229.

Bidelman, W. P. 1956. *Vistas in Astronomy*, Vol. 2, A. Beer, Ed., Pergamon, London, p. 1428.

Bidelman, W. P. and P. C. Keenan 1951. *Astrophys. J.* **114**, 473.

Blanc-Vaziaga, M.-J., G. Cayrel, and R. Cayrel 1973. *Astrophys. J.* **180**, 871.

Burbidge, E. M. and G. R. Burbidge 1957. *Astrophys. J.* **126**, 357.

Burbidge, E. M., G. R. Burbidge, W. A. Fowler, and F. Hoyle 1957. *Rev. Mod. Phys.* **29**, 547.

Cayrel, G. and R. Cayrel 1963. *Astrophys. J.* **137**, 431.

Cayrel, R. and G. Cayrel de Strobel 1966. *Annu. Rev. Astron. Astrophys.* **4**, 1.

Cayrel, R. and J. Jugaku 1963. *Ann. Astrophys.* **26**, 495.

Cayrel de Strobel, G. 1966. *Ann. Astrophys.* **29**, 413.

Climenhaga, J. L. 1966. *Colloquium on Late-Type Stars*, M. Hack, Ed., Osservatorio Astronomico, Trieste, p. 54.

Cocke, C. L., A. Stark, and J. C. Evans 1973. *Astrophys. J.* **184**, 653.

Corliss, C. H. and W. R. Bozman 1962. *Nat. Bur. Stand. Monogr.* **53**.

Corliss, C. H. and B. Warner 1964. *Astrophys. J. Suppl.* **8**, 395.

Cowley, C. R. and M. F. Aller 1972. *Astrophys. J.* **175**, 477.

Cowley, C. R. and A. P. Cowley 1964. *Astrophys. J.* **140**, 713.

Cowley, C. R. and M. R. Hartoog 1972. *Astrophys. J. Lett.* **178**, L9.

Crossley, R. J. S. 1969. *Adv. At. Mol. Phys.* **5**, 237.

Danziger, I. J. 1970. *Annu. Rev. Astron. Astrophys.* **8**, 161.

Davis, D. N. 1971. *Astrophys. J.* **167**, 327.

de Jager, C. 1971. *Highlights of Astronomy*, Vol. 2, Reidel, Dordrecht, Holland.

Demarque, P. and J. N. Heasley, Jr. 1971. *Astrophys. J.* **163**, 547.

Dixon, M. E. 1965. *Mon. Not. Roy. Astron. Soc.* **129**, 51.

Fuhr, J. R. and W. L. Wiese 1971. *Nat. Bur. Stand. Spec. Publ.* **320**, Suppl. 1.

Garstang, R. H. 1971. *Topics in Modern Physics*, W. E. Brittin and H. Odabasi, Eds., Colorado Associated University Press, Boulder, Colorado, p. 153.

Garz, T., H. Holweger, M. Kock, and J. Richter 1969. *Astron. Astrophys.* **2**, 446.

Goldberg, L., E. A. Müller, and L. H. Aller 1960. *Astrophys. J. Suppl.* **5**, 1.

Gordon, C. P. 1968. *Astrophys. J.* **153**, 915.

Greenstein, J. L. 1948. *Astrophys. J.* **107**, 151.

Grevesse, N. and J. P. Swings 1970. *Sol. Phys.* **13**, 19.

Griffin, R. 1971. *Mon. Not. Roy. Astron. Soc.* **155**, 139.

Herbig, G. H. 1964. *Astrophys. J.* **140**, 702.

Herstmonceux, 1967. *Observatory* **87**, 193.

Hill, P. W. 1960. *Mon. Not. Roy. Astron. Soc.* **129**, 137.

Hubenet, H. 1960. *Rechh. Astron. Obs. Utr.* **16**, 1.

Hubenet, H. 1966. *Abundance Determinations in Stellar Spectra, IAU Symposium No. 26*, Academic, London.

Huber, M. C. E. and W. H. Parkinson 1972. *Astrophys. J.* **172**, 229.

Keenan, P. C. 1942. *Astrophys. J.* **96**, 101.

Keenan, P. C. 1958. *Handb. Phys.* **50**, 93.

Klinglesmith, D. A., K. Hunger, R. C. Bless, and R. L. Millis 1970. *Astrophys. J.* **159**, 513.

Menzel, D. H. 1936. *Astrophys. J.* **84**, 462.

Meyer, P. 1969. *Annu. Rev. Astron. Astrophys.* **7**, 1.

Michaud, G. 1970. *Astrophys. J.* **160**, 641.

Mihalas, D. and G. R. Athay 1973. *Annu. Rev. Astron. Astrophys.* **11**, 187.

Minnaert, M. and G. F. W. Mulders 1931. *Z. Astrophys.* **2**, 165.

Minnaert, M. and C. Slob 1931. *Proc. K. Acad. Amst.* **34**, 542.

Müller, E. A. 1966. *Abundance Determinations in Stellar Spectra, IAU Symposium No. 26*, H. Hubenet, Ed., Academic, London, p. 171.

Münch, G. 1968. *Nebulae and Interstellar Matter*, B. M. Middlehurst and L. H. Aller Eds., University of Chicago Press, Chicago, p. 365.

Myerscough, V. P. 1968. *Astrophys. J.* **153**, 421.

Pagel, B. E. J. 1966. *Abundance Determinations in Stellar Spectra, IAU Symposium No. 26*, H. Hubenet, Ed., Academic, London, p. 272.

Pagel, B. E. J. 1973. *Space Sci. Rev.* **15**, 1.

Pannekoek, A. and B. B. van Albada 1946. *Publ. Astron. Inst. Univ. Amst.* **6**, Part 2.

Parsons, S. B. 1967. *Astrophys. J.* **150**, 263.

Parsons, S. B. 1970. *Astrophys. J.* **159**, 951.

Peat, D. W. and A. C. Pemberton 1968. *Mon. Not. Roy. Astron. Soc.* **140**, 21.

Peery, B. F. Jr. 1971. *Astrophys. J. Lett.* **163**, L1.

Phillips, J. G. 1966. *Appl. Opt.* **5**, 549.

Pottasch, S. R. 1964. *Mon. Not. Roy. Astron. Soc.* **128**, 73.

Ross, J. and L. Aller 1968. *Astrophys. J.* **153**, 235.

Sargent, W. L. 1964. *Annu. Rev. Astron. Astrophys.* **2**, 297.

Sargent, W. L. W., K. M. Strom, and S. E. Strom 1969. *Astrophys. J.* **157**, 1265.

Smith, M. A. 1974. *Astrophys. J.* **192**, 623.

Sneden, C. 1973. *Astrophys. J.* **184**, 839.

Spinrad, H. and W. Luebke 1970. *Bull. Am. Astron. Soc.* **2**, 2.

Spinrad, H. and B. J. Taylor 1969. *Astrophys. J.* **157**, 1279.

Strom. S. E., O. Gingerich, and K. M. Strom 1968. *Observatory* **88**, 160.

Strömgren, B. 1940. *Festschrift für Elis Strömgren*, Munksgaard, Copenhagen, p. 218.

Struve, O. and C. T. Elvey 1934. *Astrophys. J.* **79**, 409.

Suess, H. E. 1964. *Proc. Nat. Acad. Sci. Wash.* **52**, 387.

Suess, H. E. and H. C. Urey 1956. *Rev. Mod. Phys.* **28**, 53.

Taylor, R. J. 1967. *Quart. J. Roy. Astron. Soc.* **8**, 313.

Traub, W. A. and F. L. Roesler 1971. *Astrophys. J.* **163**, 629.

Traving, G. 1966. *Abundance Determinations in Stellar Spectra, IAU Symposium No. 26*, H. Hubenet, Ed., Academic, London, p. 213.

Unsöld, A. 1955. *Physik der Sternatmosphären*, 2nd ed., Springer-Verlag, Berlin.

Unsöld, A. O. J. 1970. *Publ. Astron. Soc. Pac.* Leaflet Nos. 491 and 492.

Unsöld, A. 1971. *Phil. Trans. Roy. Soc. Lond.* **270**, 23.

van der Held, E. F. M. 1931. *Z. Phys.* **70**, 508.

Van't Veer-Menneret, C. 1963. *Ann. Astrophys.* **26**, 289.

Vardya, M. S. 1970. *Annu. Rev. Astron. Astrophys.* **8**, 87.

Voigt, W. 1912, *Münch. Ber.* 603.

Wallerstein, G. 1969. *Astrophys. J.* **158**, 607.

Wallerstein, G. 1971. *Publ. Astron. Soc. Pac.* Leaflet No. 500.

Wallerstein, G. 1973. *Annu. Rev. Astron. Astrophys.* **11**, 115.

Wallerstein, G. and P. S. Conti 1969. *Annu. Rev. Astron. Astrophys.* **7**, 99.

Warner, B. 1964. *Mon. Not. Roy. Astron. Soc.* **127**, 413.

Warner, B. 1965. *Mon. Not. Roy. Astron. Soc.* **129**, 263.

Weidemann, V. 1955. *Z. Astrophys.* **36**, 101.

Weiss, A. W. 1970. *Nucl. Inst. Meth.* **90**, 121.

Withbroe, G. L. 1971. *The Menzel Symposium on Solar Physics, Atomic Spectra, and Gaseous Nebulae*, K. B. Gebbie, Ed., *Nat. Bur. Stand. Spec. Publ.* **353**, 127.

Wolf, S. C. and N. D. Morrison 1972. *Astrophys. J.* **175**, 473.

Wright, K. O. 1948. *Publ. Dom. Astrophys. Obs. Victoria* **8**, 1.

Wrubel, M. H. 1949. *Astrophys. J.* **109**, 66.

Wrubel, M. H. 1954. *Astrophys. J.* **119**, 51.

Zappala, R. R. 1972. *Astrophys. J.* **172**, 57.

Zirin, H. 1968. *Astrophys. J.* **154**, 799.

CHAPTER FIFTEEN

THE MEASUREMENT OF STELLAR TEMPERATURES AND RADII

The topics of stellar temperatures and radii are placed together in this chapter because they are closely related. Let us denote the flux emitted from the stellar surface by $\mathcal{F}_\nu$ and the flux we measure at the earth F_ν. Now let us consider two concentric spheres. The inner sphere represents the star of radius R. The outer sphere has a radius r, which is the distance from us to the star. Then by conservation of energy* we have

$$4\pi r^2 F_\nu = 4\pi R^2 \mathcal{F}_\nu. \tag{15-1}$$

We can relate this equation to the effective temperature by integrating over frequency,

$$\int_0^\infty F_\nu \, dv = \left(\frac{R}{r}\right)^2 \sigma T_{\text{eff}}^4 \tag{15-2}$$

It is clear that this expresses an intimate relation among the absolute flux measured at the earth, the angular radius of the star, and the effective temperature.

The linear radius follows by combining the angular size with the distance. If we use the conventional units of parsecs for distance, seconds of arc for the angular radius or diameter, θ_R or θ_D, and solar radii ($R_\odot = 6.96 \times 10^{10}$ cm) for R, then

$$R = 215 \, \theta_R r$$
$$= 108 \, \theta_D r.$$

---

* We assume no interstellar extinction.

The most commonly available distances come from parallax measurements, although the distances to some stars are known because the star belongs to a moving cluster. Eclipsing binaries yield linear radii without using distances, and instead the distances are needed to convert the linear radii to angular radii should they be used in an expression like eq. 15-2.

The concept of a stellar radius is predicated on the thinness of the photosphere compared to the distance from the center of the star to the photosphere. The depth of formation for all continuum radiation in the visible region of the spectrum is sensibly constant in this context. The same statement cannot be made of the full electromagnetic spectrum. The sun, for example, is much larger at some radio wavelengths because the "surface" of emission is in the corona.

The effective temperature is likewise a limited concept. It is meant to describe the total power per square centimeter coming from an average surface element of the star. We immediately run into conceptual difficulties where nonuniform behavior exists. If the star is nonspherical or has a surface brightness dependent on position (perhaps the star has sunspot structure, chemical patches, or shows gravity darkening due to rotation), then T_{eff} depends upon the orientation of the star to the line of sight.

We shall put these worries aside and spend the first part of the chapter considering stellar radii explicitly. Following this, the effective temperature scale is considered. The chapter ends with the description of those methods of measuring the temperature that are purely spectroscopic.

SPECKLE PHOTOMETRY

It is common knowledge that the sun is the only star for which we can see the surface and resolve the disc. The basis of this statement rests on the assumption that atmospheric seeing sets the minimum size of the stellar image at ~ 1 sec more or less, depending on the nightly meteorological conditions. The diffraction limit of a large telescope is much smaller. For example, $\lambda/D \sim 4000 \text{ Å}/4 \text{ m} \sim 0\overset{''}{.}02$. The catalog of Wesselink et al. (1972) indicates a significant number of stars have angular diameters larger than 10^{-2} sec of arc. So it would be possible to resolve the angular size for some of these stars if the blurring by the atmosphere could be avoided.

It has recently been realized that the instantaneous seeing disc in the focal plane of a big telescope often consists of a large number of superimposed diffraction modified images of the star itself (Labeyrie 1970, Korff et al. 1972, Liu and Lohmann 1973) (see Fig. 15-1). Apparently the atmosphere consists of a relatively small number of refracting elements that redirect portions of the wave front across the beam accepted by the telescope aperture. The

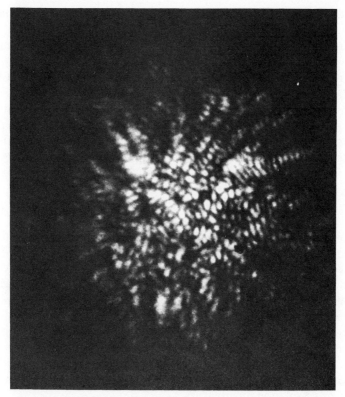

Fig. 15-1 The seeing disc speckle pattern for Vega taken in visible light with the 5 m Hale telescope. Exposure time was 10^{-2} sec. (Courtesy of D. Y. Gezari.)

intensity and position of each component of this seeing pattern is a strong function of time. Averaging over even a few seconds of time removes all trace of the individual image components and we see the usual blurred seeing disc. The rapid recording of the composite image is helped by using image intensifiers. The wavelength band in which the light is recorded must be kept small since the scale of the diffraction is proportional to wavelength.

Each image component is the convolution of the telescope instrumental profile* with the true stellar disc intensity distribution. The problem of disentangling the individual overlapping images is easily solved by taking the Fourier transform. Each speckle of the pattern must have the same amplitude transform, so it is acceptable to use a scheme that adds all the transforms together. Analog techniques have generally been used (Gezari

---

* Sometimes dominated by diffraction but also including optical aberrations, and so on.

et al. 1972, KenKnight 1972) but digital reduction is beginning to make its appearance (Irwin 1975).

The instrumental profile is obtained in the same way by measuring a star with an angular diameter much smaller than $\sim 10^{-2}$ sec of arc. Reconstruction of the Fourier amplitudes can be done in the usual way (see Chapter 12).

The dimensions for half a dozen stars were measured by Gezari et al. (1972) and Bonneau and Labeyrie (1973). Future work should enlarge the list considerably. The same stars were also measured with the Michelson interferometer as discussed in the next section.

THE INTERFEROMETERS

The phase coherent Michelson stellar interferometer (Michelson and Pease 1921) and the intensity interferometer (Hanbury Brown et al. 1967) have both been used to obtain angular sizes of stars. If we view the apparent disc of a star as an aperture in a screen (the dark sky), it is immediately clear that there is a diffraction pattern associated with this aperture. We have seen with diffraction gratings, for example, how the diffraction pattern is the Fourier transform of the aperture intensity distribution. If the stellar discs showed no limb darkening, these transforms would be Bessel functions with first zero at $x = 1.22\lambda/\theta_D$ where x is a linear coordinate measured at the earth. The interferometers are devices for measuring these transforms (also called visibility or correlation functions).

The apparent disc of the star is symmetric about the line of sight and we can view the transform as centered on one of the two apertures of the interferometer. As the second aperture of the interferometer is made to have different separation, x, from the first, the transform is recorded as a function of x. At that x where no transform amplitude is recorded, we have the condition $x = 1.22\lambda/\theta_D$, or a similar expression for models incorporating limb darkening, and hence θ_D (see also Jennison 1961).

The Michelson device has yielded only seven angular diameters (Pease 1931) all for K and M stars above the main sequence. The maximum separation of the beams was 7.3 m, limiting the resolution to $0\overset{''}{.}017$ or larger.

The only intensity interferometer in existence has a beam separation of up to 188 m. The smallest angular diameter that it has been used to measure is $0\overset{''}{.}00042$. The angular diameters of 32 stars measured with this instrument are given by Hanbury Brown et al. (1974). The errors on the angular diameters range from 2 to 14% with an average value of 6%. About one-quarter of the stars on their list have parallaxes of sufficient accuracy to allow a linear radius to be calculated. Unfortunately the large number of photons needed to obtain significant correlations between the interferometer beams precludes the measurement of all but the brightest stars until a much larger instrument can be constructed.

LUNAR OCCULTATIONS

The first positive results in using the moon to measure angular sizes of stars was obtained for α Sco (Cousins and Guelke 1953, Evans et al. 1953, Evans 1957). A value of 0″.041 was found in good agreement with the interferometer results.

The moon is used as a knife edge that moves across the beam of light coming from the star. The diffraction pattern of the occultation is recorded as a function of time. The time coordinate is converted to an angular coordinate using the angular velocity of the moon relative to the earth. Detailed analysis of deviations in the diffraction pattern from the point source (i.e., totally unresolved star) pattern gives the angular size of the star. Excellent background material is given by Nather and Evans (1970). Theoretical diffraction patterns have been computed by Morbey (1972). Many recent measurements include those of Berg (1970), Dunham et al. (1973), Herbison–Evans et al. (1971), Lasker et al. (1973), Morbey and Hutchings (1971), Nather et al. (1970), Taylor (1966), and de Jager (1973).

ECLIPSING BINARIES

About five dozen stellar radii have been reliably measured in eclipsing spectroscopic binary systems. No distance measurement is needed to obtain the stellar dimensions in absolute units. The photometry gives the ratio of radii to orbital semimajor axis, while the radial velocity measurements supply the semimajor axis times the sine of the angle of inclination (but $i \doteq 90°$ since the system is eclipsing). The data are given by Kopal (1955), Harris et al. (1963), Snowden and Koch (1969), Barnes et al. (1968), and Popper (1957, 1965, 1967, 1970, 1974). A plot of these radii as a function of $B - V$ is shown in Fig. 15-2. Although there is large scatter, the average main sequence star behavior is well defined for the early type stars.

It is important to compare the orbital dimension to the size of the stars to assess the possibility of tidal interaction and mass exchange. Binaries showing small separations usually have anomalous spectra. But binaries with large separation have small radial velocity variations and the individual spectral lines are difficult or impossible to separate, which means the orbital analysis is correspondingly difficult or impossible to perform. We must bear

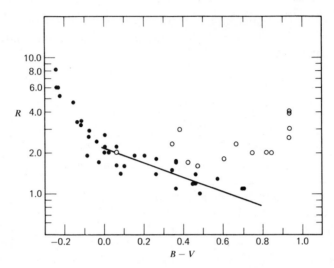

Fig. 15-2   Radii of eclipsing spectroscopic binaries are shown as a function of their $B - V$ color index. Solid symbols are for main sequence stars. The line is from Fig. 15-3.

in mind that even though the selected stars are thought to be well determined, they represent a *highly* selected group that might differ from stars in general with regard to their radius–spectral type behavior. But as yet no evidence has been found to support such fears and we generally assume that these binaries can be used as standards of comparison or for calibration work.

THE SURFACE BRIGHTNESS METHOD

A photometric method that is old in concept has been brought to fruition by Wesselink (1969) and Wesselink et al., (1972). Stellar surface brightness is calibrated as a function of $B - V$ using primarily the known angular diameters from the intensity interferometer measurements. Surface brightness is assumed to be a single valued function of $B - V$. The method is physically based on the fundamental concept that the apparent brightness of a star is proportional to its surface brightness and the solid angle it subtends at the observer. The conversion from angular to linear dimensions is made through the 1963 calibration of M_V with MK spectral type done by Blaauw. Angular and linear dimensions are given for over 2000 stars by Wesselink et al. (1972).

It is difficult to evaluate the accuracy of the method, but Wesselink et al. (1972) estimate the uncertainty in their angular diameters to be a few per cent. The surface brightness method, while powerful and easy to use, depends heavily on the calibration using the interferometrically measured radii.

THE BOLOMETRIC FLUX METHOD

This photometric method for measuring a stellar radius depends on a knowledge of the total or bolometric flux of the star. Suppose we take eq. 15-2 and write

$$R = r \left[\frac{\int_0^\infty F_v \, dv}{\sigma T_{\text{eff}}^4} \right]^{1/2}. \tag{15-3}$$

We assume r is obtained from a trigonometric parallax. The trick in using eq. 15-3 properly is to avoid circular reasoning. The situation can be seen more readily if we write eq. 15-3 completely in terms of flux

$$\begin{aligned} R &= r \left[\frac{\int_0^\infty F_v \, dv}{\int_0^\infty \mathcal{F}_v \, dv} \right]^{1/2} \\ &= r \left[\frac{a_1 \int_0^\infty F_v W(v) \, dv}{a_2 \int_0^\infty \mathcal{F}_v W(v) \, dv} \right]^{1/2} \end{aligned} \tag{15-4}$$

where a_1 and a_2 are the bolometric factors of eq. 10-6 in Chapter 10 and $W(v)$ is the window in which absolute flux measurements are actually made. We see immediately that we must be exactly consistent in handling the radiation outside $W(v)$ to ensure that $a_1 = a_2$. The bolometric correction applied to the measured stellar flux must be the same one used in computing the effective temperature scale. It does not matter if it is an observed or a theoretical bolometric correction since it cancels out.

Logically it would be more sensible to define a pseudoeffective temperature according to

$$\int_0^\infty \mathcal{F}_v W(v) \, dv \equiv \sigma t_{\text{eff}}^4.$$

Then

$$R = r \left[\frac{\int_0^\infty F_v W(v) \, dv}{\sigma t_{\text{eff}}^4} \right]^{1/2}$$

and the problem of the unseen flux is avoided.

The circular reasoning may enter if, for example, the relation between $T_{\rm eff}$ and $B - V$ is used to obtain $T_{\rm eff}$ for the star. The calibration of $T_{\rm eff}$ as a function of $B - V$ is done in its purest form using stellar radii, as we see in a few pages. Furthermore, in most of the $T_{\rm eff}$-$(B - V)$ calibrations, bolometric corrections are incorporated. The unwary researcher may use a completely different bolometric correction to obtain $\int_0^\infty F_\nu \, d\nu$. It may be more appropriate in these cases to estimate a radius directly from the R versus $B - V$ plot of other stars.

The bolometric flux method is perhaps the most commonly used method, examples being Terashita and Matsushima (1969) and Heintze (1973).

As we see later in this chapter, the effective temperature calibration ultimately rests on some absolute flux calibration. Instead of using eq. 15-3 or 15-4 it is more direct to obtain the radius from an absolute flux calibration without going through the intermediary of an effective temperature.

RADII FROM ABSOLUTE FLUX

This method (Gray 1967, 1968a) is based on the absolute flux calibration discussed in Chapter 10. Starting with the fundamental relation expressed in eq. 15-1, we write

$$R = r\left(\frac{F_\nu}{\mathscr{F}_\nu}\right)^{1/2}$$

or

$$R = 4.43 \times 10^7 r\left(\frac{F_\nu}{\mathscr{F}_\nu}\right)^{1/2} \tag{15-5}$$

for R in units of solar radii and the distance r in parsecs. Photometric observations are used to determine the absolute apparent flux F_ν and model photospheres are used to compute the absolute flux at the star's surface, $\mathscr{F}_\nu$. The model that most closely represents the star being studied is chosen on the basis of energy distribution—primarily a fit of the model's Paschen continuum to the observed Paschen continuum.* The model photospheres explicitly relate the slope of the Paschen continuum to the absolute flux, a relation that is very obvious in Fig. 10-9. When a model continuum has been correctly fitted to the observed energy distribution, the ratio $F_\nu/\mathscr{F}_\nu$ is independent of ν. The square root of this flux ratio is the angular radius of the star in radians.

No use is made of the concept of effective temperature or of bolometric

---

* It would also be possible to choose the model from any other suitable features in the spectrum. Then F_ν and $\mathscr{F}_\nu$ need be observed and computed, respectively, at only one wavelength.

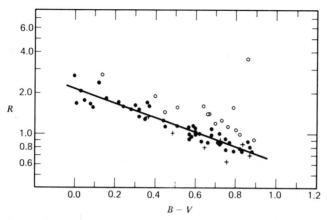

Fig. 15-3 Photometric radii measured with the absolute flux method are shown plotted against $B - V$: ($\bullet$) main sequence stars; (+) subdwarfs, ($\bigcirc$) stars above the main sequence; the line corresponds to eq. 15-6. The calibration in Table 10-1 was used in the reduction.

correction. No reliance is placed on a guess about the flux in the IR or UV that cannot be easily measured. On the other hand, the method depends heavily upon the absolute calibration—both the shape and the true flux at 5556 Å.

Radii for several dozen normal stars have been measured using the method. In Fig. 15-3 we see that the logarithm of the radius as a function of color index shows a remarkably linear relation for the main sequence. Such a relation can be written

$$\log R = 0.333 - 0.528(B - V) \tag{15-6}$$

The scatter of the points about this relation indicates an average probable error of 8 % for the radii.

The same method has been applied to white dwarfs by Shipman (1972) and to giants by Parsons (1970).

Now let us turn from the question of radii to that of temperatures.

EFFECTIVE TEMPERATURES FROM ABSOLUTE FLUX

The relation expressed in eq. 15-2 gives

$$T_{\text{eff}} = \frac{(\int_0^\infty F_v \, dv/\sigma)^{1/4}}{(R/r)^{1/2}}. \tag{15-7}$$

The angular size is obtained from one of the fundamental methods described earlier and the absolute flux is measured over the entire spectrum. Because of the terrestrial atmospheric filtering, $\int_0^\infty F_v \, dv$ is measured in pieces, with the

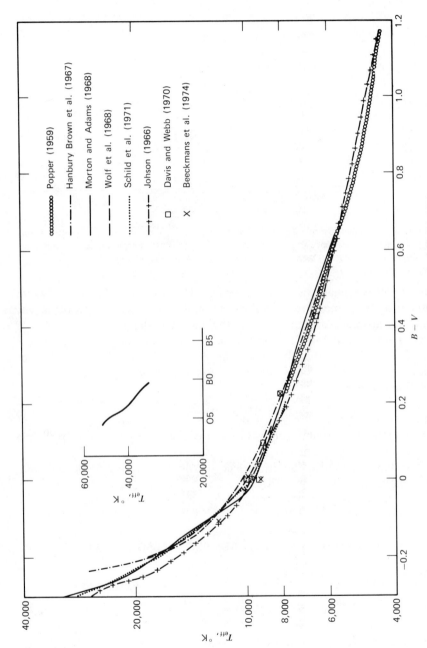

Fig. 15-4 The effective temperature calibrations for main sequence stars. Data from Conti (1973).

UV and IR photometry being done from rockets, satellites, or high altitude balloons. Often the UV and IR flux is measured with respect to the visible flux. This results in measured bolometric flux factors, a, as defined in eq. 10-6. The relative numbers are reduced to absolute flux by means of the absolute photometric calibration in eq. 10-5. The results for several individual stars according to Davis and Webb (1970) and Beeckmans et al. (1974) are given in Fig. 15-4 along with several smoothed T_{eff} calibrations.

A second approach is to ratio eq. 15-7 for the star and the sun and use the absolute flux calibration for the sun. That is,

$$T_{eff} = \frac{T_\odot (\int_0^\infty F_v \, dv / \int_0^\infty F_v^\odot \, dv)^{1/4}}{(\theta/\theta_\odot)^{1/2}}. \tag{15-8}$$

This method was used in most work until recently because the absolute fluxes for stars were less well established than they were for the sun. Equation 15-8 is often put in logarithmic form,

$$\log T_{eff} = \log T_\odot - 0.1 \, \Delta M_V + 0.1 \, \Delta BC - 0.5 \log \left(\frac{\theta}{\theta_\odot}\right)$$

where we have used ΔM_V for the difference in absolute visual magnitude and ΔBC for the difference in bolometric correction (Figs. 10-20 and 10-21) between the star and the sun. The absolute flux calibration enters through $T_\odot$. Popper's analysis of Sirius (Popper 1959) is a good illustration of this approach.

The relation between T_{eff} and $B - V$ is nearly independent of luminosity class, with significant differences occurring only for supergiants and for M stars. The results of Johnson (1966) are given in Fig. 15-5.

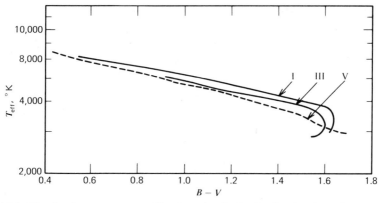

Fig. 15-5 The effective temperature calibrations showing luminosity class dependence according to Johnson (1966).

Although systematic errors may still affect some of the calibrations, a carefully executed measurement of a star's effective temperature results in random errors of about $\pm 5\%$ or less.

EFFECTIVE TEMPERATURES FROM MODEL PHOTOSPHERES

If one knows enough about the temperature (or source function) distribution(s), and the absorption coefficients including the UV and IR, it is possible to compute $\int_0^\infty \mathscr{F}_\nu \, d\nu$ and obtain the effective temperature for the model. The next step is to somehow associate particular models with individual real stars, thereby assigning a model's T_{eff} to each star of interest. Clearly the value of T_{eff} found in this way depends strongly on how good the models are. The association of the model with a star can be made in several ways but remains one of the uncertain parts of a T_{eff} calibration of this sort.

Hanbury Brown et al. (1967) have used (essentially) eq. 15-1, obtaining (R/r) from their interferometer measurements and F_ν from the photometry of Willstrop (1965). The value of $\mathscr{F}_\nu$ so derived was used to select the proper model and hence T_{eff} from the tabulation by Mihalas (1966). The temperature calibration obtained in this way is also shown in Fig. 15-4.

A second means of choosing the model is to match the slope of the Paschen continuum as computed and observed. This approach has been used by Schild et al. (1971) and Wolf et al. (1968), for example. These calibrations have been added to Fig. 15-4. The close agreement of the result of Wolf et al. with that of Hanbury Brown et al. above is in part due to both groups using the same set of model photospheres. The agreement does show that the models are consistent between their absolute flux and the slope of the Paschen continuum.

In the calibration by Morton and Adams (1968), several different sets of models were used. The spectral type and color index to be assigned to each model were chosen according to the size of the Balmer discontinuity, the temperature sensitivity of which is discussed later in the chapter.

Conti (1973) has reviewed and redone the T_{eff} calibration for O stars using the model predictions of Auer and Mihalas (1972) chosen on the basis of the He I $\lambda 4471$/He II $\lambda 4541$ ratio. The insert in Fig. 15-4 shows the relation derived by Conti. Spectral type is used as abscissa since the $B - V$ color index loses all sensitivity to temperature in this range and the problem of interstellar reddening is avoided.

For the cooler stars, the slope of the relation in Fig. 15-4 shows that a step of ~ 0.01 magnitudes in $B - V$ corresponds to a change in effective temperature of $\sim 1\%$.

It is clear that any temperature sensitive spectrum feature could be used to choose a model. At this stage it is interesting to reflect on the fact that the spectrum depends on $T(\tau)$, or its counterpart in the full non-LTE case, and not upon T_{eff}. In many studies of stellar spectra the actual value of T_{eff} is irrelevant. It makes sense to consider the temperature measurement of stars from their spectra in this more general frame of mind. This we do in the remainder of the chapter.

Several of the temperature sensitive spectral features are also gravity dependent. These and other pressure indicators are elaborated on in Chapter 16 and we choose to minimize discussion of the pressure dependence in this chapter.

THE PASCHEN CONTINUUM

One of the best measures of the temperature of a star is the slope of the Paschen continuum (we assume no interstellar reddening is present). It is relatively independent of the spectral lines and unlikely to be affected by the complications of non-LTE. The dependence on surface gravity is very small as we have seen in Chapter 10. The change in slope for the model continua shown in Fig. 10-9 is given in Fig. 10-11. The more rapid the change in slope with temperature, the better the thermometer we have. We find from Fig. 10-11,

$$\frac{\partial \log (\mathscr{F}_{4000}/\mathscr{F}_{7000})}{\partial \log T_{\text{eff}}} \doteq 2.3$$

for main sequence models having $T_{\text{eff}} \lesssim 10{,}000°$K. Taken literally, it implies that if we can measure the continuum slope between $\lambda 4000$ and $\lambda 7000$ to 2.3%, then the temperature is established to $\pm 1\%$. In practice, using the whole Paschen continuum, the temperature can be established with an internal consistency of ± 2 or 3%.

The loss of sensitivity for high temperatures occurs rather abruptly, starting at $T_{\text{eff}} \sim 10{,}000°$K as shown in Fig. 10-11. We expect to lose leverage when the peak of the star's energy distribution lies well away from the Paschen continuum. In the cooler stars the line absorption becomes more and more a problem and the temperature error is correspondingly worse. Measurements in the red and infrared become increasingly important for cooler stars.

When scaled solar models are matched to observed Paschen continua, a value of the scaling factor, S_0, can be associated with the empirical classification parameters of the star such as color index and spectral type. This type of temperature calibration is analogous to the effective temperature calibration in Fig. 15-4. The observed relation between S_0 and $B - V$ is shown in

Fig. 15-6 as defined by several dozen normal stars. For $(B - V) < 0.60$ the linear relation

$$\log S_0 = 0.243 - 0.405(B - V) \qquad (15\text{-}9a)$$

describes the behavior. For $(B - V) > 0.70$ we find instead

$$\log S_0 = 0.268 - 0.405(B - V). \qquad (15\text{-}9b)$$

The scatter around these lines corresponds to a probable error of $\pm 2\%$. As with T_{eff}, a step of 0.01 magnitude in $B - V$ corresponds to a change in $S_0 \sim 1\%$.

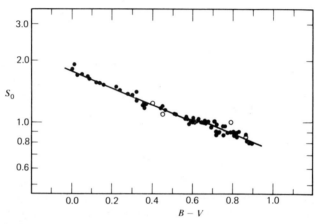

Fig. 15-6 S_0 versus $B - V$ determined from energy distributions based on the Vega calibration in Table 10-1. The line corresponds to eq. 15-9. ($\bullet$) main sequence stars; ($\bigcirc$) stars above the main sequence.

The depth of formation for the Paschen continuum is characterized by $\tau_0 \sim 1$, so it is really the temperatures of the deeper layers of the photosphere that are being measured and not T_{eff}. At this point we review the following thoughts. The continuum and the lines, particularly the low excitation lines, have depths of formation that are sufficiently different to allow a certain amount of decoupling between them. This was illustrated in Chapter 14 when the influence of the temperature distribution on the curve of growth was considered. It is conceivable, especially for chemically anomalous stars, that the Paschen continuum slope cannot be automatically translated into line excitation temperatures and so on, using standard models of the photosphere. Nevertheless the Paschen continuum measurements should be viewed as important pieces of information because they *do* establish the scale of $T(\tau)$ in the deeper layers of the star to be analyzed.

THE BALMER JUMP

The temperature sensitivity of the Balmer jump is useful in the hotter stars. But in the A and F stars the Balmer jump is also pressure sensitive and additional information must be used to resolve the ambiguity. We saw in Figs. 10-9, 10-10, and 10-11 the behavior of the Balmer jump as a function of temperature. The Balmer discontinuity, $D \equiv \log \mathscr{F}_+ / \mathscr{F}_-$ where $+$ and $-$ refer to wavelength longer and shorter than 3647 Å, is plotted in Fig. 10-11. For models between about G0 and A5, the rate of change $\partial D / \partial \log T_{\text{eff}}$ is between 2.3 and 2.8, depending on the surface gravity. An error of 10% in $\mathscr{F}_+ / \mathscr{F}_-$ results in an error of $\sim 4\%$ in temperature if no error is introduced because of the uncertainty in surface gravity.

For higher temperatures the Balmer jump decreases monononically and $\partial D / \partial \log T_{\text{eff}} \sim 1.3$ so the temperature leverage is lower but the complication of the gravity uncertainty is gone.

In practice it is not trivial to measure the Balmer jump because the strong hydrogen lines blend together at the series limit causing distortion in the Paschen continuum. But there is no serious problem in fitting the Paschen continuum toward longer wavelengths and the Balmer continuum below the Balmer jump. Theoretical models do not all show the same relation between Paschen slope and Balmer jump, which leads to some ambiguity in the interpretation of the observations. The depth of formation is sufficiently different for $\mathscr{F}_+$ and $\mathscr{F}_-$ that the size of the Balmer jump depends on the temperature gradient (see Fig. 9-13). Scaled solar models show a smaller gradient than, for instance, the models of Mihalas (1966). The result is that the scaled solar models show a smaller Balmer discontinuity. Observational evidence that the larger gradient is more correct comes from the fact that consistency between the Paschen continuum slope and the Balmer jump is not as good when using models with low temperature gradients (see Fig. 10-18). Deviations from LTE generally produce marginally larger flux in the Balmer continuum (see Chapter 10), which, if anything, increases the argument for the steeper source function gradient.

Examples of temperature measurement using the Balmer discontinuity are found in the work of Morton and Adams (1968) and Chalonge and Divan (1952). A similar method has been used by Gray (1968b). In stars as cool as G0, the Balmer jump is obliterated by the many metal lines.

THE HYDROGEN LINES

The temperature sensitivity of H_γ is illustrated in Fig. 13-8. Other Balmer lines show a similar behavior. Below about 8000°K effective temperature the rapid change of line strength with temperature and the null gravity dependence make these lines very useful. An estimate of the temperature sensitivity can be obtained from Fig. 13-8b. We compute $\partial \log W / \partial \log T_{eff} \sim 6$. The leverage is good, but it is exceedingly difficult to measure a true equivalent width for a hydrogen line and the advantages of profiles should be used whenever possible (see discussion in Chapter 12). Different theoretical formulations of the Stark profile lead to considerable range in interpretation. In practice the hydrogen lines give a precision comparable to that afforded by the Paschen continuum or the Balmer jump.

When $T_{eff} > 8000°K$ an independent surface gravity measurement is necessary to uniquely identify the temperature from the Balmer lines. The gradient $\partial \log W / \partial \log T_{eff}$ near 10,000°K is ~ -2 but gradually comes closer to zero with increasing T_{eff}.

Cayrel and Cayrel (1963) used H_α in ε Vir (G9 III) to help establish its temperature. The consistency of H_γ and the Paschen continuum as temperature indicators was investigated by Searle and Oke (1962).* In more recent investigations such as that by Schild et al. (1971) good agreement is found between continuum and hydrogen lines as used to interpret the spectra of B and A stars.

METAL LINES AS TEMPERATURE INDICATORS

A line's sensitivity to temperature can vary widely as we have seen in Chapter 13. In solar type stars the lines of neutral metals are useful because they are independent of pressure. It is necessary to use ratios of line strengths for one element in order to eliminate the dependence on abundance. Lines of the same ion but with widely different excitation potential would make suitable ratios, that is, they would give a significant temperature dependence. The enhancement of the line strength by microturbulence or other desaturating mechanisms can only be avoided by going to weak lines. Two problems are introduced by this restriction. First, it is difficult to find measurable lines of widely different excitation potential both of which are weak. Second, to reach a small percentage error on the equivalent width of a weak line is not yet possible. A 1% flux measurement error translates into a 10% error in central depth for a line with a central depth of 10%.

---

* Note that the recalibration of Vega's energy distribution has occurred in the meantime and more advanced hydrogen line broadening theories have come into use.

A ratio can also be made between lines from two ionization stages of the same element. This ratio can be sensitive to temperature but is also pressure dependent. In some cases the surface gravity is known from independent data, so pressure ceases to be a variable. Peterson and Scholz (1971) have used the ratios of C II/C III and C III/C IV in their study of O stars with limited success.

In place of individual line ratios, many lines can be brought to bear by use of the curves of growth as a temperature measuring device. In one form of the technique (see Cowley and Cowley 1966, for example) the excitation temperature is arbitrarily adjusted to minimize the scatter in the empirical curve of growth. The more standard technique (Aller 1963) is to plot separate curves for small intervals in excitation potential—the shift between curves being proportional to $\theta_{ex} \chi$ according to eq. 14-5. (The curve of growth examples cited here assume a single layer model atmosphere, but the concept can be extended to more complex models where θ_{ex} is a function of depth.)

Certainly the full potential of lines as temperature indicators has yet to be exploited. Progress has been uncertain because of the many physical variables, questions of uniqueness, deviations from LTE, and aerodynamic phenomena.

REFERENCES

Aller, L. H. 1963. *The Atmospheres of the Sun and Stars*, 2nd ed., Ronald, New York, p. 379.

Auer, L. H. and D. Mihalas 1972. *Astrophys. J. Suppl.* **24**, 193.

Barnes, R. C., D. S. Hall, and R. H. Hardie 1968. *Publ. Astron. Soc. Pac.* **80**, 69.

Beeckmans, F., D. Macau, and D. Malaise 1974. *Astron. Astrophys.* **33**, 93.

Berg. R. A. 1970. *Bull. Am. Astron. Soc.* **2**, 182.

Blaauw, A. 1963. *Basic Astronomical Data*, K. Aa. Strand, Ed., University of Chicago Press, Chicago.

Bonneau, D. and A. Labeyrie 1973. *Astrophys. J. Lett.* **181**, L1.

Cayrel, G. and R. Cayrel 1963. *Astrophys. J.* **137**, 431.

Chalonge, D. and L. Divan 1952. *Ann. Astrophys.* **15**, 201.

Conti, P. S. 1973. *Astrophys. J.* **179**, 181.

Cousins, A. W. J. and R. Guelke 1953. *Mon. Not. Roy. Astron. Soc.* **113**, 776.

Cowley, C. R. and A. D. Cowley 1966. *Abundance Determinations in Stellar Spectra, IAU Symposium No. 26*, H. Hubenet, Ed., Academic, London, p. 43.

Davis, J. and R. J. Webb 1970. *Astrophys. J.* **159**, 551.

de Jager, C. 1973. *Highlights in Astronomy*, Vol. 2, Reidel, Dordrecht, Holland.

Dunham, D. W., D. S. Evans, E. C. Silverberg, and J. R. Wiant 1973. *Astron. J.* **78**, 199.

Evans, D. S. 1957. *Astron. J.* **62**, 83.

Evans, D. S., J. C. R. Heydenrych, and J. D. N. van Wyk 1953. *Mon. Not. Roy. Astron. Soc.* **113**, 781.

Gezari, D. Y., A. Labeyrie, and R. V. Stachnik 1972. *Astrophys. J. Lett.* **173**, L1.

Gray, D. F. 1967. *Astrophys. J.* **149**, 317.

Gray, D. F. 1968a. *Astron. J.* **73**, 769.

Gray, D. F. 1968b. *Astrophys. J. Lett.* **153**, L113.

Hanbury Brown, R., J. Davis, and L. R. Allen 1974. *Mon. Not. Roy. Astron. Soc.* **167**, 121.

Hanbury Brown, R., J. Davis, L. R. Allen, and J. M. Rome 1967. *Mon. Not. Roy. Astron. Soc.* **137**, 393.

Harris, D. L., K. Aa. Strand, and C. E. Worley 1963. *Basic Astronomical Data*, K. Aa. Strand, Ed., University of Chicago Press, Chicago, p. 273.

Heintze, J. R. W. 1973. *Problems of Calibration of Absolute Magnitudes and Temperatures of Stars, IAU Symposium No. 54*, B. Hauck and B. E. W. Westerlund, Eds., Reidel, Dordrecht, Holland, p. 231.

Herbison-Evans, D., R. Hanbury Brown, J. Davis, and L. E. Allen 1971. *Mon. Not. Roy. Astron. Soc.* **151**, 161.

Irwin, J. B. 1975. *Sky and Telescope* **49**, 164.

Jennison, R. C. 1961. *Fourier Transforms and Convolutions for the Experimentalist*, Pergamon, New York.

Johnson, H. L. 1966. *Annu. Rev. Astron. Astrophys.* **4**, 193.

KenKnight, C. E. 1972. *Astrophys. J. Lett.* **176**, L43.

Kopal, Z. 1955. *Ann. Astrophys.* **18**, 379.

Korff, D., G. Dryden, and M. G. Miller 1972. *Opt. Commun.* **5**, 187.

Labeyrie, A. 1970. *Astron. Astrophys.* **6**, 85.

Lasker, B. M., S. B. Bracker, and W. E. Kunkel 1973. *Publ. Astron. Soc. Pac.* **85**, 109.

Liu, C. Y. C. and A. W. Lohmann 1973. *Opt. Commun.* **8**, 372.

Michelson, A. A., and F. G. Pease 1921. *Astrophys. J.* **53**, 249.

Mihalas, D. 1966. *Astrophys. J. Suppl.* **13**, 1.

Morbey, C. L. 1972. *Publ. Dom. Astrophys. Obs. Victoria* **14**, 45.

Morbey, C. L. and J. B. Hutchings 1971. *Publ. Astron. Soc. Pac.* **83**, 156.

Morton, D. C. and T. F. Adams 1968. *Astrophys. J.* **151**, 611.

Nather, R. E. and D. S. Evans 1970. *Astron. J.* **75**, 575, 583, 963.

Nather, R. E., M. M. McCants, and D. S. Evans 1970. *Astrophys. J. Lett.* **160**, L181.

Parsons, S. B. 1970. *Astrophys. J.* **159**, 951.

Pease, F. G. 1931. *Ergeb. exakten Naturwiss. Berlin* **10**, 84.

Peterson, D. M. and M. Scholz 1971. *Astrophys. J.* **163**, 51.

Popper, D. M. 1957. *J. Roy. Astron. Soc. Can.* **51**, 51.

Popper, D. M. 1959. *Astrophys. J.* **129**, 647.

Popper, D. M. 1965. *Astrophys. J.* **141**, 126.

Popper, D. M. 1967. *Annu. Rev. Astron. Astrophys.* **5**, 85.

Popper, D. M. 1970. *Mass. Loss and Evolution in Close Binaries, IAU Colloquium* 6, K. Gyldenkerne and R. M. West, Eds., Copenhagen University Observatory Publications, p. 13.

Popper, D. M. 1974. *Astrophys. J.* **188**, 559.

Schild, R., D. M. Peterson, and J. B. Oke 1971. *Astrophys. J.* **166**, 95.

Searle, L. and J. B. Oke 1962. *Astrophys. J.* **135**, 790.

Shipman, H. L. 1972. *Astrophys. J.* **177**, 723.

Snowden, M. S. and R. H. Koch 1969. *Astrophys. J.* **156**, 667.

Taylor, J. H. Jr. 1966. *Nature* **210**, 1105.

Terashita, Y. and S. Matsushima 1969. *Astrophys. J.* **156**, 203.

Wesselink, A. J. 1969. *Mon. Not. Roy. Astron. Soc.* **144**, 297.

Wesselink, A. J., K. Paranya, and K. DeVorkin 1972. *Astron. Astrophys. Suppl.* **7**, 257.

Willstrop, R. V. 1965. *Mem. Roy. Astron. Soc.* **69**, 83.

Wolf, S. C., L. V. Kuhi, and D. Hayes 1968. *Astrophys. J.* **152**, 871.

CHAPTER SIXTEEN

THE MEASUREMENT
OF PHOTOSPHERIC PRESSURE

Precise spectroscopic measurement of the photospheric pressure is difficult to achieve. There are no spectrum features that have a striking sensitivity to pressure as the lines and continuum do for temperature. Continuum measurements over the Balmer jump can be used with the A and F stars to measure the electron pressure. A comparison of lines formed by neutral atoms to lines of ions also gives a measure of electron pressure. Finally, the lines strong enough to show wings (including hydrogen) often have the wing strength dependent on the pressure through the van der Waals and Stark broadening. In each case there is also a temperature dependence. Either the temperature must be established securely before executing the pressure analysis or a simultaneous pressure–temperature solution must be made.

The electron pressure is connected to the gas pressure according to the relations given in Chapter 9. The surface gravity controls the scale of the pressure distributions, for example, according to eq. 9-11 and 9-12. Therefore we usually assume that the depth dependence of P_e and P_g are fixed by the chemical composition, the temperature distribution, and hydrostatic equilibrium. What we mean by determining the atmospheric pressure then is the absolute scale involved. This is conveniently described by the surface gravity, g.

In addition to the gas pressure, contributions to the hydrostatic support may come from radiation pressure, magnetic fields, and turbulence. Any g value obtained by the usual analysis of a stellar spectrum is the effective gravity needed in the hydrostatic equation. In extreme cases it may differ from GM/R^2 because of these other pressures.

Stellar masses have been measured from time to time by spectroscopic means. The procedure is to fix g from the spectroscopic data, obtain the radius from independent measurements, and $M = R^2(g/g_\odot)$ follows. As we see later in this chapter, it is hard to measure surface gravity with sufficient precision to make such a mass calculation very reliable. But in some situations any estimate of mass is welcome.

It is an empirical fact that the luminosity of a star is inversely related to the surface gravity. Pressure effects are often referred to as luminosity effects. Those pressure effects that are large enough to see at classification dispersion are used to assign the luminosity class to the spectrum.

We now discuss the spectroscopic determination of surface gravity, describing the methods and estimating the precision. The chapter closes with a summary of the binary star data and the best determined values of surface gravity.

THE CONTINUUM AS A PRESSURE INDICATOR

The visible continuum sensitivity to surface gravity is shown in Figs. 10-10 and 10-11. The Balmer jump is the only feature sufficiently sensitive to g to make a useful tool. The gradient $\partial D/\partial \log g$ (where as in the last chapter, $D \equiv \log F_+/F_-$ is the Balmer discontinuity) is shown in Fig. 16-1. The larger

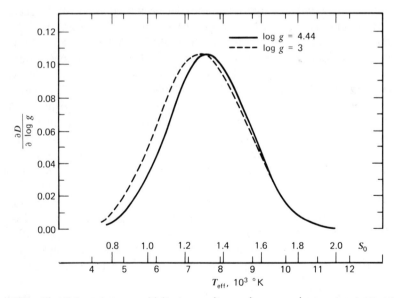

Fig. 16-1 The Balmer jump sensitivity to gravity reaches a maximum at a temperature $\sim 7500°K$. (Data from Fig. 10-11.)

$\partial D/\partial \log g$, the better D is as a pressure indicator. As shown in the figure, the maximum occurs at $7500°K$ with a value somewhat over 0.1. So even at its best, the leverage is about one twenty-fifth of that found for the temperature sensitivity. If we can measure F_+/F_- to 5%, then g is fixed to $\pm 50\%$, assuming no uncertainty in the temperature.

The usefulness of this pressure indicator does not extend to stars as cool as one might guess from Fig. 16-1. The Balmer jump is essentially masked by the multitude of metal lines that appear already in the late F stars ($\sim 6500°K$). The Balmer jump in the solar spectrum is measured only with great difficulty (Houtgast 1968). In addition, the abundances of the electron donors affect the size of the Balmer discontinuity in the cooler stars. But in the hotter stars having anything like normal chemical composition, the electrons come primarily from hydrogen and the details of the chemical composition become irrelevant. An example can be seen in Böhm-Vitense and Szkody (1974).

Several investigators have made use of the Balmer jump as a gravity indicator, including Bless (1960), Searle and Sargent (1964), Newell et al. (1969), Breger and Kuhi (1970), to cite a few interesting examples.

THE HYDROGEN LINES

Pressure sensitivity in the hydrogen lines has been used as one of the classical luminosity indicators in early type stars (Abt et al. 1968, Morgan et al. 1943). Empirical calibrations were done by Petrie (1953, 1965), Petrie and Maunsell (1950), Petrie and Lee (1966), and Crawford (1973), for example. Attempts at quantitative interpretation began early in the century with the work of Hulbert (1924), Russell and Stewart (1924), and especially Elvey and Struve (1930). The more recent application of model photospheres was fostered by Underhill (1951). In the years following, the development of our understanding of the hydrogen line broadening has led to many revisions and readjustments of the deduced numbers. Some of the most up to date applications can be seen in the papers by Strom and Peterson (1968), Olson (1968, 1969, 1975), Schild et al. (1971), and Peterson and Shipman (1973).

Calculated hydrogen line profiles are shown in Fig. 16-2. The gravity dependence of the equivalent width is shown in the insert (see also Fig. 13-8b). We find the equivalent width gradient is $\partial \log W/\partial \log g \cong \frac{1}{4}$ for $T_{\text{eff}} > 10^4 °K$. That is, an error of 10% in measured equivalent width translates into a 40% error in the surface gravity. Somewhat better precision can be attained by fitting the profiles.

Dufton (1972) investigates some of the merits of Balmer lines as pressure indicators and finds g can be fixed to $\sim 25\%$. A similar precision is found by

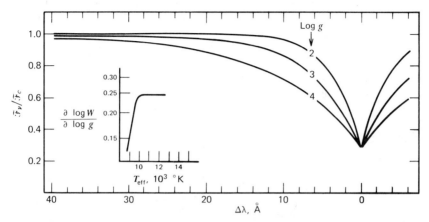

Fig. 16-2 Hγ line profiles computed from scaled $T(\tau_0)$ models having $S_0 = 1.70$. The gravity dependence is shown. The insert shows the equivalent width gravity dependence as a function of temperature. Compare with Fig. 13-8b.

Olson (1975). The quoted uncertainty refers to the internal consistency with which g can be obtained. When uncertainties in the hydrogen line absorption coefficient and the model photosphere are included, the error is much larger.

OTHER STRONG LINES

Lines like the Ca II H and K lines, Ca I $\lambda 4227$, Na I D lines, and Mg b lines show strong pressure broadened wings in the spectra of the cooler stars. These lines persist over a considerable range in temperature, which makes them useful for the task at hand. Let us consider the Na I D lines 5890 and 5896 as in Fig. 16-3. The equivalent width sensitivity can be written $\partial \log W/\partial \log g \sim \frac{1}{4}$ and it diminishes toward low surface gravities where the pressure broadening becomes smaller. From mid-G spectral type and cooler, the D line doublet shows significant blending, so it is advantageous to fit the profiles rather than the equivalent widths. The type of fit that can be made is seen in Fig. 16-4. At the low resolution (0.27 Å) of the observations nothing definitive can be said about the cores, except that they are stronger than the LTE model predicts. The wings, that is, the pressure sensitive parts, agree with the Voigt function theory to within the $\pm 2\%$ error in the measurements. The precision of the fit amounts to $\pm 20\%$ in g assuming all the other physical variables are known without error.

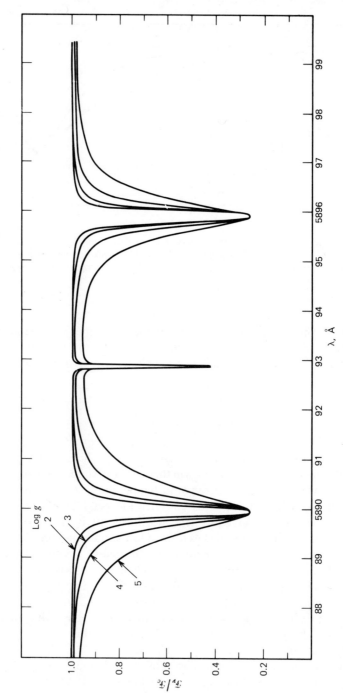

Fig. 16-3 The surface gravity dependence of the Na D lines is shown for models using the solar $T(\tau_0)$.

382

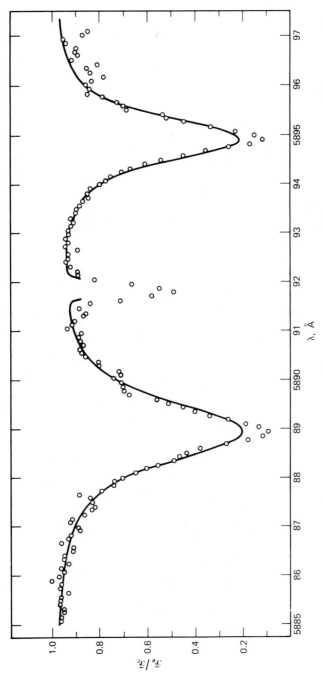

Fig. 16-4 The observed (○) D lines in γ Cep are fit with theoretical profiles (——). The computed profiles have been convolved with the instrumental profile. Some of the small depressions are real lines in the stellar spectrum. Others are telluric features that have not been fully corrected for and there is also the inevitable noise. The telluric line wavelength scale is shown.

Unfortunately temperature and the chemical composition are physical variables of some significance in this regard. For example, a few per cent change in the temperature of the model used in the analysis causes a few hundred per cent change in the deduced surface gravity. The leverage of the chemical composition is less marked, but, in view of Figs. 13-14, we see it is both the abundance of the species causing the strong line under study and the abundances of the electron donors that are involved. Unambiguous use of strong lines in a cool star requires an abundance analysis.

Examples of papers dealing with strong lines as pressure indicators are those of Vardya and Böhm (1965), where the pressure in HD 95735 has been investigated using Ca I 4227, and of Curtis and Jefferies (1967), who measured the pressure in the solar photosphere using the Na D lines. An extensive photographic investigation of K stars has been published by Cayrel de Strobel (1966) in which the Na D and Mg b lines were valuable tools. Lutz et al. (1973) have documented the pressure sensitivity of the H and K lines.

THE WEAKER LINES

The measurement of pressure by means of weak or modestly strong lines involves the comparison of two stages of ionization. Line ratios involving lines from neutral species and often unrelated ionic species are used for luminosity classification in stellar spectra. In quantitative analysis it is desirable to use lines of the same element to reduce the complications of chemical composition. But because the ionic line strength really depends on P_e and only indirectly on g (as shown in Chapter 13), the abundance of the electron donors still affects the situation. As with the strong lines, the weak lines can be used to obtain g in a rigorous way only when some details of the chemical composition are known. The equivalent widths of weak ionic lines in the cooler stars show $\partial \log W / \partial \log g \sim -0.3$ (as in Table 14-1 and Fig. 14-2). This pressure dependence enters through κ_v and so it is quite independent of the particular line and its atomic parameters.

Only a few investigations have dealt quantitatively with individual line ratios, for example, Wright and Jacobson (1966) and Oke (1959), while more frequently many lines have been used in the form of curves of growth. In the case of a solar type star, the curve of growth for the neutral species is used to fix the abundance of that element. The surface gravity of the model is then adjusted to ensure that the curve of growth for the ionic species gives the same abundance. With a precision in fitting the abscissa of the curve of growth being ~ 0.1 dex, g can be measured to about a factor of 2. Illustrations of this technique can be seen in Koelbloed (1967), Cayrel de Strobel (1968), Hardorp and Scholz (1968), Parsons (1967), and Conti and Strom (1968).

THE GRAVITY–TEMPERATURE DIAGRAM

The spectroscopic pressure measurement in virtually every case is compromised by the much greater temperature sensitivity of the same features. Further, the problem lies not only with the analysis of equivalent widths but remains even when line profiles are used. An observed line can be fit with theoretical lines based on a wide range of models in which differences in g are compensated by adjustment of T_{eff}. The sets of (g, T_{eff}) points generated in this way define a locus in the g–T_{eff} diagram, as shown in Fig. 16-5. The locus has a finite width associated with it representing the errors of the solution. It is clear, upon a moment's reflection, that uniqueness is not lost but requires additional information for its recovery. Any other spectrum feature can be used to map a second relation on the diagram that will intersect with the first relation. The point of intersection defines the unique surface gravity and temperature for the star. The more nearly orthogonal the loci, the more precisely the intersection is established.

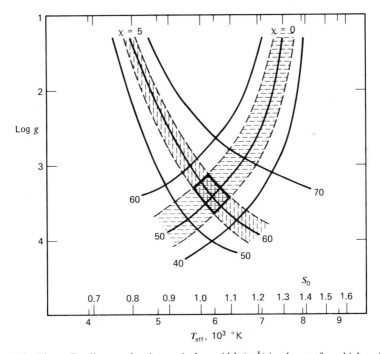

Fig. 16-5 The g–T_{eff} diagram showing equivalent width (mÅ) isochrones for a high and low excitation Fe II line. An error range is indicated that fixes the solution to the domain included in the shaded box.

   The logical next step is to incorporate a multitude of spectroscopic features, Paschen continuum, Balmer jump, hydrogen lines, strong lines, weak lines of various excitation and ionization potential. In the ideal case all of these many loci intersect at one and the same point within the error widths. Differences from the ideal case may arise from an inadequate model atmosphere, errors in the data, and so on.

   But we may not be finished with the analysis if other physical parameters have not been eliminated as variables. Such parameters as turbulence, chemical abundances, and rapid rotation can alter the isochrones in Fig. 16-5. In the most general case, each variable generates a coordinate. The concept of intersection of loci still holds but cannot be visualized geometrically unless pairs (or at most triplets) of coordinates are considered at a time. The large number of physical variables makes it difficult to distinguish between errors in the fitting of the model to the star and inadequacies in the basic model itself.

   Application of the $g-T_{\text{eff}}$ diagram can be seen in the papers by Traving (1955), Van't Veer-Menneret (1963), Mihalas (1964), Kodaira (1964), Hardorp (1966), Rodgers (1967), Newell et al. (1969), Klinglesmith et al. (1970), Peterson and Scholz (1971), Hundt (1972), and Klinglesmith (1972) among others.

THE HELIUM ABUNDANCE

The role of helium in shaping stellar spectra is in many respects equivalent to that of surface gravity. Only in the hot stars can the helium abundance be measured in the usual way. In stars cooler than A0, the helium spectrum is no longer visible. Helium in these cooler stars interacts by increasing the mean molecular weight of the atmospheric material, so the mass absorption coefficient decreases with increasing helium abundance. This results in larger pressures at a given optical depth. The approximate analytical relations are given by eq. 9-17.

   There is this uncertainty then: the helium abundance is a free parameter in cool stars that affects the spectroscopically deduced value of surface gravity. How large is the effect? We can estimate it by returning to the derivation of eq. 9-17. There we saw that to a reasonable approximation $[1 + 4A(\text{He})]$ and g behaved in a similar way. It follows that the fractional changes are related by

$$\frac{\partial g}{g} = \frac{\partial[1 + 4A(\text{He})]}{1 + 4A(\text{He})} = \frac{4 \, \partial A(\text{He})}{1 + 4A(\text{He})}.$$

If we take $A(\text{He}) = 0.10$ and the uncertainty in this value to be ± 0.05, then $dg/g = \pm14\%$ or ±0.06 in log g. In many investigations this is a negligibly small uncertainty, but in the better determined cases it may present a problem for which there appears to be no solution.

SURFACE GRAVITY FROM VISUAL BINARIES

Several visual binaries have definitive orbits leading to well determined masses when the parallax is large and well determined.* The mass can be combined with a radius determination to give g. There are only a few systems where the separation of the stellar images in the focal plane of the telescope is large enough to allow precise photometry and spectroscopy to be done on each of the components separately. Table 16-1 lists the visual binaries for

Table 16-1 Surface Gravities of Visual Binaries

Star	Sp	$B - V$	π	$g \pm$ p.e.†	log g
α CMa A	A1 V	0.01	0″.377	2.0E4 $\pm$ 13%	4.31
γ Vir A	F0 V	0.35	0.099	1.6E4 $\pm$ 15%	4.19
α CMi A	F5 IV–V	0.41	0.285	1.4E4 $\pm$ 14%	4.14
ζ Sco AB*	F5 IV	0.45	0.036	1.3E4 $\pm$ 17%	4.11
η Cas A	G0 V	0.57	0.170	2.6E4 $\pm$ 14%	4.42
HD 158614*	G8 IV–V	0.73	0.052	3.1E4 $\pm$ 25%	4.50
ζ Boo A	G8 V	0.75	0.148	2.8E4 $\pm$ 14%	4.45
70 Oph A	K0 V	0.82	0.203	2.4E4 $\pm$ 14%	4.39

* Unresolved double. The value of g assumes identical components.
† The notation p.e. stands for probable error.

which g has been measured. The radii have been found by matching the measured and model energy distributions and applying eq. 15-5. The masses are derived from Harris et al. (1963), Worth and Heintz (1974), and Worley (1963). The precision of these surface gravities is very high, $\sim \pm15\%$, making these stars good standards for model atmosphere comparisons.

THE ECLIPSING BINARY DATA

Masses and radii obtained from eclipsing spectroscopic binaries are completely independent of any stellar atmosphere analysis. No parallax measurement is needed—the absolute scale of the system is given by the velocity

* There are some systematic uncertainties even in definitive cases though. One of these is the mass ratio correction. (see Feierman 1971).

Table 16-2   Surface Gravities of Eclipsing Binaries

Star	Sp	$B - V$*	log g	Star	Sp	$B - V$*	log g
Y Cyg	09.5 V	(−0.24)	4.14	YZ Cas	A3	(0.09)	4.0
	09.5 V	(−0.24)	4.14	EE Peg	A4	(0.12)	4.15
AO Cas	09.5	(−0.24)	3.45	TX Her	A5	0.26	4.13
AH Cep	B0 V	(−0.24)	4.05		F0	0.34	4.23
VV Ori	B1 V	(−0.14)	3.95	WW Aur	A6 V	0.15	4.14
	B6	—	—		A7 V	0.20	4.12
u Her	B2 V	(−0.19)	4.00	ZZ Boo	F2 V	0.39	4.19
	B5	(−0.14)	3.55		F2 V	0.39	4.20
σ Aql	B3 V	(−0.18)	4.00	EI Cep	—	0.38	3.76
AG Per	B3	(−0.18)	4.15		—	0.34	3.93
λ Tau	B3 + A3	—	3.65	RS CVn	F4 IV	0.42	4.11
AR Cas	B3	(−0.18)	4.0		K0 IV	(0.93)	3.38
RS Vul	B4 V	(−0.16)	3.85	Z Her	F4 IV	0.47	4.12
Z Vul	B4 V	(−0.16)	3.83		K0 IV	(0.93)	3.65
	A2 III	(0.06)	4.20	VZ Hyd	F5 V	0.45	4.37
U Oph	B5 V	(−0.14)	4.10		F5 V	0.48	4.49
	B6 V	(−0.12)	4.12	HR 7484	F5 V	0.46	4.27
ζ Phe	B6 V	(−0.12)	4.16		F5 V	0.46	4.40
	A0 V	(0.00)	4.31	WZ Oph	G0 V	0.57	4.26
U Cep	B6 V	(−0.12)	4.20		G0 V	0.57	4.19
AR Aur	B8 V	−0.09	4.29	UV Leo	G0 V	0.70	4.34
	B9.5 V	−0.03	4.34		G2 V	0.70	4.34
U Sgr	B8	(−0.07)	3.8	AR Lac	G2 IV	0.60	4.20
RX Her	B9.5 V	−0.04	4.12		K0	—	3.63
	A1 V	0.02	4.20	WW Dra	G2 IV	(0.66)	3.86
v451 Oph	A0 V	0.00	4.02		K0 IV	(0.93)	3.40
	A0 V	0.06	4.12	SS Boo	G5 IV	(0.74)	3.79
δ Lib	A0 IV–V	(0.00)	3.85		G8 IV	(0.84)	3.76
AS Eri	A0 V	0.06	4.24	RZ Cnc	K1 III	(1.14)	2.85
	K0 V	(0.82)	3.14		K4 III	(1.46)	1.95
AI Dra	A1	(0.03)	4.2	ζ Aur	K4 II	(1.54)	0.95
TU Cam	A1	(0.03)	3.6		B7 V	—	—
RZ Cas	A2 V	(0.06)	4.25	31 Cyg	K4 Ib	(1.58)	0.84
β Aur	A2 V	0.04	4.05		B4 V	—	—
CM Lac	A2 V	0.10	4.31	YY Gem	M1e V	(1.46)	4.64
	A8 V	0.26	4.32		M1e V	(1.46)	4.64
v477 Cyg	A3 V	0.08	4.40				
	F5 V	0.36	4.48				

Source.   Popper (1957, 1965, 1967, 1970, 1974), Olson (1968, 1969), Snowden and Koch (1969), Barnes et al. (1968), and Harris et al. (1963).
* Values in parentheses are estimated from the spectral type.

measurements. This group of stars forms a testing ground for our spectral interpretation procedures because, as with the visual binaries, the surface gravity is known with high precision.

The spectrophotometry can be difficult because we need to isolate the light of one star from that of the other. During eclipse, when the larger star is in front, we see only the light from the larger star. Observation at other phases can be combined with eclipse measurements to obtain the spectrum from the smaller star alone. In some systems the components are identical and the problem of separating the spectra reduces to making allowance for the radial velocity shifts. The stars having dependable measurements are listed in Table 16-2.

The eclipsing binaries have been used by Olson (1975) to test the hydrogen line calculations. From a detailed comparison of hydrogen line profiles, he finds evidence that the formulation of the absorption coefficient by Edmonds et al. (1967) agrees less well with the orbital solutions than the formulations by Griem (1967) or Vidal et al. (1970).

THE VARIATION OF SURFACE GRAVITY ALONG THE MAIN SEQUENCE

The best determined g values from binary stars (Tables 16-1 and 16-2) are plotted as a function of $B - V$ in Fig. 16-6. The surface gravity is seen to vary by only a small amount along the main sequence. The straight line in the

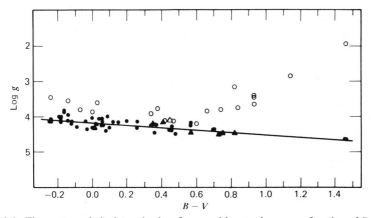

Fig. 16-6 The most precisely determined surface gravities are shown as a function of $B - V$. (Where $B - V$ was not available, the spectral type was converted to $B - V$ using Fig. 1-5.) Values for main sequence stars: (●) spectroscopic eclipsing binaries; (▲) visual binaries. The line is given by eq. 16-1.

figure has the equation

$$\log g = 4.17 + 0.38(B - V) \tag{16-1}$$

and is seen to represent the main sequence data well within the scatter. The average deviation from the line is 0.089 dex. If we assume no cosmic scatter, a probable error of 0.075 dex or $\pm 20\%$ is implied. The luminosity class IV stars are systematically lower in surface gravity by about a factor of 2.

Surface gravities obtained from eq. 16-1 are useful as starting approximations in model computations and as a reference against which spectroscopic gravity measurements can be compared.

REFERENCES

Abt, H. A., A. B. Meinel, W. W. Morgan, and J. W. Tapscott 1968. *An Atlas of Low Dispersion Grating Stellar Spectra.* Kitt Peak National Observatory, Tucson, Arizona.

Barnes, R. C., D. S. Hall, and R. H. Hardie 1968. *Publ. Astron. Soc. Pac.* **80**, 69.

Bless, R. C. 1960. *Astrophys. J.* **132**, 532.

Böhm-Vitense, E. and P. Szkody 1974. *Astrophys. J.* **193**, 607.

Breger, M. and L. V. Kuhi 1970. *Astrophys. J.* **160**. 1129.

Cayrel de Strobel, G. 1966. *Ann. Astrophys.* **29**, 413.

Cayrel de Strobel, G. 1968. *Ann. Astrophys.* **31**, 43.

Conti, P. S. and S. E. Strom 1968. *Astrophys. J.* **154**, 975.

Crawford, D. L. 1973. *Problems of Calibration of Absolute Magnitudes and Temperatures of Stars,* IAU Symposium No. 54, B. Hauck and B. E. Westerlund, Eds., Reidel, Dordrecht, Holland, p. 93.

Curtis, G. W. and J. T. Jefferies 1967. *Astrophys. J.* **150**, 1061.

Dufton, P. L. 1972. *Astron. Astrophys.* **18**, 335.

Edmonds, F. N. Jr., H. Schlüter, D. C. Wells, III 1967. *Mem. Roy. Astron. Soc.* **71**, 271.

Elvey, C. T., and O. Struve 1930. *Astrophys. J.* **72**, 277.

Feierman, B. H. 1971. *Astron. J.* **76**, 73.

Griem, H. R. 1967. *Astrophys. J.* **147**, 1092.

Hardorp, J. 1966. *Z. Astrophys.* **63**, 137.

Hardorp, J. and M. Scholz 1968. *Z. Astrophys.* **69**, 350.

Harris, D. L. III, K. Aa. Strand, and C. E. Worley 1963. *Basic Astronomical Data,* K. Aa. Strand, Ed., University of Chicago Press, Chicago, p. 273.

Houtgast, J. 1968. *Sol. Phys.* **3**, 47.

Hulbert, E. O. 1924. *Astrophys. J.* **59**, 177.

Hundt, E. 1972. *Astron. Astrophys.* **21**, 413.

Klinglesmith, D. A. III 1972. *Astrophys. J.* **171**, 79.

Klinglesmith, D. A. III, K. Hunger, R. C. Bless, and R. L. Millis 1970. *Astrophys. J.* **159**, 513.

Kodaira, K. 1964. *Z. Astrophys.* **59**, 139.

Koelbloed, D. 1967. *Astrophys. J.* **149**, 299.

Lutz, T. E., I. Furenlid, and J. H. Lutz 1973. *Astrophys. J.* **184**, 787.

Mihalas, D. 1964. *Astrophys. J.* **140**, 885.

Morgan, W. W., P. C. Keenan, and E. Kellman 1943. *An Atlas of Stellar Spectra*, University of Chicago Press, Chicago.

Newell, E. B., A. W. Rodgers, and L. Searle 1969. *Astrophys. J.* **156**, 597.

Oke, J. B. 1959. *Astrophys. J.* **130**, 487.

Olson, E. C. 1968. *Astrophys. J.* **153**, 187.

Olson, E. C. 1969. *Publ. Astron. Soc. Pac.* **81**, 97

Olson, E. C. 1975. *Astrophys. J.* Supplement **29**, 43.

Parsons, S. B. 1967. *Astrophys. J.* **150**, 263.

Peterson, D. M., and M. Scholz 1971. *Astrophys. J.* **163**, 51.

Peterson, D. M., and H. L. Shipman 1973. *Astrophys. J.* **180**, 635.

Petrie, R. M. 1953. *Publ. Dom. Astrophys. Obs. Victoria* **9**, 251.

Petrie, R. M. 1965. *Publ. Dom. Astrophys. Obs. Victoria* **12**, 317.

Petrie, R. M. and E. K. Lee 1966. *Publ. Dom. Astrophys. Obs. Victoria* **12**, 435.

Petrie, R. M. and C. D. Maunsell 1950. *Publ. Dom. Astrophys. Obs. Victoria* **8**, 253.

Popper, D. M. 1957. *J. Roy. Astron. Soc. Can.* **51**, 51.

Popper, D. M. 1965. *Astrophys. J.* **141**, 126.

Popper, D. M. 1967. *Annu. Rev. Astron. Astrophys.* **5**, 85.

Popper, D. M. 1970. *Mass Loss and Evolution in Close Binaries, IAU* Colloquium No. 6, K. Gyldenkerne and R. M. West, Eds., Copenhagen University Observatory Publication, p. 13.

Popper, D. M. 1974. *Astrophys. J.* **188**, 559.

Rodgers, A. W. 1967. *Observatory* **87**, 127.

Russell, H. N. and J. Q. Stewart 1924. *Astrophys. J.* **59**, 197.

Schild, R., D. M. Peterson, and J. B. Oke 1971. *Astrophys. J.* **166**, 95.

Searle, L. and W. L. W. Sargent 1964. *Astrophys. J.* **139**, 793.

Snowden, M. S. and R. H. Koch 1969. *Astrophys. J.* **156**, 667.

Strom, S. E. and D. M. Peterson 1968. *Astrophys. J.* **152**, 859.

Traving, G. 1955. *Z. Astrophys.* **36**, 1.

Underhill, A. B. 1951. *Publ. Dom. Astrophys. Obs. Victoria* **8**, 357 and 385.

Van't Veer-Menneret, C. 1963. *Ann. Astrophys.* **26**, 289.

Vardya, M. S. and K. H. Böhm 1965. *Mon. Not. Roy. Astron. Soc.* **131**, 89.

Vidal, C. R., J. Cooper, and E. W. Smith 1970. *J. Quant. Spectrosc. Radiat. Transfer* **10**, 1011.

Worley, C. E. 1963. *Publ. U.S. Nav. Obs.*, second series, Vol. 18, part III.

Worth, M. D. and W. D. Heintz 1974. *Astrophys. J.* **193**, 647.

Wright, K. O. and T. V. Jacobson 1966. *Publ. Dom. Astrophys. Obs. Victoria* **12**, 373.

CHAPTER SEVENTEEN

STELLAR ROTATION

The existence of stellar rotation is something we might expect based on our experience with nature. It is rare when a "free" object like a star has zero angular momentum. If the angular momentum is zero for some object, our natural inclination is to look for a braking mechanism to account for the unusual circumstance. Since we believe stars form out of clouds, and clouds collide with one another and are subject to other torques like those induced by galactic rotation, we would predict rather large stellar rotation rates on the basis of the conservation of angular momentum. The largest observed rotation rates are near the "break-up" velocity, which means that the gravitational acceleration at the equator is comparable to the centripetal force needed to retain the equatorial material as part of the star.

The sun shows rotation but it is very slow (1.9 km/sec equatorial velocity). Such slowness is typical of cooler stars. The sun also shows a small differential rotation with the equator having an angular velocity about 25% larger than higher latitudes $\sim 75°$.

The effects of rotation on the continuous spectrum are small except for rotation rates very near the break-up velocity. The spectral lines, however, are strongly changed by the relative Doppler shifts of the light coming from different parts of the stellar disc. In fact, one of the early confirmations of rotation came from eclipsing binary observations consisting of radial velocity measurements as a function of time (Rossiter 1924, McLaughlin 1924, Shajn and Struve 1929). The orbital radial velocity is expected to go through zero during eclipse. But the observed curve shows a steep rise between first and second contact and the inverse behavior between third and fourth contact. Figure 17-1 illustrates the effect. Part way through eclipse the observable portion of the star being eclipsed has a net velocity due to rotation that is away from the observer. The part of any absorption line that we observe at this phase then appears red shifted. Upon egress the blue shifted portion of the profile is seen. The resolution on the spectrograms is usually too low for changes in line shape to be discerned, but the line appears to show a shift in position that the observer translates into radial velocity.

Rotation is difficult to separate from *macro*turbulence (the term used to describe large scale gas velocities in the star's atmosphere), but modern observational technique is now reaching the precision needed to separate them. From most observations the line width can be measured but its cause cannot be unambiguously assigned to rotation or to macroturbulence. Still, the line widths of several angstroms in some stars indicate rather high velocities—up to several hundred kilometers per second. To explain such high velocities as macroturbulence seems out of the question.

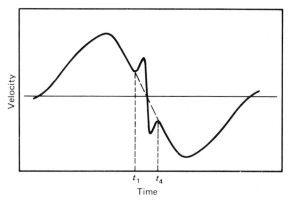

Fig. 17-1 The limb effect seen in eclipsing binaries is illustrated. First contact occurs at t_1, fourth at t_4. The anomolous behavior between t_1 and t_4 is attributed to stellar rotation.

The Doppler line broadening due to rotation depends upon the orientation of the axis of rotation relative to the line of sight. Unless there is some preferential orientation of axes, we expect a wide range in line broadening corresponding to the maximum for equator-on objects and to zero for the pole-on objects. The observations are in agreement with this except in the case of supergiants where the minimum broadening seems to be distinctly greater than zero. This minimum is interpreted as the residual macroturbulent broadening in these stars. Random orientation of rotation axes seems to be true even in star clusters.

Several catalogs of rotation velocities have been compiled. The older list of Boiarchuk and Kopylov (1958) has been used in many discussions. The

1964 catalog of Boiarchuk and Kopylov gives weighted rotation values for 2558 stars. Two newer catalogs have been compiled by Bernacca and Perinotto (1970–1971) and Uesugi and Fukuda (1970).

THE ROTATION PROFILE

The line profile we expect from a rotating star has a characteristic shape that we now show how to calculate. The star is assumed to be spherical and rotate as a rigid body. We choose the coordinate arrangement as shown in Fig. 17-2.

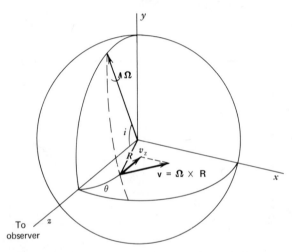

Fig. 17-2   The rotation axis of the star is inclined at an angle i to the line of sight. The y axis is chosen so Ω lies in the yz plane. For some arbitrary point on the stellar surface at an angle θ from the line of sight, the velocity is $\Omega \times R$ where R is the stellar radius to the point. The z component of velocity gives the Doppler shift.

The axis of rotation lies in the yz plane. The xy plane is perpendicular to the line of sight along the z axis. The angle of inclination, i, is measured from the z axis to the axis of rotation. We call the angular velocity of rotation Ω. Then the linear velocity of any point on the surface of the star is

$$\mathbf{v} = \Omega \times \mathbf{R}$$

where $\times$ means vector product and $\mathbf{R}$ is the radius vector as shown in Fig. 17-2. The Doppler effect arises from the z component,

$$v_z = x\Omega \sin i \qquad (17\text{-}1)$$

and

$$\Delta\lambda = \frac{v_z}{c}\lambda = \left(\frac{\lambda\Omega \sin i}{c}\right)x. \qquad (17\text{-}2)$$

This is an important relation because it shows that all elements on the stellar surface having the same x coordinate also show the same wavelength displacement, $\Delta\lambda$, due to rotation. We can then imagine the stellar disc divided into strips as shown in Fig. 17-3. $\Delta\lambda$ is constant along each strip. The largest shift occurs at the limbs where $x = R$ and has a value

$$\Delta\lambda_L = \left(\frac{\lambda\Omega \sin i}{c}\right)R$$

$$= \frac{\lambda}{c}\, v \sin i \qquad (17\text{-}3)$$

where v is the equatorial rotation velocity.

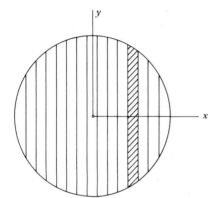

Fig. 17-3   The apparent disc of the star (z axis straight out of the page) can be thought of as a series of strips each having a unique radial velocity according to eq. 17-1.

The flux for such a star is still given by eq. 5-5,

$$\mathcal{F}_v = \oint I_v \cos\theta \, d\omega,$$

but I_v is Doppler shifted according to eq. 17-2. We take $d\omega = dA/R^2$ where dA is an increment of surface area on the star of radius R. Then the increment of area on the apparent disc of the star, using the coordinates of Fig. 17-2, is

$$dx\, dy = dA \cos\theta$$

so that

$$\mathcal{F}_v = \iint \frac{I_v \, dx \, dy}{R^2} \tag{17-4}$$

where the integration is carried over the apparent stellar disc.

The most general procedure is to calculate I_v at many points on the disc (using a model photosphere) and obtain $\mathcal{F}_v$ numerically. This scheme has the advantages that limb darkening is automatically taken into account and one can incoporate center to limb variations in the line profile. Underhill (1968), Hutchings (1972), Stoeckley and Mihalas (1973), Stoeckley and Morris (1974), and Collins (1974) have used this numerical approach.

The analytical development is less general and proceeds as follows. Let us define the line profile at any point on the disc to be

$$H(\lambda) = \frac{I_v}{I_c}, \tag{17-5}$$

that is, the ratio of intensity at any point in the spectrum to the continuum intensity. The flux profile in a nonrotating star would then be written

$$\frac{\mathcal{F}_v}{\mathcal{F}_c} = \frac{\oint H(\lambda) I_c \cos \theta \, d\omega}{\oint I_c \cos \theta \, d\omega}, \tag{17-6}$$

and if $H(\lambda)$ were independent of position on the disc, eq. 17-6 would reduce to $\mathcal{F}_v/\mathcal{F}_c = H(\lambda)$. In the case of a rotating star we have in place of eq. 17-6

$$\frac{\mathcal{F}_v}{\mathcal{F}_c} = \frac{\oint H(\lambda - \Delta\lambda) I_c \cos \theta \, d\omega}{\oint I_c \cos \theta \, d\omega}. \tag{17-7}$$

Even if $H(\lambda)$ were independent of position on the disc, $H(\lambda - \Delta\lambda)$ is not since $\Delta\lambda$ is proportional to x according to eq. 17-2. In the sun center to limb variations in $H(\lambda)$ are ~ 10–50% or $\Delta\lambda \lesssim 10^{-1}$ Å. By comparison, rotation in some stars can give line broadening as large as $\Delta\lambda \sim 5$ Å. Therefore as a first approximation we neglect the change in $H(\lambda)$ with the y coordinate and the only change we consider in the x coordinate is the translation $\Delta\lambda = \text{const} \cdot x$. With this in mind, we transform eq. 17-7 into

$$\frac{\mathcal{F}_v}{\mathcal{F}_c} = \frac{\int H(\lambda - \Delta\lambda) I_c (dx \, dy/R)}{\oint I_c \cos \theta \, d\omega}$$

$$= \frac{\int_{-\Delta\lambda_L}^{\Delta\lambda_L} H(\lambda - \Delta\lambda) \int_{-y_1}^{y_1} I_c (dy/R)(d\Delta\lambda/\Delta\lambda_L)}{\oint I_c \cos \theta \, d\omega}. \tag{17-8}$$

Equations 17-2 and 17-3 have been used to transform dx to $d\Delta\lambda$ and y_1 has a value

$$y_1 = (R^2 - x^2)^{1/2} = R\left[1 - \left(\frac{\Delta\lambda}{\Delta\lambda_L}\right)^2\right]^{1/2}.$$

Now we define $G(\Delta\lambda)$ to be*

$$G(\Delta\lambda) \equiv \frac{1}{\Delta\lambda_L} \frac{\int_{-y_1}^{y_1} I_c\, dy/R}{\oint I_c \cos\theta\, d\omega} \quad \text{for} \quad |\Delta\lambda| \le \Delta\lambda_L$$

$$\equiv 0 \quad \text{for} \quad |\Delta\lambda| > \Delta\lambda_L. \tag{17-9}$$

Then eq. 17-8 can be written

$$\frac{\mathcal{F}_v}{\mathcal{F}_c} = \int_{-\infty}^{\infty} H(\lambda - \Delta\lambda)G(\Delta\lambda)\, d\Delta\lambda$$

$$= H(\lambda)*G(\lambda). \tag{17-10}$$

This is an interesting and powerful result. It says that as long as $H(\lambda)$ has the same shape over the disc of the star, we can take the flux profile for a non-rotating star and convolve it with the "rotation profile," $G(\lambda)$, to obtain the rotationally broadened flux profile. The larger the rotation rate, the wider $G(\lambda)$ is compared to $H(\lambda)$, and the more the flux profile approaches $G(\lambda)$.

Usually $G(\lambda)$ given in eq. 17-9 can be evaluated without excessive error by assuming a limb darkening law for I_c of the form

$$I_c = I_c^0[(1 - \varepsilon) + \varepsilon \cos\theta], \quad I_c^0 = \text{constant}. \tag{17-11}$$

The limb darkening (see Fig. 9-2) coefficient, ε, is a slowly varying function of λ and is considered constant over a line profile. The analytical integration of $\oint I_c \cos\theta\, d\omega$ then gives $\pi I_c^0(1 - \varepsilon/3)$. In the numerator of $G(\lambda)$ we have

$$\int_{-y_1}^{y_1} I_c \frac{dy}{R} = 2I_c^0 \int_0^{y_1} [(1 - \varepsilon) + \varepsilon \cos\theta] \frac{dy}{R}$$

$$= \frac{2I_c^0(1 - \varepsilon)y_1}{R} + 2\varepsilon I_c^0 \int_0^{y_1} \cos\theta \frac{dy}{R}.$$

Using $\cos\theta = [R^2 - (x^2 + y^2)]^{1/2}/R$ as shown in Fig. 17-4 and the fact that $\int (A^2 - y^2)^{1/2}\, dy = \frac{1}{2}[y(A^2 - y^2)^{1/2} + A^2 \sin^{-1}(y/A)]$ we obtain

$$\int_{-y_1}^{y_1} I_c \frac{dy}{R} = 2I_c^0(1 - \varepsilon)\left[1 - \left(\frac{\Delta\lambda}{\Delta\lambda_L}\right)^2\right]^{1/2} + \frac{1}{2}\pi\varepsilon I_c^0\left[1 - \left(\frac{\Delta\lambda}{\Delta\lambda_L}\right)^2\right]^{1/2}$$

---

* $G(\Delta\lambda)$ will often be abbreviated $G(\lambda)$ when no ambiguity results.

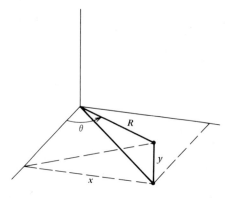

Fig. 17-4 Cosine θ can be written in terms of the x, y, and R coordinates of the point on the surface of the star.

and finally,

$$G(\Delta\lambda) = \frac{2(1 - \varepsilon)[1 - (\Delta\lambda/\Delta\lambda_L)^2]^{1/2} + \frac{1}{2}\pi\varepsilon[1 - (\Delta\lambda/\Delta\lambda_L)^2]}{\pi\Delta\lambda_L(1 - \varepsilon/3)}$$

$$= c_1\left[1 - \left(\frac{\Delta\lambda}{\Delta\lambda_L}\right)^2\right]^{1/2} + c_2\left[1 - \left(\frac{\Delta\lambda}{\Delta\lambda_L}\right)^2\right] \qquad (17\text{-}12)$$

where the constant terms have been lumped together in c_1 and c_2. Again we recall that $\Delta\lambda_L = (\lambda v \sin i)/c$. Figure 17-5 is a graph of the two parts of

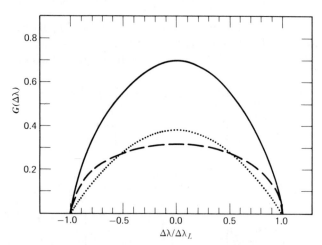

Fig. 17-5 The rotation profile $G(\Delta\lambda)$ according to the analytical solution in eq. 17-12: (-----) the first term (no limb darkening); ($\cdots\cdots$) the second term; (——) the sum for $\varepsilon = 0.6$.

$G(\Delta\lambda)$ and their sum (for $\varepsilon = 0.6$). Of course, if $\varepsilon = 0$, which means a uniformly illuminated stellar disc, the second part of $G(\Delta\lambda)$ is zero. The first term alone is the equation for an ellipse.

The theoretical line profile as a function of $v \sin i$ is shown in Fig. 17-6. Notice how the original profile becomes "washed out." The equivalent width of the line is conserved according to theorem 6 in Chapter 2. This is also clear from physical considerations. Since I_v as emitted at any spot on the stellar surface is a complete spectrum independent of the rest of the star, Doppler shifting each I_v spectrum cannot cause the total absorption in a line to increase or decrease.

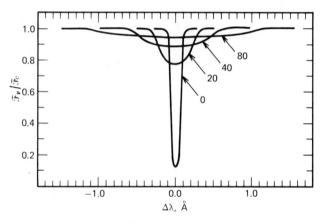

Fig. 17-6 Rotationally broadened profiles according to eqs. 17-10 and 17-12. The profiles are labeled with $v \sin i$. The wavelength is 4243 Å and the line has an equivalent width of 100 mÅ.

We also see that as $v \sin i$ becomes large such that $G(\lambda)$ is much wider than $H(\lambda)$, the assumption of $H(\lambda)$ not changing shape over the disc no longer matters. The shape of the flux profile is then $G(\lambda)$ given by eq. 17-12 essentially independent of $H(\lambda)$ and subject only to the validity of the limb darkening law assumed in eq. 17-11.

Conversely the choice of $H(\lambda)$ becomes important for the slower rotator. Sometimes it is to advantage to use for $H(\lambda)$ a profile computed from a model atmosphere, but the more common practice is to use an observed profile from a narrow lined star. The instrumental profile is automatically accounted for in such a procedure (at least approximately) provided the same spectrograph is used for the whole investigation.

MEASUREMENT OF ROTATION

The quantity $v \sin i$ can be established for a star by comparing its spectral line profiles to computed rotation profiles of the type discussed in the last section. The lines chosen should be free of strong pressure broadening and yet must be strong enough so the rotationally broadened profile can still be seen and measured. He I $\lambda 4471$ is used extensively for hot stars even though the forbidden component at $\lambda 4470$ Å is blended with it. He I $\lambda 4388$ is also used for hot stars. The Mg II $\lambda 4481$ line is nearly ideal for late B through mid-A stars as long as the rotational broadening is large compared to the 0.2 Å splitting of this doublet. In later types Fe I $\lambda 4405$, $\lambda 4473$, $\lambda 4476$, $\lambda 4549$, and Fe II $\lambda 4508$ have been measured. With the advent of red sensitive photocathodes and photographic emulsions, we can expect lines in the red to come into use, especially for cooler stars.

Detailed fitting of profiles has been done by several workers (e.g., Elvey 1929, Struve 1930, Slettebak 1954, Abt 1957). An example is shown in Fig. 17-7. $\lambda 4508$ in ν Per (F5 II) is matched with a rotation profile having $v \sin i = 46$ km/sec.

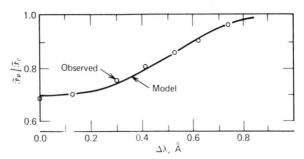

Fig. 17-7   The profile of Fe II 4508 in ν Per fit with a rotation profile having $v \sin i = 46$ km/sec. (Adapted from Abt 1957 with permission from the Astrophysical Journal, Copyright 1957 by the University of Chicago.)

Although agreement was found between theory and observation, it was soon realized that the resolution on most spectrograms was sufficient to determine only the lower Fourier components of the profile. Furthermore, $G(\lambda)$ scales with $v \sin i$ so that only one parameter *needs* to be measured. Consequently the width of the profile has come to be adopted as the single parameter used to measure $v \sin i$. The saving of labor resulting from using just the line width makes statistical studies of rotation feasible. After a series of standards has been set up, the width can be estimated by eye directly from the spectrograms using the standards for comparison. Higher precision can be obtained by measuring the half width of the line from a microphotometer tracing of the spectrogram or from photoelectric scans of the spectrum.

The stars measured by Slettebak (1954, 1955, 1956) and Slettebak and Howard (1955) are often adopted as standards. More recently Slettebak et al. (1975) have defined a standard system incorporating photoelectric line measurements. A plot of half width versus $v \sin i$ for these standards facilitates the reduction of new rotation velocities. Figure 17-8 shows such a standard curve. The measured half width is read into the curve to obtain $v \sin i$ for any nonstandard star. The standard curve becomes nonunique for very high rotation rates because the profile depends on the angle of inclination. Some restraint should be used in the interpretation of line widths as the result of rotation alone, especially if we suspect the star of showing other types of line broadening that we have not explicitly taken into account.

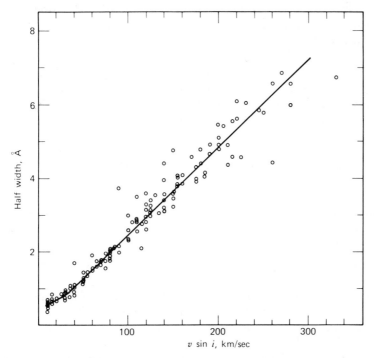

Fig. 17-8 The relation between the half width of the line profile and $v \sin i$ is sensibly linear until the instrumental profile becomes significant compared to the rotation broadening (lower left). Data from Slettebak et al. (1975).

Carroll noticed some time ago (Carroll 1933a, 1933b) that the Fourier transform analysis would be useful in view of eq. 17-10. He obtained the following expression for the transform of $G(\lambda)$ when $\varepsilon = 0.6$,

$$g(\sigma) = \frac{J_1(\Delta\lambda_L\sigma)}{\Delta\lambda_L\sigma} - \frac{3\cos\Delta\lambda_L\sigma}{2(\Delta\lambda_L\sigma)^2} + \frac{3\sin\Delta\lambda_L\sigma}{2(\Delta\lambda_L\sigma)^3} \qquad (17\text{-}13)$$

in which J_1 stands for the first order Bessell function. The important point he made use of is that $g(\sigma)$ has zero amplitude at certain Fourier frequencies given by

$$\Delta\lambda_L\sigma_1 = 0.660$$
$$\Delta\lambda_L\sigma_3 = 1.162$$
$$\Delta\lambda_L\sigma_3 = 1.661 \qquad (17\text{-}14)$$
$$\vdots \qquad \vdots$$

The multiplication of $g(\sigma)$ by the transform of $H(\lambda)$ (according to the transform of eq. 17-10) may add zeros but will not change the position of the zeroes at $\sigma_1, \sigma_2, \ldots$ given above. Furthermore, if the rotation is reasonably large, the first two or three zeroes of $g(\sigma)$ may occur at σ values well below the cutoff resulting from the transform of $H(\lambda)$. Clearly then, the plan of Carroll was to find $\sigma_1, \sigma_2, \ldots$ from the transform of the observed profile and by using eqs. 17-14 find $\Delta\lambda_L$, which is equivalent to $v \sin i$ via eq. 17-3.

Carroll's method was applied to i Ori, ζ Per, λ Ori, and β Per by Carroll and Ingram (1933) and to α Boo by Wilson (1969). The advantages of the method are difficult to obtain in practice without good photometric precision.

FOURIER ANALYSIS OF THE ROTATION PROFILE

High photometric precision can now be obtained using photoelectric line scanners and it is natural to extend Carroll's method from a location of zeroes into a fitting of $g(\sigma)$ to the transform of the observed line profile. Figure 17-9 shows the important sidelobe character in the transform of $G(\lambda)$. The first sidelobe has an amplitude of about one-tenth of the main lobe and is negative in sign. The higher order sidelobes alternate in sign and diminish in amplitude as shown in the figure. The scale of the rotation profile is proportional to $v \sin i$ so of course the scale of $g(\sigma)$ is inversely proportional to $v \sin i$. When $v \sin i$ is large, there is some advantage in using $\log \sigma$ as abscissa. Then $g(\sigma)$ has a unique shape for all $v \sin i$ not near the break-up value and can be matched to data plotted on the same coordinates by simple translations in both directions. If we chose to plot $g(\sigma)$ for $v \sin i = 1$ km/sec, the $\log \sigma$ difference between $g(\sigma)$ and the data gives $\log (v \sin i)$ directly. In Fig. 17-10

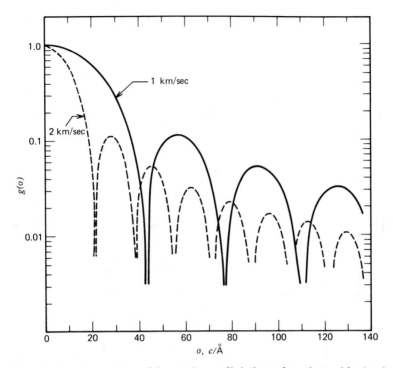

Fig. 17-9 The Fourier transform of the rotation profile is drawn for $v \sin i = 1$ km/sec (——) and $v \sin i = 2$ km/sec (-----).

the data of α Peg is analyzed in this way. A $v \sin i$ of 133 km/sec is indicated. Inclusion of the thermal profile of $\lambda 4481$ with $g(\sigma)$ results in the dashed curve.

Identification of the sidelobes in $g(\sigma)$ can be used to distinguish rotational from macroturbulent broadening—a point we consider in more detail in Chapter 18. And it is for those cases where rotation does not completely dominate the broadening that the Fourier analysis is of particular value.

Let us consider the combination of rotation with the thermal (plus microturbulence) profile as shown in Fig. 17-11. When $v \sin i$ is small (2 km/sec example), the rotation profile transform filters the thermal profile giving the curve labeled A. But when $v \sin i$ is larger (16 km/sec example), it is the thermal profile that filters $g(\sigma)$ to give the curve labeled B. In either case we can reconstruct the filtered function if we know the shape of the filter. The reconstructed function can then be analyzed to obtain additional information about the star. Specifically, the other line shaping mechanisms can be studied.

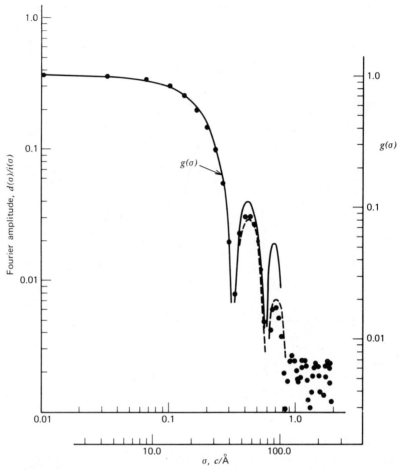

Fig. 17-10 A log–log plot of $g(\sigma)$ is matched to the data (points) in the transform of $\lambda4481$ (α Peg) by translation in both coordinates. The left coordinate grid is for the data. The right coordinate grid is for $g(\sigma)$ having $v \sin i = 1$ km/sec. The abscissas differ by a factor of $133 = v \sin i$ for α Peg. Including the thermal profile improves the fit in the side lobes (----).

    There is also a less obvious point that enhances the value of the Fourier analysis. The noise distribution has a relation to the signal that is different from that of the λ domain. Typically the noise is white. Then, because the rotation profile is characterized by certain frequencies that are much stronger than others (i.e., the central lobe and sidelobe structure), the signal to noise ratio becomes high at these frequencies. In other words, the "characteristics" of the rotation profile are concentrated to certain bands of σ, making them easier to identify.

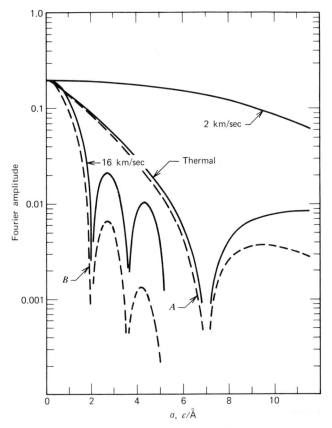

Fig. 17-11 The thermal profile is combined with the two rotation profiles having 2 and 16 km/sec for $v \sin i$. The results are the dashed curves labeled A and B.

We started out to apply Fourier analysis to measure the rotation of a star. What has emerged is not only a precise method for obtaining $v \sin i$ but a more general philosophy for dissecting a spectral line according to its broadening mechanisms. These thoughts are taken up again in the next chapter. Now we turn our attention to the general and statistical results of the rotation measurements.

STATISTICAL CORRECTION FOR AXIAL PROJECTION

Evidence for random orientation of rotation axes is found in the lack of correlation between $v \sin i$ and galactic latitude or longitude and in the frequency distribution of $v \sin i$ for homogeneous groups of stars. Under the assumption of random orientation, it is possible to convert the average $v \sin i$ characterizing a group of stars to an average rotation velocity $\bar{v}$.

Let us consider Fig. 17-12 where it is shown that all orientations of axes with the angle i relative to the line of sight are represented by the shaded ring. The fraction of randomly oriented axes between i and $i + di$ is proportional to the solid angle subtended by the ring,

$$P(i) \, di = 2\pi \sin i \, di.$$

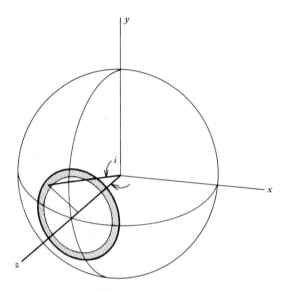

Fig. 17-12 For randomly oriented rotation axes, the fraction at any angle of inclination i is proportional to the area of the shaded ring.

The average $\sin i$ is then given by

$$\langle \sin i \rangle = \frac{\int_0^{\pi/2} P(i) \sin i \, di}{\int_0^{\pi/2} P(i) \, di} = \frac{\pi}{4}.$$

This statistical result can be utilized to convert, for example, the average measured $v \sin i$ for A0V to A2V stars to an average rotational velocity $\bar{v}$, but in the remainder of the chapter we continue to use the uncorrected values, $v \sin i$, leaving the removal of $\langle \sin i \rangle$ to the reader should he find it desirable to do so. A more complete discussion of statistical corrections is given by Bernacca and Perinotto (1974).

OBSERVED ROTATION OF MAIN SEQUENCE STARS

The behavior of rotation along the main sequence is remarkable. Figure 17-13 shows the variation as a function of spectral type. The rotation is large for early type stars and drops rapidly in the region of F stars and finally reaches values below the resolution of most investigations about G0. The circles are the means of $v \sin i$ in the spectral interval. The dispersion about the mean is indicated by the vertical bars. The smooth curve denotes the general run of the observations and no one knows if the deviations of individual points from the mean trend are significant.

This behavior is not clearly understood, although most workers in the area seem to favor the braking mechanism of Schatzman (1962). He proposes a dissipation of angular momentum by a coupling between a star's mass ejection and its magnetic field. Presumably cooler main sequence stars have stronger stellar winds and therefore stronger rotational braking (see also Weber and Davis 1967).

Two other interesting phenomena show up beginning at about ~F2 spectral type and may be related to the rotation behavior. One is the occurrence of convection in the envelopes of stars cooler than F2, which may be the cause of the stellar wind. The other is the start of the chromospheric activity as seen in the Ca II H and K emission cores. Wilson (1966)

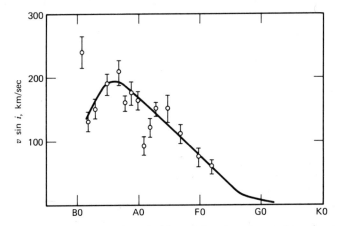

Fig. 17-13 Average observed rotation velocities as a function of spectral type for single main sequence field stars (data points from Abt 1970 based on the data of Slettabak and Howard 1955). The error bars indicate the standard deviation of the measurements about the mean. The solid line represents the overall trend indicated by the points and is similar to the smooth relation given by Abt and Hunter (1962).

has given strong evidence that the HK emission is stronger in younger stars. Kraft (1967) has shown the rotation velocity to be correlated with the H and K emission. This implies that for stars cooler than F2 the rate of rotation diminishes with age. The time scale is of the order of 10^9 years or less (Skumanich 1972, Blanco et al. 1974).

OBSERVED ROTATION ABOVE THE MAIN SEQUENCE

The observations of Slettebak (1955), Herbig and Spalding (1955), and others gives some indication of the rotation rate of field stars above the main sequence. The average rotation velocity for luminosity classes III and IV is shown in Fig. 17-14. We see that for the stars cooler than about A3 the giant stars show higher rotation than luminosity classes V stars. This can be qualitatively understood in terms of a stellar evolution phenomenon. The rapid rotators from the hot end of the main sequence evolve to luminosity III–IV in later spectral types. To be sure, the increase in radius as the star moves off the main sequence reduces its rotation rate, but the drop in rotation *along* the main sequence is so steep that the evolving star still has the larger rotation even with its reduced rotation rate. Several angular momentum calculations have been done including those of Sandage (1955), Abt (1957, 1958), and Oke and Greenstein (1954).

Rotation of higher luminosity stars is not as well determined. The measurements do indicate considerably smaller average velocities than obtained for

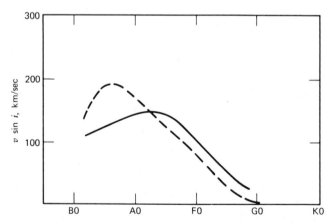

Fig. 17-14   The behavior of rotation in luminosity classes III and IV (———) based on the work of Slettebak (1955, 1970). For comparison the main sequence relation of Fig. 17-11 is shown (----).

the main sequence, with individual studies done by Slettebak (1956), Oke and Greenstein (1954), Abt (1957, 1958), Kraft (1969), and Huang and Struve (1953).

ROTATION IN OPEN STAR CLUSTERS

The rotational velocity distribution varies widely from one cluster to another. Some show higher rotation than the field stars while others show the opposite. No clear picture of the age sequence has yet emerged and while the effects of evolution must certainly be taken into account, they seem insufficient to explain all the observations.

Most clusters and associations that have been measured show a sharp rise at the start of the distribution, reach a maximum, and then decline smoothly on the faint end. There are also exceptions to this general behavior as seen in Fig. 17-15. This figure is a composite of many people's data as compiled by Kraft (1970). The spectral types of many of the cluster stars are unknown so the absolute magnitude has been chosen as the abscissa. The data generally include all luminosity classes lumped together. On the faint end, the field star distribution seems to be a lower bound for the cluster distributions. Evidence for this point has also been put forward by Bernacca and Perinotto (1974).

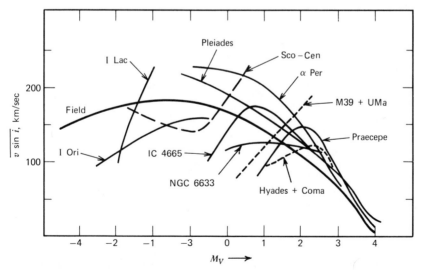

Fig. 17-15   Rotation velocity distributions are shown for several clusters (adapted from Kraft 1970 with permission of the University of California Press).

The frequency of binaries in a cluster is inversely related to the mean rotation velocities of the cluster stars (Abt 1970). The frequency of occurrence of Ap stars in clusters may also cause anomalies because the Ap stars show slower rotation than normal stars (e.g., Conti 1965, Abt et al. 1972).

RAPID ROTATION

In the B stars the most rapid rotators are found to come close to or possibly reach the theoretical break-up velocities. The relations according to Slettebak are shown in Fig. 17-16. Hardorp and Strittmatter (1968) and Stoeckley (1968a) have shown how more detailed theoretical line profile calculations lead one to deduce somewhat large rotation velocities from the observed profiles. It is logical then to associate the existence of emission line B stars with rapid rotation. Mass is presumably leaving the star near the equator and

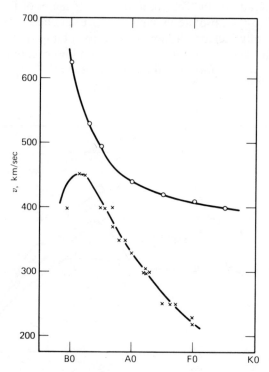

Fig. 17-16 The fastest rotators are shown by ×. It is assumed that sin $i \doteq 1$ for these stars. The theoretical breakup velocities (top curve) approach the observed relation most closely in the B star range (data from Slettabak 1966).

is seen as a "disc" around the star. Several models have been proposed by McLaughlin (1961), Limber (1970), Huang (1972), and Marlborough (1969). Reviews on Be stars are given by Underhill (1960, 1961).

Changes in the continuous spectrum have been computed by Collins (1966), Faulkner et al. (1968), and Sweet and Roy (1953). When a star rotates very rapidly it loses its spherical shape and the effects of so-called gravity darkening (von Zeipel 1924, Eddington 1926) become important. Such observational parameters as $B - V$, M_v, T_{eff}, and so on become a function of the aspect angle, i. Scatter in the zero age main sequence may in part be due to rotation as discussed by Kraft and Wrubel (1965) and Roxburgh and Strittmatter (1965a, 1965b, 1966). Observational evidence of these effects is still inconclusive (see Bless and Code 1972). The reader may also find useful the review by Strittmatter (1969).

The effects of rapid rotation on spectral absorption lines has been considered by Stoeckley (1968a, 1968b), Collins (1968, 1974), and Hardorp and Strittmatter (1968). Stoeckley has analyzed photoelectric data for five rapid rotators and has shown how v and i can be separated. Very precise measurements are needed in order to obtain definitive conclusions of this type.

ROTATION IN BINARY STARS

Angular momentum problems take on special interest when dealing with binary systems where the orbital parameters can be considered along with the rotation. McNally (1965) was the first to point out that the angular velocities, as opposed to linear velocities, show a maximum at spectral type A5. Figure 17-17 shows the situation. The angular velocity falls rapidly on both sides of A5. Plavec (1970) has carried the result one step farther by noting that the observed average period of rotation and the shortest possible orbital period (where both stars fill the critical Roche lobe) are very nearly

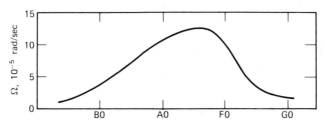

Fig. 17-17   The angular velocity is seen to reach a maximum in mid-A spectral type (data from McNally 1965).

Table 17-1 A Comparison of Rotational with Orbital Periods

Spectral Type	P_{rot}	P_{orbit} (minimum)
O5	4.8	4.0
B0	1.9	1.8
B5	1.0	1.1
A0	0.73	0.78
A5	0.56	0.53
F0	0.73	0.40
F5	2.42	0.39
G0	4.55	0.36

identical for A5 stars and earlier (see Table 17-1). No hypothesis has yet been advanced to explain this relationship.

The question of rotation and orbital synchronism has been looked at theoretically by several investigators, some of the most recent being Huang (1966), Zahn (1966) and Kopal (1968). Synchronism is expected as a result of tidal interaction. The observations of eclipsing binaries by Koch et al. (1965), Olson (1968), and Plavec (1970) show that synchronous rotation does hold for most systems, but some individual stars appear to rotate faster than the synchronized rate. The mean rotation rates of these binary components is lower than main sequence field stars as illustrated in Fig. 17-18.

The substantially lower rate of rotation seen for these binaries is probably a result of the tidal braking induced upon each component by the other.

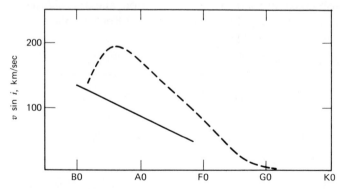

Fig. 17-18 The rotation velocities of eclipsing binaries (———) according to Plavec (1970) are substantially slower than normal main sequence stars (-----).

A detailed study of 116 visual binary systems, where the separation between the components is large, shows normal rotational behavior for their position in the HR diagram (Slettebak 1963, Weis 1974).

CONCLUSIONS

Many photographic line profile observations, as described in this chapter, have given us the basis for the overview just presented. Many details remain to be fitted into the picture. Photoelectrically measured profiles that are now becoming available will allow not only greater precision and resolution in measuring $v \sin i$ but may also allow us to measure differential rotation, to separate i from v for the rapid rotators, and to distinguish rotation from turbulence. In the next chapter, we consider turbulence and some methods for disentangling turbulence from rotation.

REFERENCES

Abt, H. A. 1957. *Astrophys. J.* **126**, 503.

Abt, H. A. 1958. *Astrophys. J.* **127**, 658.

Abt, H. A. 1970. *Stellar Rotation*, IAU Colloquium, Columbus, Ohio 1969, A. Slettebak, Ed., Reidel, Dordrecht, Holland, p. 193.

Abt, H. A., F. H. Chaffee, and G. Suffolk 1972. *Astrophys. J.* **175**, 779.

Abt, H. A. and J. H. Hunter 1962. *Astrophys. J.* **136**, 381.

Bernacca, P. L. and M. Perinotto 1970–1971. *Contributi dell'Osservatorio Astrofisico dell' Universita di Padova in Asiago*, Nos. 239 and 249.

Bernacca, P. L. and M. Perinotto 1974. *Astron. Astrophys.* **33**, 443.

Blanco, C., S. Catalano, E. Marilli, and M. Rodono 1974. *Astron. Astrophys.* **33**, 257.

Bless, R. C. and A. D. Code 1972. *Annu. Rev. Astron. Astrophys.* **10**, 208.

Boiarchuk, A. A. and I. M. Kopylov 1958. *Sov. Astron. AJ*, **2**, 752.

Boiarchuk, A. A. and I. M. Kopylov 1964. *Crimean Astrophys. Obs. Publ.* **31**, 44.

Carroll, J. A. 1933a. *Mon. Not. Roy. Astron. Soc.* **93**, 478.

Carroll, J. A. 1933b. *Mon. Not. Roy. Astron. Soc.* **93**, 680.

Carroll, J. A. and L. J. Ingram 1933. *Mon. Not. Roy. Astron. Soc.* **93**, 508.

Collins, G. W. II 1966. *Astrophys. J.* **146**, 914.

Collins, G. W. II 1968. *Astrophys. J.* **151**, 217.

Collins, G. W. II 1974. *Astrophys. J.* **191**, 157.

Conti, P. 1965. *Astrophys. J.* **142**, 1594.

Eddington, A. S. 1926. *The Internal Constitution of the Stars*, Dover, New York, p. 282.

Elvey, C. T. 1929. *Astrophys. J.* **70**, 141.

Faulkner, J., I. Roxburgh, and P. A. Strittmatter 1968. *Astrophys. J.* **151**, 203.

Hardorp, J. and P. Strittmatter 1968. *Astrophys. J.* **153**, 465.

Herbig, G. H. and J. F. Spalding, Jr. 1955. *Astrophys. J.* **121**, 118.

Huang, S.-S. 1966. *Annu. Rev. Astron. Astrophys.* **4**, 35.

Huang, S.-S. 1972. *Astrophys. J.* **171**, 549.

Huang, S.-S. and O. Struve 1953. *Astrophys. J.* **118**, 463.

Hutchings, J. B. 1972. *Publ. Dom. Astrophys. Obs. Victoria* **14**, 59.

Koch, R. H., E. C. Olson, and K. M. Yoss 1965. *Astrophys. J.* **141**, 955.

Kopal, Z. 1968. *Astrophys. Space Sci.* **1**, 179, 284, 411.

Kraft, R. P. 1967. *Astrophys. J.* **150**, 551.

Kraft, R. P. 1969. *Stellar Astronomy*, H.-Y. Chiu, R. L. Warasila, and J. L. Remo, Eds., Gordon and Breach, New York, p. 317.

Kraft, R. P. 1970. *Spectroscopic Astrophysics*, G. H. Herbig, Ed., University of California Press, Berkeley, Calif., p. 385.

Kraft, R. P. and M. H. Wrubel 1965. *Astrophys. J.* **142**, 703.

Limber, D. N. 1970. *Stellar Rotation*, IAU Colloquium, Columbus, Ohio 1969, A. Slettebak, Ed., Dordrecht, Holland, p. 274.

Marlborough, J. M. 1969. *Astrophys. J.* **156**, 135.

McLaughlin, D. B. 1924. *Astrophys. J.* **60**, 22.

McLaughlin, D. B. 1961. *J. Roy. Astron. Soc. Can.* **55**, 13 and 73.

McNally, D. 1965. *Observatory* **85**, 166.

Oke, J. B. and J. L. Greenstein 1954. *Astrophys. J.* **120**, 384.

Olson, E. C. 1968. *Publ. Astron. Soc. Pac.* **80**, 185.

Plavec, M. 1970. *Stellar Rotation*, IAU Colloquium, Columbus, Ohio 1969, A. Slettebak, Ed., Reidel, Dordrecht, Holland, p. 133.

Rossiter, R. A. 1924, *Astrophys. J.* **60**, 15.

Roxburgh, I. and P. Strittmatter 1965a. *Mon. Not. Roy. Astron. Soc.* **133**, 1.

Roxburgh, I. and P. Strittmatter 1965b. *Z. Astrophys.* **63**, 15.

Roxburgh, I. and P. Strittmatter 1966. *Mon. Not. Roy. Astron. Soc.* **133**, 345.

Sandage, A. 1955. *Astrophys. J.* **122**, 263.

Schatzman, E. 1962. *Ann. Astrophys.* **25**, 18.

Shajn, G. and O. Struve 1929. *Mon. Not. Roy. Astron. Soc.* **89**, 222.

Skumanich, A. 1972. *Astrophys. J.* **171**, 565.

Slettebak, A. 1954. *Astrophys. J.* **119**, 146.

Slettebak, A. 1955. *Astrophys. J.* **121**, 653.

Slettebak, A. 1956. *Astrophys. J.* **124**, 173.

Slettebak, A. 1963. *Astrophys. J.* **138**, 118.

Slettebak, A. 1966. *Astrophys. J.* **145**, 126.

Slettebak, A. 1970. *Stellar Rotation*, IAU Colloquium, Columbus, Ohio, 1969, A. Slettebak, Ed., Reidel, Dordrecht, Holland, p. 3.

Slettebak, A., G. W. Collins, P. B. Boyce, and N. M. White 1975. *Astrophys. J.* (in press).

Slettebak, A. and R. F. Howard 1955. *Astrophys. J.* **121**, 102.

Stoeckley, T. R. 1968a. *Mon. Not. Roy. Astron. Soc.* **140**, 121.

Stoeckley, T. R. 1968b. *Mon. Not. Roy. Astron. Soc.* **140**, 141.

Stoeckley, T. R. and D. Mihalas 1973. *Boulder National Center for Atmospheric Research Publication NCAR-TN/STR-84* (June) or Publication No. 32, Astronomy Department, Michigan State University.

Stoeckley, T. R. and C. S. Morris 1974. *Astrophys. J.* **188**, 579.

Strittmatter, P. 1969. *Annu. Rev. Astron. Astrophys.* **7**, 681.

Struve, O. 1930. *Astrophys. J.* **72**, 1.

Sweet, P. A. and A. E. Roy 1953. *Mon. Not. Roy. Astron. Soc.* **113**, 701.

Uesugi, A. and I. Fukuda 1970. *Contrib. Inst. Astrophys. Kwasan Obs. Univ. Kyoto*, No. 189.

Underhill, A. B. 1960. *Stellar Atmospheres*, J. L. Greenstein, Ed., University of Chicago Press, Chicago, p. 411.

Underhill, A. B. 1961. *Nuovo Cimento Suppl.* **22**, Series 10, 69.

Underhill, A. B. 1968. *Bull. Astron. Inst. Neth.* **19**, 526.

von Zeipel, H. 1924. *Mon. Not. Roy. Astron. Soc.* **84**, 665.

Weber, E. J. and L. Davis, Jr. 1967. *Astrophys. J.* **148**, 217.

Weis, E. W. 1974. *Astrophys. J.* **190**, 331.

Wilson, A. 1969. *Mon. Not. Roy. Astron. Soc.* **144**, 325.

Wilson, O. 1966. *Astrophys. J.* **144**, 695.

Zahn, J.-P. 1966. *Ann. Astrophys.* **29**, 313, 489, 565.

CHAPTER EIGHTEEN

TURBULENCE IN STELLAR PHOTOSPHERES

Motions of the photospheric gas that are on a large scale compared to atomic dimensions but on a small scale compared to the size of the star are called turbulence (the term being introduced by Rosseland in 1928). Turbulence alters the line profiles in a stellar spectrum through the Doppler displacement. We have little or no evidence concerning turbulence other than this line of sight component of the motion as seen averaged over the disc of the star. As a direct result of this limitation, uniqueness in interpretation of the line profile may be impossible to establish. Nevertheless, with reasonable assumptions it is possible to obtain *some* information on the velocity distribution of the photospheric material.

We would also like to know the geometry of the motion, whether we are dealing with "bubbles" of material, local density fluctuation caused by eddies, or wave motion in the material, or perhaps prominence-like structures in the atmosphere. Knowledge of this geometry is necessary if we are to construct meaningful models of the photosphere that take into account turbulence. If the geometry is not known, the transfer equation cannot be solved, that is, model atmosphere construction is stopped at the fundamental level. At the present state of development, the sun is the only star for which some idea of the geometry can be formulated. In dealing with *stellar* spectra, it has been necessary to make asymptotic type approximations: (1) the size of the turbulent element is small compared to the mean free path of the photon or (2) the size of the turbulent element is large compared to the mean free path of the photon. These limiting cases go by the names microturbulence and macroturbulence, respectively. If we assume one or both of these approximations is applicable to the star we wish to study, then the geometry transfer equation problem can be handled and we can proceed with an analysis to obtain the velocity distributions or, in less favorable cases, the characteristic widths of the velocity distributions.

Whether or not any of the velocity fields in stellar atmospheres is related to the turbulence studied in aerodynamics is still an open question. The appearance of aerodynamic turbulence depends on the so-called Reynold's number being greater than $\sim 10^3$ where the Reynold's number is defined by the product of density, average flow velocity, and linear dimension across the tube containing the flow divided by the viscosity. If one can rightfully assume the solar granulation size can be associated with the tube dimension, then the Reynold's number calculated for the solar photosphere is $> 10^{11}$ (Minnaert 1953) and the condition for turbulent flow is definitely met. Additional references dealing with aerodynamic turbulence in the astronomical context are Chandrasekhar (1949, 1952, 1955), Tandberg-Hanssen (1967), Huang and Struve (1960), and especially Thomas (1961).

In the material presented in this chapter, we neglect any change in ionization, excitation, or temperature that might be caused by the turbulent velocities. The increase in pressure caused by collision between microturbulent cells is $\frac{1}{2}\rho v^2$ (ρ is the gas density, v the mean turbulence velocity) and in main sequence stars it is usually small compared to the gas pressure (see Vernazza et al. 1973, for example). In giant stars it could be much more important.

Several modes of macromotion are unambiguously observed in the sun. There is also evidence for macroturbulence in other stars. On the other hand, some investigators have strongly criticized the concept of microturbulence— not only as turbulence but the very existence of small scale velocities. Worrall and Wilson (1972, 1973) summarize many of the uncertainties in the area. They particularly emphasize the ad hoc manner in which microturbulence is introduced to explain the enhanced equivalent widths on the flat part of the curve of growth or the center to limb variations in solar line profiles. Wilson and Guidry (1974) have even shown how inhomogeneous thermal structure may explain some of the observed line broadening. But still, the undisputed existence of chromospheres (in the sun and probably in stars as well judging by the H and K line emission cores) and large scale turbulence are evidence for nonthermal motion, some of which is likely to appear on a small geometrical scale.

Part of the study of turbulence deals with the cause of the turbulence and the energy content and transfer of the turbulence field (see Hern 1974, for instance). Although the evidence is partially circumstantial, one can put forward a simple picture where convection in the stellar envelope induces turbulent motions in the overlying photosphere (Schwartzschild 1961, Frazier 1966). These motions generate mechanical waves that propagate outward, turning into shock waves at sufficiently low densities. The energy released in the shock heats the high atmosphere causing the temperature inversion. The H and K line emission cores may be coupled to convection and turbulent motion in this way. Further, mass loss by means of stellar winds may be caused by or enhanced by turbulence.

FROM VELOCITY TO SPECTRUM

The turbulence velocity distribution must first be projected onto the line of sight to obtain the distribution of Doppler shifts. The observed Doppler shifts are spacially averaged over the disc of the star and temporally averaged over the observing time.

Let us consider the simple model of radial prominence-type motion. It is clear that at the limb ($\theta = 90°$) no Doppler shift is introduced, while at the disc center ($\theta = 0°$) the full velocity enters. For purposes of illustration, suppose we specify the velocity distribution to be

$$N(v)\, dv = \frac{1}{\sqrt{\pi}\, v_0}\, e^{-(v/v_0)^2}\, dv \tag{18-1}$$

where $N(v)\, dv$ is the fraction of material having velocities in the range v to $v + dv$ and for this model v is allowed only along stellar radii. We assume $N(v)$ is statistically uniform over the stellar surface and applies to the material in any small portion of the surface.

We take the projection of all velocities onto the line of sight and write

$$\Delta\lambda = \frac{\lambda}{c}\, v \cos\theta$$

$$\beta \equiv \frac{\lambda}{c}\, v_0 \cos\theta \equiv \beta_0 \cos\theta.$$

Then

$$N(\Delta\lambda)\, d\Delta\lambda = \frac{1}{\sqrt{\pi}\, \beta}\, e^{-(\Delta\lambda/\beta)^2}\, d\Delta\lambda$$

$$= \frac{1}{\sqrt{\pi}\, \beta_0 \cos\theta}\, e^{-(\Delta\lambda/\beta_0 \cos\theta)^2}\, d\Delta\lambda \tag{18-2}$$

follows directly from the velocity distribution. Notice that β, the width parameter, is a function of θ, while β_0 is a constant. At the center of the disc, $N(\Delta\lambda)$ reflects $N(v)$ directly, but away from the center the Doppler distribution becomes narrower. At the limb $N(\Delta\lambda)$ is a δ function. This example illustrates the general way in which $N(v)$ and $N(\Delta\lambda)$ are related.

The isotropic case comes up so frequently that it is appropriate to mention it at this stage. In such a case the distributions $N(v)$ and $N(\Delta\lambda)$ are identical because there is no θ dependence by definition. A frequently made assumption is that the velocity distribution is Gaussian. This assumption may not be true, but as yet the observational evidence about velocity distributions is so meager that no decision can be made against it.

MICROTURBULENCE IN LINE COMPUTATIONS

When the line of sight penetrates through many cells of motion in the atmosphere, the velocity distribution of the cells molds the line profile in the same way the particle velocity distribution does. This is the reason for defining microturbulence as having element sizes that are small compared to the mean free path of a photon. In Chapter 11 we handled the particle velocity distribution by incorporating it into the line absorption coefficient. We now let α' represent the line absorption coefficient without micro-turbulence (eq. 11-55). Then the absorption coefficient broadened by micro-turbulence can be written as the convolution

$$\alpha = \alpha' * N(\Delta\lambda) \tag{18-3}$$

and incorporated directly into the photospheric flux calculations using eq. 13-6.

Should $N(v)$ be an isotropic Gaussian function of the form

$$N(v)\,dv = \frac{1}{\sqrt{\pi}\,\xi}\,e^{-(v/\xi)^2}\,dv,$$

then the convolution of $N(v)$ with α' can be included by a trivial modification of the thermal broadening part of α' which is also Gaussian (eq. 11-48). The convolution of Gaussians is a new Gaussian with a dispersion parameter given by

$$v^2 = v_0^2 + \xi^2.$$

Therefore we simply write

$$\Delta\lambda_D = \frac{\lambda}{c}\left(\frac{2kT}{m} + \xi^2\right)^{1/2}$$

$$\Delta v_D = \frac{v}{c}\left(\frac{2kT}{m} + \xi^2\right)^{1/2}$$

(18-4)

in place of eq 11-50 or 11-51 and automatically include this special type of isotropic turbulence. In practice this procedure has been used universally (Struve and Elvey, 1934, originally introduced this formulation) and the parameter ξ is referred to as *the* microturbulence. The name classical microturbulence has also been assigned to this case especially if ξ is independent of depth in the atmosphere. The extension to $\xi = \xi(\tau)$, that is, a depth dependent microturbulence, has been invoked to explain the dependence of ξ on excitation potential. The observations are discussed later in the chapter.

A second extension of classical microturbulence, made with solar studies in mind, was advanced by Allen (1949), developed by Waddell (1958), and incorporated into a multitude of recent papers. It is the hypothesis of non-isotropic microturbulence. Led by the granulation flow patterns, one postulates separate radial and tangential velocity distributions given by

$$N(v_r)\,dv_r = \frac{1}{\xi_r\sqrt{\pi}}\,e^{-(v_r/\xi_r)^2}\,dv_r \quad \text{and} \quad N(v_t) = \frac{1}{\xi_t\sqrt{\pi}}\,e^{-(v_t/\xi_t)^2}\,dv_t.$$

Then, following the same path that led to eq. 18-2, the wavelength distributions are

$$N_r(\Delta\lambda)\,d\Delta\lambda = \frac{1}{\xi_r'\sqrt{\pi}\cos\theta}\,e^{-(\Delta\lambda/\xi_r'\cos\theta)^2}\,d\Delta\lambda$$

(18-5)

and

$$N_t(\Delta\lambda)\,d\Delta\lambda = \frac{1}{\xi_t'\sqrt{\pi}\sin\theta}\,e^{-(\Delta\lambda/\xi_t'\sin\theta)^2}\,d\Delta\lambda$$

(18-6)

in which $\xi_r' = \xi_r\lambda/c$ and $\xi_t' = \xi_t\lambda/c$. If we allow each element in the radial stream to possess the full distribution of tangential velocities (or vice versa) then the complete wavelength distribution is given by

$$N(\Delta\lambda)\,d\Delta\lambda = N_r(\Delta\lambda) * N_t(\Delta\lambda)\,d\Delta\lambda$$

$$\equiv \frac{1}{\xi'\sqrt{\pi}}\,e^{-(\Delta\lambda/\xi')^2}\,d\Delta\lambda$$

(18-7)

where $\xi' = \xi\lambda/c$ and

$$\xi^2 = \xi_r^2 \cos^2 \theta + \xi_t^2 \sin \theta$$

or since $\sin^2 \theta = 1 - \cos^2 \theta$,

$$\xi^2 = \xi_t^2 - (\xi_t^2 - \xi_r^2) \cos^2 \theta. \tag{18-8}$$

This equation is then combined with eq. 18-4, making α in eq. 18-3 a function of θ. Line profiles can be computed for various positions on the solar disc and comparisons made with observations.

An alternative anisotropic model is to assume the radial and tangential distributions do not mix but rather the turbulent streaming is either radial *or* tangential, depending on where the material is relative to the granulation cell pattern. Then in place of eq. 18-7 the distribution is

$$N(\Delta\lambda)\, d\Delta\lambda = [a_r N_r(\Delta\lambda) + a_t N_t(\Delta\lambda)]\, d\Delta\lambda \tag{18-9}$$

where a_r and a_t are the fractions that specify how much material is in each mode of micromotion. This microturbulence can be convolved with α in the usual way. Since there is a summation in $N(\Delta\lambda)$, α is the sum of "radial" and "tangential" absorption coefficients, which again make α a function of θ.

The anisotropic model leading to eq. 18-8 has been applied with some success in explaining the center to limb variations of solar profiles. It has also generated considerable controversy. No one has yet applied such a model to other stars. Without the center to limb observations, it is difficult to establish values for more than one microturbulence parameter.

MACROTURBULENCE IN LINE COMPUTATIONS

When a physical portion of the stellar atmosphere is large enough so that photons remain in the same moving mass element from the time they are created until they escape from the star, we have the case of macroturbulence. In order to have line broadening by macroturbulence, the individual spectra from many of these elements must be viewed simultaneously. This is exactly the case for a star where the light from the full stellar disc is seen integrated together. Each macroelement gives a complete spectrum that is shifted by the Doppler shift corresponding to the velocity of the element. In this respect macroturbulence and rotation are similar. In the solar spectrum, if a long slit is used, one can see the Doppler displacements of the individual macroelements as illustrated in Fig. 18-1. The spectrum in Fig. 18-1 is really a composite of many individual spectra stacked one above the other according to the projection of the macroelements along the spectrograph slit. Measurements of these "wiggly lines" as they are called lead to interesting results as discussed later in the chapter.

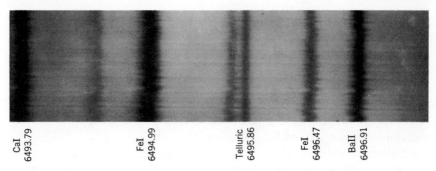

CaI
6493.79

FeI
6494.99

Telluric
6495.86

FeI
6496.47

BaII
6496.91

Fig. 18-1   Wiggly lines in the solar spectrum. (Courtesy of McMath-Hulbert Observatory.)

Let us imagine the surface of a star subdivided into domains within which there are many macroelements. We can specify the velocity distribution in a general way by defining $\Theta(v)\, dv$ to be the fraction of the emitting area in the velocity range v to $v + dv$. In the usual way we expect the intensity spectrum coming from each of the small portions on the disc to be given by

$$I_v = I_v^0 * \Theta(\Delta\lambda)$$

in which I_v^0 is the intensity spectrum without macroturbulence and where $\Theta(v)$ has been converted to $\Theta(\Delta\lambda)$ as discussed above. The equivalent width of the line is conserved since Θ is normalized to unit area. The linear width of the line grows at the expense of the depth.

The flux, being the integral of I_v over the disc, is then given by

$$\mathcal{F}_v = \oint I_v^0 * \Theta(\Delta\lambda) \cos\theta \, d\omega, \qquad (18\text{-}10)$$

which is eq. 5-5 adapted to the present case. Equation 18-10 is very general and is the way in which stellar line profiles can be computed.

The isotropic case is especially useful in understanding some of the methods of analysis described later in the chapter and is a first order model in its own right. If $\Theta(\Delta\lambda)$ is isotropic, there is no dependence on ω and eq. 18-10 becomes

$$\mathcal{F}_v = \Theta(\Delta\lambda) * \oint I_v^0 \cos\theta \, d\omega. \qquad (18\text{-}11)$$

This result is very interesting because it says that the effect of isotropic macroturbulence on a stellar spectrum can be simulated by convolving the macroturbulent velocity distribution, Θ, with the flux spectrum of a star without macroturbulence. This behavior for macroturbulence is reminiscent of the convolution derived for rotation (eq. 17-10).

The line profiles expressed in I_ν^0 may have microturbulence broadening already, so eq. 18-10 or 18-11 can be used to compute the spectrum for a model atmosphere having both large scale and small scale turbulence.

Beyond the isotropic model one might again postulate radial and tangential macroturbulence as being more in accord with the solar granulation flow—or perhaps pure radial flow if one imagines a star covered with "prominences" that rise and fall. These and other models can be incorporated through eq. 18-10. An example is given at the end of the next section.

THE AFFECT OF TURBULENCE ON LINE PROFILES

Turbulent motions always broaden the spectral lines. Consider the flux integral $\mathcal{F}_\nu = \oint I_\nu \cos \theta \, d\omega$ for the case of microturbulence. For the weak line I_ν mimics the line absorption coefficient (by arguments similar to those leading to eq. 13-19) which, as we see in eq. 18-3, is a convolution between the microturbulence velocity distribution and the thermal profile. For the weak line then, the flux profile has a form identical to eq. 18-10. We conclude that macroturbulence cannot be distinguished from microturbulence by studying weak lines. The effect of all dimensional scales of turbulence on the weak line profile is the same.

This conclusion makes sense from another point of view as well. If the line is weak the line opacity if small in all parts of the profile, which implies that the mean free path of the photon is comparable to the thickness of the line forming region of the photosphere. In such a case *all* the turbulent elements must be smaller than the photon mean free path length. Each absorbed photon exhibits the Doppler shift for the turbulence velocity of the material where it is absorbed. If we compare two photons that differ by $\Delta\lambda$ in the profile, we cannot tell if they were affected by different turbulent elements along the same line of sight (micro) or if they were affected by different turbulent elements at different spots on the stellar disc (macro). In fact, if both macro- and microturbulence are active in the star, then, from eq. 18-3, the weak line form of eq. 18-10 can be written

$$\mathcal{F}_\nu = \text{const} \oint \Theta(\Delta\lambda) * N(\Delta\lambda) * \alpha \cos \theta \, d\omega.$$

Any velocity distribution or turbulence broadening parameter that we determine from a weak line refers to the combined effects of macro- and microturbulence.

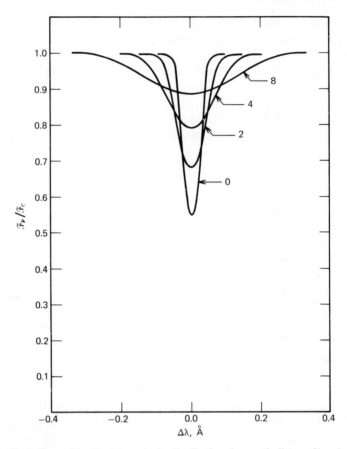

Fig. 18-2 Turbulence with a Gaussian velocity distribution changes the line profile as indicated. Velocity dispersions are shown in kilometers per second.

Figure 18-2 shows theoretical weak line profiles for turbulence having an isotropic Gaussian velocity distribution. The profiles become broader and shallower with increasing turbulence in keeping with the expected behavior of convolutions. If we write the Doppler profile as (eq. 11-53)

$$\alpha = \frac{\sqrt{\pi}\,e^2}{mc}\,f\,\frac{\lambda^2}{c}\,\frac{1}{\Delta\lambda_D}\,e^{-(\Delta\lambda/\Delta\lambda_D)^2}$$

and then compute $d\alpha/d\Delta\lambda_D$, we find

$$\frac{d\alpha}{d\Delta\lambda_D} = \frac{\sqrt{\pi}\,e^2}{mc^2}\,f\lambda^2\left(\frac{1}{\Delta\lambda_D}\right)^2\left[2\left(\frac{\Delta\lambda}{\Delta\lambda_D}\right)^2 - 1\right]e^{-(\Delta\lambda/\Delta\lambda_D)^2}$$

which is positive for $\Delta\lambda > \Delta\lambda_D/\sqrt{2}$ and negative for $\Delta\lambda < \Delta\lambda_D/\sqrt{2}$. Thus, if we may use the simple relation in eq. 18-4, we can expect α to increase with ξ for $\Delta\lambda > \Delta\lambda_D/\sqrt{2}$ but decrease with ξ for $\Delta\lambda < \Delta\lambda_D/\sqrt{2}$. The equivalent width of the weak lines is conserved.

The situation for stronger lines is quite different and it is by the study of stronger lines that stellar micro- and macroturbulence can be separated. The broadening of the line caused by microturbulence results from the broadening of α. Absorption can then occur over a greater wavelength range because of the turbulence induced Doppler shifts. If a line is not weak, then by definition saturation effects are present. Broadening α reduces the amount of saturation, allowing the equivalent width to increase. We have already seen this effect in the curves of growth of Figs. 14-4 and 14-8. Figure 18-3 shows the theoretical behavior in the profile.

Small values of ξ cause a squaring up of the profile. This is shown in Fig. 18-3 by the profiles labeled 0 km/sec and 2 km/sec. Further increase in ξ results in complete desaturation and a profile reflecting the velocity distribution (8 km/sec). A nonisotropic or non-Gaussian velocity distribution would produce a different profile but the general desaturation behavior would still go on.

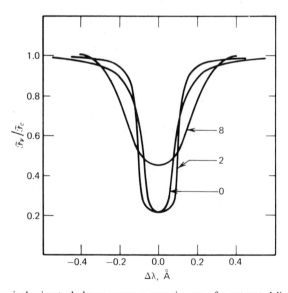

Fig. 18-3 Classical microturbulence causes a squaring-up of a saturated line profile. Large values of microturbulence may desaturate the line. ξ is in kilometers per second. (Adapted from Gray 1973 with permission from the Astrophysical Journal, Copyright 1973 by the University of Chicago.)

The change in profile with *macro*turbulence depends exclusively on the velocity distribution as seen integrated over the stellar disc according to eq. 18-10. The only general statement one can make is that a stellar spectrum in which all the lines are shallow is indicative of macroturbulence or rotation. Weak lines and strong lines alike are broadened and shallowed in the same way. The effect of isotropic macroturbulence on the line profiles is described by eq. 18-11. It is obvious that the profiles reflect the shape of $\Theta(\Delta\lambda)$. But let us consider the anisotropic model for macroturbulence. Suppose we postulate pure radial and tangential motion with a Gaussian distribution of macro-velocities (as in eq. 18-9 but now for macroturbulence). Then $\Theta(\Delta\lambda)$ is given by

$$\Theta(\Delta\lambda) = A_T \Theta_T(\Delta\lambda) + A_R \Theta_R(\Delta\lambda)$$

$$= \frac{A_T}{\sqrt{\pi}\,\zeta_T \sin\theta} e^{-(\Delta\lambda/\zeta_T \sin\theta)^2} + \frac{A_R}{\sqrt{\pi}\,\zeta_R \cos\theta} e^{-(\Delta\lambda/\zeta_R \cos\theta)^2} \qquad (18\text{-}12)$$

where A_T and A_R are the fractional areas having tangential and radial motions. From eq. 18-10 we have

$$\mathcal{F}_\nu = 2\pi A_T \int_0^{\pi/2} \Theta_T * I_\nu \sin\theta \cos\theta \, d\theta + 2\pi A_R \int_0^{\pi/2} \Theta_R * I_\nu \sin\theta \cos\theta \, d\theta.$$

$$(18\text{-}13)$$

It would be possible to include differences in temperature between tangential and radial flows and so on by computing I_ν appropriately, but instead let us keep the simplest case where even center to limb changes are ignored. Then I_ν can be factored out giving

$$\mathcal{F}_\nu = 2\pi A_T I_\nu * \int_0^{\pi/2} \Theta_T \sin\theta \cos\theta \, d\theta + 2\pi A_R I_\nu * \int_0^{\pi/2} \Theta_R \sin\theta \cos\theta \, d\theta.$$

In this way we are led to evaluate the integrals

$$\frac{1}{\sqrt{\pi}\zeta_T} \int_0^{\pi/2} e^{-(\Delta\lambda/\zeta_T \sin\theta)^2} \cos\theta \, d\theta \quad \text{and} \quad \frac{1}{\sqrt{\pi}\zeta_R} \int_0^{\pi/2} e^{-(\Delta\lambda/\zeta_R \cos\theta)^2} \sin\theta \, d\theta.$$

By substitution of $u = (\zeta_T \sin\theta)/\Delta\lambda$ in the first integral and $u = (\zeta_R \cos\theta)/\Delta\lambda$ in the second, we obtain

$$\mathcal{F}_\nu = 2\pi A_T I_\nu * \frac{\Delta\lambda}{\sqrt{\pi\zeta_T^2}} \int_0^{\zeta_T/\Delta\lambda} e^{-1/u^2} \, du + 2\pi A_R I_\nu * \frac{\Delta\lambda}{\sqrt{\pi\zeta_R^2}} \int_0^{\zeta_R/\Delta\lambda} e^{-1/u^2} \, du.$$

$$(18\text{-}14)$$

In other words, the radial and tangential components behave the same way. Figure 18-4 shows the shape of the profile given by the above integrals. Notice how much more pointed the profile is compared to a Gaussian. This

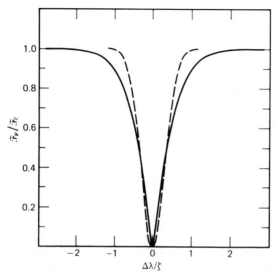

Fig. 18-4   The profile for the radial–tangential model (——) is more pointed than a Gaussian (-----) of the same half width.

triangulation results from the radial streaming or the tangential streaming or their sum.

Several other velocity distributions can be proposed, but we shall not go beyond the previous examples. The reader is referred to the investigation of Huang and Struve (1953) in which several other simple velocity distributions are considered. The effect of sound waves on line profiles has been considered by Eriksen and Maltby (1967).

TECHNIQUES FOR MEASURING TURBULENCE

Microturbulence has been measured for many stars in the sense that the parameter ξ in eq. 18-4 has been estimated from a fit of measured equivalent widths to theoretical curves of growth. Baschek and Reimers (1969), Chaffee (1970), Smith (1971), and Andersen (1973) have studied the behavior of ξ along the main sequence using the curve of growth as their tool. But usually ξ is a by-product of an abundance analysis. In most of these cases no attempt is made at finding the velocity distribution, and the simple isotropic, homogeneous, Gaussian microturbulence is assumed in computing the theoretical curves of growth.

There are several reasons to doubt the validity of the measured micro-turbulent ξ values. These reasons all have one common point—the height of the saturation portion of the curve of growth depends upon more than the microturbulence. There are at least the other desaturating mechanisms of Zeeman splitting and hyperfine structure. In addition, f value errors, the temperature distribution, damping due to Stark and van der Waals inter-actions, and non-LTE effects change the flat part of the curve of growth. (Additional discussion can be seen in Gray and Evans 1973, in Holweger 1973, and in Smith 1974.) These mechanisms need to be carefully accounted for if the curve of growth ξ values are to be taken seriously. A study of curve growth derived ξ's may be, in reality, a study of temperature gradients as they differ from star to star!

With these ambiguities in mind, refer back to Figs. 13-17, 14-4, and 14-8 for quantitative examples. The precision in reading ξ from a curve of growth fit is about $\frac{1}{2}$ km/sec.

Line profile measurements can potentially be used to detect and measure microturbulence in accordance with the results portrayed in Fig. 18-3. Several investigators have attempted to use the half width as a measure of the turbulent broadening, for example, Smith (1973), Day et al. (1973), Bonsack and Culver (1966), van den Heuvel (1963), Huang (1952), and Huang and

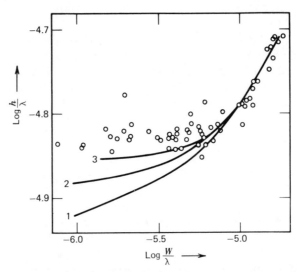

Fig. 18-5 The curve of line width. The half width is denoted by h (ordinate), the equivalent width by W (abscissa). Solid lines are models with Gaussian macro- (ζ) and micro- (ξ) turbulence: (1) $\xi = 1.8$, $\zeta = 0.0$, (2) $\xi = 0.0$, $\zeta = 2.0$, (3) $\xi = 1.5$, $\zeta = 1.5$. (Adapted from Elste 1967 with permission from the Astrophysical Journal, Copyright 1967 by the University of Chicago.)

Struve (1952). One tool for handling the half widths is the curve of line width (first presented by Huang 1952) which consists of a plot of half width as a function of equivalent width. For weak lines, where macroturbulence is indistinguishable from microturbulence, one expects the line to grow primarily in depth as the equivalent width increases, making the half width nearly independent of equivalent width. With the onset of saturation the growth in equivalent width depends more and more on the half width of the line, making the two widths approach a direct proportionality relation. In principle this characteristic shape can be used to obtain a value for ξ and a velocity parameter for macroturbulence as well. One example of this process as applied to the sun by Elste (1967) is shown in Fig. 18-5. Elste finds $\xi = 1.5$ and a Gaussian macroturbulence dispersion of $\zeta = 1.5$ km/sec. As Elste points out, and as can be seen from Fig. 18-3, the half width is a poor choice of parameter because its sensitivity to microturbulence is not large. The widths at three-quarter and one-quarter depths have been proposed as more sensitive parameters. Successful application to stellar spectra has yet to be made (see Slettebak 1956, and Elste 1967).

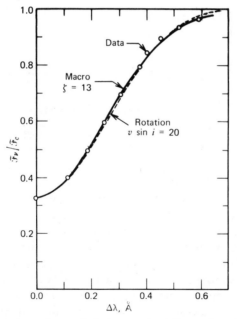

Fig. 18-6 The comparison of macroturbulence and rotation as hypotheses to explain the observed line profiles (○). Both theoretical profiles include 6.4 km/sec microturbulence. (From Abt 1958 with permission from the Astrophysical Journal, Copyright 1958 by the University of Chicago.)

The only way *macro*turbulence can be detected is from the profiles. Furthermore, even at relatively high resolution (~ 0.1 Å) only the lower Fourier components of the profile are not attenuated by the spectrograph. As a result, rotational broadening and macroturbulent broadening may not be separable one from the other. Slettebak (1956), Abt (1958), and Kraft (1969) have emphasized this point. Figure 18-6 shows the calculations of Abt comparing radial Gaussian macroturbulence and a rotationally broadened profile. The profiles have similar widths when the macroturbulent Gaussian dispersion is equal to $\frac{2}{3}v \sin i$. The ambiguity is compounded by the modest photometric accuracy afforded by the photographic process.

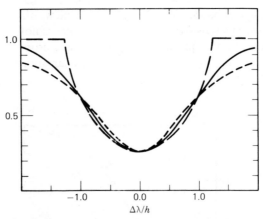

Fig. 18-7 Curves: (– – –) $G(\lambda)$; (———) Gaussian; (- - - -) dispersion profile. If it were not for the blurring of other line broadening mechanisms, accurate measurement would allow them to be distinguished from each other.

In Fig. 18-7 we see a comparison of a rotation profile, a Gaussian profile, and a dispersion profile all normalized to the same half width and central depth. It is clear that with photometric precision of $\sim 1\%$, as attainable with photoelectric line scanners, it would be possible to distinguish between them. The clear differences we see in Fig. 18-7 can become smaller as the blurring of the thermal profile, microturbulence, and the spectrograph all take their toll. We might also expect macroturbulence and rotation to occur together in the same star, further compromising the uniqueness of the shapes.

A far more comprehensive approach than half or three-quarter widths can be used to exploit the profile boxiness induced by microturbulence and make use of the small differences between rotation and macroturbulence. That approach is to extend the Fourier analysis begun in the last chapter.

FOURIER ANALYSIS FOR MACROTURBULENCE

Let us start with eq. 18-11 where we have an explicit convolution between the macroturbulence velocity distribution and the flux profile. In the Fourier domain we have the products of the transforms that give rise to filtering with close analogy to the rotation analysis of the previous chapter. But in place of $g(\sigma)$ we have the transform of $\Theta(\Delta\lambda)$ and because the macroturbulence

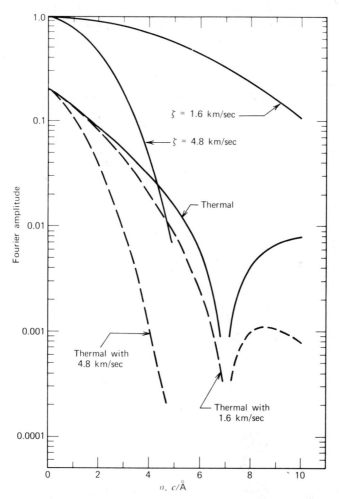

Fig. 18-8 The filtering of the thermal profile on $\zeta = 1.6$ and 4.8 km/sec Gaussian dispersion macroturbulence is shown. The dashed curves are the result. (Compare Fig. 17-11.)

velocities are typically 10^{-1} or 10^{-2} as large as the rotation velocities we looked at in Chapter 17, the interaction of the broadening mechanisms is more balanced with none of them in complete domination.

Let us consider the example in Fig. 18-8. Although an isotropic Gaussian macroturbulence of dispersion ζ is used in the figure, the general behavior remains the same regardless of the form of $\Theta(\Delta\lambda)$. We think of the thermal profile transform as the filter. When the characteristic width of $\Theta(\Delta\lambda)$ exceeds the width of the thermal profile (the $\zeta = 4.8$ km/sec curve), we can reconstruct the macroturbulence transform with the usual division by the filter. We then have obtained an important result, namely, the transform of $\Theta(\Delta\lambda)$ and hence $\Theta(\Delta\lambda)$. We have measured the macroturbulence velocity distribution.

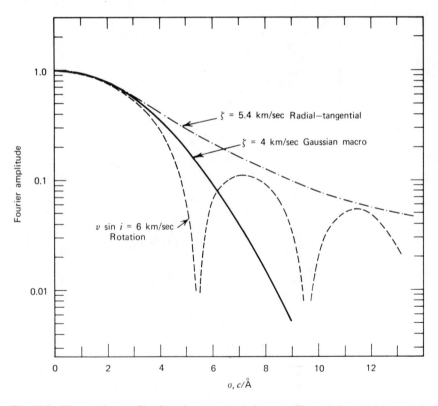

Fig. 18-9 The rotation profile, Gaussian macroturbulence profile, and the radial-tangential profile transforms are compared. Differences occur only at the higher Fourier frequencies ($\sigma \gtrsim 3$ $c/\text{Å}$ in this example).

The amount of filtering by the thermal profile differs from star to star and in some cases we can expect the high frequency components to be lost from $\Theta(\Delta\lambda)$. In the most severe situations (like the $\zeta = 1.6$ km/sec case in Fig. 18-8), it is possible to obtain only a characteristic width for $\Theta(\Delta\lambda)$.

Should the line broadening arise from rotation rather than macroturbulence, the transform will show sidelobes as in Figs. 17-9, 17-10, and 17-11. Very few macroturbulence velocity distributions produce such strong sidelobes and in this way macroturbulence can be distinguished from rotation. In Fig. 18-9 we compare the rotation and Gaussian macroturbulence transforms for $\zeta = \frac{2}{3} v \sin i$. Contrary to the analysis of the profiles in the λ domain, there is little similarity in the transforms. Even the main lobe shows a characteristic difference. At the same time it can readily be seen how low spectral resolution filters out the distinguishing sidelobes. The additional filtering of a thermal (plus microturbulent) profile could further reduce the high frequency amplitudes. This explains explicitly the reason for the macroturbulence–rotation ambiguity in the λ domain.

Now let us turn to some real data. The $\lambda 4481$ profile of α Cyg is analyzed in Fig. 18-10. The dots show the data after removal of the instrumental profile. The triangles denote the residual transform after removal of the thermal plus microturbulence profile. Rotation does not explain the residual transform because the main lobe has the wrong shape. But a radial tangential (eq. 18-14) macroturbulence model with $\zeta = 40$ km/sec fits very closely the residual transform from the main lobe. (See Gray 1975 for additional information.) The thermal plus microturbulence sidelobe structure is not visible because of the filtering by the macroturbulence.

The situation for β Ori is more interesting (Fig. 18-11). Symbols like those of Fig. 18-10 are used to denote the data transform and the first residual after removal of the thermal profile. Both the isotropic Gaussian and the radial-tangential macroturbulence fail to explain the situation (triangles). Rotation fairs better. By using the *first zero* in the data transform, we position $g(\sigma)$ and find $v \sin i = 42$ km/sec. Dividing $g(\sigma)$ into the first residual transform results in the second residual transform indicated by the $\times$'s. This is the macroturbulence transform. The macroturbulence is adequately represented by an isotropic Gaussian function with $\zeta = 17$ km/sec or the radial tangential model with $\zeta = 23$ km/sec. We cannot distinguish between them.

The macroturbulence analysis is particularly simple for the isotropic case described by eq. 18-11 and other forms such as eq. 18-14 where an explicit convolution occurs. In the examples of α Cyg and β Ori we find evidence that these are reasonable approximations. In more complicated situations recourse must be made to eq. 18-10. Then we can no longer derive $\Theta(\Delta\lambda)$ directly, but instead models must be constructed so that the convolution $\Theta(\Delta\lambda) * I_\nu$ can be done before integration over the visible disc.

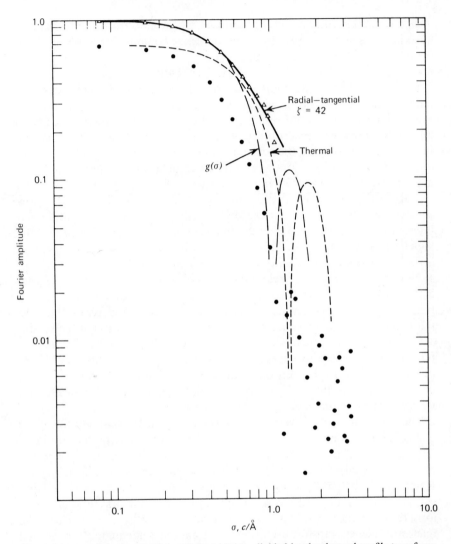

Fig. 18-10 (●) Data for λ4481 in α Cyg; (△) data divided by the thermal profile transform. Rotation (−−−) and radial-tangential macroturbulence (——) are shown fitted to the data. The latter works; the former does not.

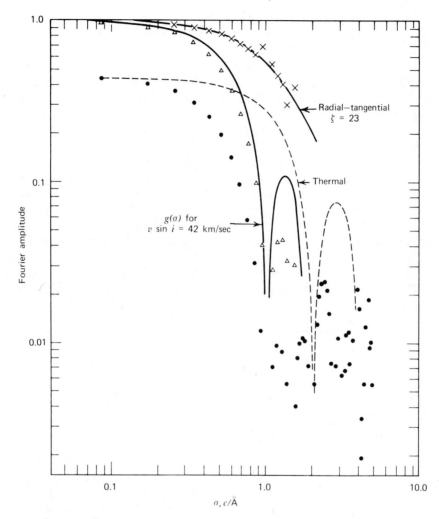

Fig. 18-11 (●) Data for $\lambda 4481$ in β Ori; (△) data after removal of the thermal profile. The rotation profile transform is positioned *by the zero*, making $g(\sigma)$ slightly above and to the right of the triangles. Removal of rotation changes each triangle into the × above it. The ×'s are the transform of $\Theta(\Delta\lambda)$ and are well represented by eq. 18-12 with dispersion $\zeta = 23$ km/sec.

FOURIER ANALYSIS FOR MICROTURBULENCE

Fourier analysis for microturbulence can best be done on saturated lines and must generally rely heavily on models because the convolution with $N(\Delta\lambda)$ is buried in the flux profile computation. Weak lines offer a simpler analysis and under the assumption of an isotropic $N(v)$, we can deduce $N(\Delta\lambda)$ the same way we did for macroturbulence. The only serious problem is that we cannot know that we have not measured the macroturbulence instead. In saturated lines we find the characteristics introduced by microturbulence and that cannot be due to macroturbulence or rotation. These characteristics are seen, for the special case of classical microturbulence, in Fig. 18-3 and result in the interesting transforms shown in Fig. 18-12. The boxiness (so obvious in the $\xi = 2$ km/sec profile) results in a strengthening of the sidelobes. A microturbulence large enough to desaturate the line (8 km/sec in the example of Fig. 18-3) causes a return toward a Gaussian line shape and the corresponding reduction in the sidelobes. It is clear that for any given star, one must seek out a saturated line of the proper strength to maximize the sidelobe effect. Should there arise an ambiguity between the sidelobes of the micro-

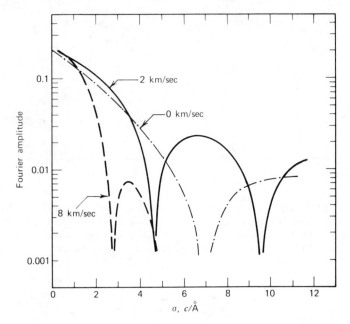

Fig. 18-12 Shown are the Fourier transforms for the profiles in Fig. 18-3. The $\xi = 2$ km/sec transform shows large sidelobes. (From Gray 1973.) (Adapted from Gray 1973 with permission from the Astrophysical Journal, Copyright 1973 by the University of Chicago.)

turbulence profile and the rotation profile, the behavior of the sidelobe as a function of lines of different strength could be used to resolve it. The sidelobe pattern for rotation is the same for all lines in a stellar spectrum while those for microturbulence are not.

A somewhat different approach was followed by Gray (1973) where, because of filtering of the instrumental profile, all sidelobe information was unavailable. The main lobe analysis resulted in values of ξ under the assumption of classical microturbulence. Figure 18-13 shows the data and the division by the instrumental profile transform for $\lambda 6065$ as measured in μ Her A.

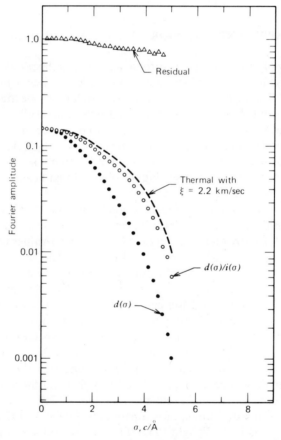

Fig. 18-13 $\lambda 6065$ in μ Her A is analyzed: ($\bullet$) the data, $d(\sigma)$, ($\bigcirc$) after removal of the instrumental profile; (-----) a thermal plus 2.2 km/sec classical microturbulence. The residual transform ($\triangle$) is the top of the main lobe of a rotation or macroturbulence transform. It corresponds to $v \sin i = 2.9$ km/sec. (Adapted from Gray 1973 with permission from the Astrophysical Journal, Copyright 1973 by the University of Chicago.)

The maximum σ attainable, beyond which the signal is lost in the noise, is about $5\ c/\text{Å}$. At the top of Fig. 18-13 we see the data after division by a theoretical profile transform having a thermal component for S_0 in eq. 9-5 of 0.87 ($T_{\text{eff}} \cong 5000°\text{K}$) and a classical microturbulent component with $\xi = 2.2$ km/sec. The residual transform can be interpreted as the main lobe of the transform of $\Theta(\Delta\lambda)$ or of $G(\Delta\lambda)$ or their product.* The value of ξ is chosen by trial and error with the condition that the residual transform must be reasonable. Values of ξ significantly different from the value shown result in residual transforms that are unphysical and show values in excess of unity (which implies negative areas of emission).

THE APPARENT BEHAVIOR OF ξ IN THE HR DIAGRAM

The summary of results on stellar turbulence, about to be presented, must be treated with some reserve. Uncertainties in models of stellar photospheres, the physics of line formation used in the analysis, limited spectral resolution, and so on all contribute to the tentative nature of the results.

The thermal velocities range from about 1 to 6 km/sec in cool stars (iron to hydrogen) and from about 2 to 16 km/sec in hot stars. The velocity of sound is ~ 6 km/sec in cool stars and ~ 16 km/sec in hot stars. These values should be kept in mind when assessing the physical significance of the microturbulence to be discussed.

Glebocki (1973) has summarized the available measurements of the microturbulence parameter ξ as derived from curve of growth analyses (375 stars; see his paper for individual references). He finds mean values of ξ for F, G, and K main sequence stars as shown in Fig. 18-14. The error bars indicate the 30% uncertainty he estimated from the internal consistency of the data. It seems fair to conclude that $\xi \sim 2$ km/sec independent of position along the main sequence. Baschek and Reimers (1969), Chaffee (1970), Smith (1971), and Andersen (1973) also investigated ξ along the main sequence and each found a slightly different behavior. In the very early spectral types there is some evidence that higher values of ξ exist. Kamp (1973), for example, finds 15 km/sec is typical of the O stars τ Sco, 10 Lac, and HD 57862.

Above the main sequence ξ remains relatively small for most stars, starting at about 2 km/sec for the main sequence, staying ~ 2 km/sec for giants, but then increasing to 6 or 7 km/sec for supergiants (Glebocki 1973, Buscombe et al. 1972, Parsons 1967, Wright 1955). These higher velocities are probably at or near the sonic value. As with the main sequence stars, there is some

---

* Much higher spectral resolution would be needed to distinguish between them. The filtering of the thermal–microturbulence profile is so strong that it may in fact be impossible to do so for this star.

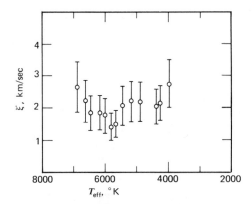

Fig. 18-14   Microturbulence along the main sequence according to data compiled by Glebocki (1973).

indication that ξ may be larger in hotter supergiants. Some remarkably large values have been reported. For instance, van Helden (1972) found $\xi = 24$ km/sec for the B3 Ia star o^2 CMa, while Voigt (1952) obtained 30 km/sec in 55 Cyg (B3 Ia also).

The value of ξ in the sun is not much better determined than in some stars. As early as 1937 Allen came to the conclusion that the extra width in the solar line profiles could be accounted for with a microturbulence having $\xi = 1.6$ km/sec. More recently Cowley and Cowley (1964) have obtained a value of 1.4 km/sec from a curve of growth analysis. Elste (1967) obtained a value near 1 km/sec from a curve of line width study, while Gray (1973) measured 1.6 km/sec from the Fe I $\lambda6065$ line profile. Most measurements of solar velocity fields are, however, interpreted in terms of depth dependence and anisotropic distributions, the subject of the next section.

THE DEPTH DEPENDENCE AND ANISOTROPY OF MICROTURBULENCE

The depth dependence of ξ has been looked at in only a few stars, most of which are supergiants. The depth dependence is probed by means of the depth of formation dependence on ionization and excitation (see Chapter 13). Wright (1946, 1947) and Wright and van Dien (1949) were the first to find such a dependence.

Usually ξ is found to increase outward. For example, Groth (1972) has obtained values for α Cyg (A2 Ia) ranging from 2 km/sec at the bottom of the atmosphere to 20 km/sec at the top. Wolf (1971) found $\xi = 2$–10 km/sec through the atmosphere of η Leo (A0 Ib). Parsons (1967) found a similar outward increase in the atmospheres of α Per (F5 Ib) and β Aqr (G0 Ib). Rosendahl (1970), on the other hand, finds ξ in α Cyg to go from 12 km/sec in the deep layers to 6 km/sec in the higher layers and then increase again in still higher layers.

Some of the giant eclipsing systems have been used to obtain ξ as a function of depth by measurement of the absorption lines during the atmospheric eclipse phases (e.g., Wilson 1948, Wright and Hesse 1969). The smaller B star shining through the atmosphere of the giant star acts as a probe. Curve of growth results seem to support ξ increasing outward.

In the sun the depth dependence of ξ in the photosphere has been obtained mostly from center to limb profile variations—a method inapplicable to other stars. Several complex models have been made in an attempt to explain the observed changes (usually an increase) in line half widths toward the limb. The models can be grouped into isotropic and anistropic classes. The isotropic models show ξ increasing with height. The anisotropic models have ξ increasing with depth. Usually separate radial and tangential components are assumed as in the example given earlier in the chapter. In the anisotropic model the line width increases toward the limb because the tangential ξ is chosen to be larger than the radial ξ. In the isotropic model the increase in line width toward the limb is a result of the depth of formation being higher toward the limb. Allen (1949) found $\xi_t = 2.8$ km/sec and $\xi_r = 1.7$ km/sec, which can be compared to Waddell's (1956) values of $\xi_t = 2.0 \pm 0.3$ km/sec and $\xi_r = 1.8 \pm 0.1$ km/sec. Other studies of solar microturbulence have been carried out by Schmalberger (1963), Zirin (1966), Müller et al. (1968), Dunn and Olson (1971), de Jager and Neven (1972), Canfield (1971), Athay and Canfield (1969), Gurtovenko and Troyan (1971), and Vernazza et al. (1973), to name a few. Some of the models can become quite elaborate as, for instance, the multicolumn nonisotropic model of Gonczi and Roddier (1971). In their model the radial and tangential components have different $\xi(\tau)$ for rising and falling material. Figure 18-15 shows the behavior of their turbulence model with depth. Lites (1973) has made a comparison between several solar microturbulence models. One general rule seems to hold in these investigations. The number of variable parameters will expand to equal the number of measured features!

It is worth mentioning that even in most non-LTE treatments microturbulence is still introduced to obtain a profile fit. Such is the case in the investigations of the solar Ce II lines (Canfield 1969), solar FeI lines (Lites 1973), and the B star spectra by Kamp (1973). Rosendahl (1973) has also given a reasonable explanation of the observed change in He I $\lambda 6678$ line strength in supergiant OB stars (an effect not predicted by the non-LTE calculations) by introducing microturbulence desaturation.

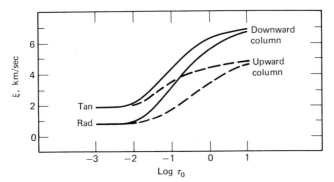

Fig. 18-15 The photospheric microturbulence model for the sun proposed by Gonczi and Roddier (1971) published with permission from Astronomy and Astrophysics.

OBSERVATIONS OF MACROTURBULENCE

In the sun we see a complex array of macromotions. The three most prominent types are associated with granulation, supergranulation, and vertical oscillations.

The granulation seen in Fig. 18-16 lies at the bottom of the photosphere and has a cell-type structure similar to that expected from convective columns. The bright portions are interpreted as the top of the hot column where the

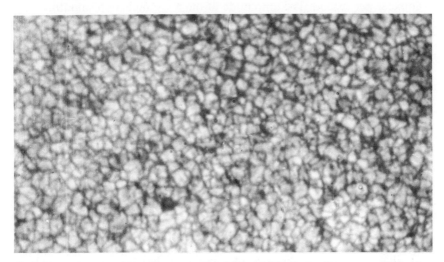

Fig. 18-16 Solar Granulation. (Courtesy of M. Schwarzschild, Project Stratoscope of Princeton University sponsored by ONR, NSF, and NASA.)

convected energy is being released as radiation (Wilson 1969a,b). The cooled gas is seen returning to the deeper layers in the dark lanes between the bright areas. The characteristic size of these cells is about 10^{-3} solar diameters. The flow velocities are of the order of 1 km/sec and center to edge temperature differences are about $10^{3}°$K. The granulation cell pattern changes in a random way with a time scale of a few minutes. The so-called convective overshoot of the granulation induces motions throughout the photosphere (Frazier 1966, 1968a, 1968b).

The Doppler shifts caused by the granulation cause the major wiggles in the wiggly line spectrograms such as that in Fig. 18-1. The line strength can also be seen to differ from cell to cell in response to the differences in temperature. The cause of the convective instability is the hydrogen ionization in the subphotosphere. This convection zone extends in depth over a few hundred kilometers.

The second type of solar macroturbulence is the supergranulation. The gas flow in a supergranule mimics that of the granule, but the size of the convective cells is about 20 times as large, the flow velocities are typically only a few tenths of a kilometer per second, and there is no measurable brightness variation across a supergranule. It is likely that the helium ionization, about 10^{3}–10^{4} km below the photosphere, causes this weak convection pattern (Simon and Leighton 1964).

Vertical oscillations are a completely different type of solar macroturbulence and consist of wave motion characterized by wavelengths of about twice the size of the granulation. The velocities are again a few tenths of a kilometer per second. The amplitude of the wave builds up smoothly, lasts for several cycles, and then decays smoothly. Fourier analysis of the time variations shows a predominant periodicity at 300 sec (Noyes and Leighton 1963, Evans et al. 1965, Frazier 1966, Edmonds and Webb 1972), but other frequencies also appear to exist (Gonczi and Roddier 1969, Frazier 1968a, 1968b, Howard 1967).

The solar macroturbulence is obviously a complicated but structured phenomenon. A detailed review of the solar velocity fields is presented by Gibson (1973).

Now, when we ask what these motions look like in the case of other stars, we must imagine the wiggly line structure and the supergranulation velocities all integrated together. A study of integrated sunlight has been done on Fe I $\lambda6065$ by Gray (1973) who finds $\zeta \sim 2$ km/sec from a Fourier analysis of the line profile. A similar value has been obtained by Ross and Aller (1972), Ross (1973), and Elste (1967) from profile fitting and the curve of line width analysis mentioned earlier.

Evidence of macroturbulence in other stars has been found mainly in supergiants. Huang and Struve (1953) showed how the profiles of lines in the spectrum of ρ Leo (B1 Ib) were more triangular than a Gaussian—certainly not what one would expect from rotation or classical microturbulence. Scholz (1972) has found a similar result for two O stars. The profile analysis of η Leo (A0 Ib) by Wolf (1971) was made internally consistent by adding 15 km/sec Gaussian macroturbulence. Similar results are obtained by Groth (1961) for α Cyg where 22 km/sec was found (compare Fig. 18-10).

The statistical result that the observed line widths in supergiants, when interpreted as rotation, have a lower bound of a few tens of kilometers per second instead of the expected zero value has been taken as evidence of macroturbulence. Slettebak (1956) has suggested the ratio of macroturbulence to rotation increases with luminosity and possibly with temperature (see also Huang and Struve 1954).

COMMENTS

Such meager results at least give a start to our understanding of the velocity fields. So much less appears to be known about stellar macroturbulence compared to microturbulence because the former requires profile measurements while the latter has been attacked by the equivalent widths. On the other hand, the results for macroturbulence are more believable. We have every reason to hope that additional measurements will allow us not only to separate microturbulence, macroturbulence, and rotation, but to measure some aspects of the velocity distributions themselves. The step beyond that may be to go past the geometrical approximations of micro and macroturbulence and include a full dimensional spectrum for stellar turbulence.

REFERENCES

Abt, H. 1957. *Astrophys. J.* **126**, 503.

Abt, H. 1958. *Astrophys. J.* **127**, 658.

Allen, C. W. 1937. *Astrophys. J.* **85**, 165.

Allen, C. W. 1949. *Mon. Not. Roy. Astron. Soc.* **109**, 343.

Andersen, P. H. 1973. *Publ. Astron. Soc. Pac.* **85**, 666.

Athay, R. G. and R. C. Canfield 1969. *Astrophys. J.* **156**, 695.

Baschek, B. and D. Reimers 1969. *Astron. Astrophys.* **2**, 240.

Bonsack, W. K. and R. B. Culver 1966. *Astrophys. J.* **145**, 767.

Buscombe, W., J. Neatrour, and E. Albert 1972. *Bull. Am. Astron. Soc.* **4**, 237.

Canfield, R. C. 1969. *Astrophys. J.* **157**, 425.

Canfield, R. C. 1971. *Sol. Phys.* **20**, 275.

Chaffee, H. T. Jr. 1970. *Astron. Astrophys.* **4**, 291.

Chandrasekhar, S. 1949. *Astrophys. J.* **110**, 329.

Chandrasekhar, S. 1952. *Mon. Not. Roy. Astron. Soc.* **112**, 475. (or 1954. *Am. Math. Mon. Suppl.* **61**, 32).

Chandrasekhar, S. 1955. *Vistas in Astronomy*, A. Beer, Ed., Pergamon, London, **1**, 344.

Cowley, C. R. and A. P. Cowley 1964. *Astrophys. J.* **140**, 713.

Day, R. W., D. L. Lambert, and C. Sneden 1973. *Astrophys. J.* **185**, 213.

de Jager, C. and L. Neven 1972. *Sol. Phys.* **22**, 49.

Dunn, A. R. and E. C. Olson 1971. *Sol. Phys.* **16**, 272.

Edmonds, F. N. Jr. and C. J. Webb 1972. *Sol. Phys.* **25**, 44.

Elste, G. H. E. 1967. *Astrophys. J.* **148**, 857.

Eriksen, G. and P. Maltby 1967. *Astrophys. J.* **148**, 833.

Evans, J. W., R. Michard, and R. Servajean 1965. *Ann. Astrophys.* **26**, 368.

Frazier, E. N. 1966. *An Observational Study of the Hydrodynamics of the Lower Photosphere*, thesis, University of California.

Frazier, E. N. 1968a. *Z. Astrophys.* **68**, 345.

Frazier, E. N. 1968b. *Astrophys. J.* **152**, 557.

Gibson, E. G. 1973. *The Quiet Sun*, NASA SP-303, Scientific and Technical Information Office, Washington, D.C., Chapter 5.

Glebocki, R. 1973. *Acta Astron.* **23**, 135.

Gonczi, G. and F. Roddier 1969. *Sol. Phys.* **8**, 255.

Gonczi, G. and F. Roddier 1971. *Astron. Astrophys.* **11**, 28.

Gray, D. F. 1973. *Astrophys. J.* **184**, 461.

Gray, D. F. 1975. *Astrophys. J.* **201**, (in press).

Gray, D. F. and J. C. Evans 1973. *J. Roy. Astron. Soc. Can.* **67**, 241.

Groth, H.-G. 1961. *Z. Astrophys.* **51**, 206.

Groth, H.-G. 1972. *Astron. Astrophys.* **21**, 337.

Gurtovenko, E. A. and V. Troyan 1971. *Sol. Phys.* **20**, 264.

Hearn, A. G. 1974. *Astron. Astrophys.* **31**, 415.

Holweger, H. 1973. *Sol. Phys.* **30**, 35.

Howard, R. 1967. *Sol. Phys.* **2**, 3.

Huang, S.-S. 1952. *Astrophys. J.* **115**, 529.

Huang, S.-S. and O. Struve 1952. *Astrophys. J.* **116**, 410.

Huang, S.-S. and O. Struve 1953. *Astrophys. J.* **118**, 463.

Huang, S.-S. and O. Struve 1954. *Ann. Astrophys.* **17**, 85.

Huang, S.-S. and O. Struve 1960. *Stellar Atmospheres*, J. L. Greenstein, Ed., University of Chicago Press, Chicago, Chapter 8.

Kamp, L. W. 1973. *Astrophys. J.* **180**, 447.

Kraft, R. P. 1969. *Stellar Astronomy*, H.-Y. Chiu, R. L. Warasila, and J. L. Remo, Eds., Gordon and Breach, New York, p. 317.

Lites, B. W. 1973. *Sol. Phys.* **32**, 283.

Minnaert, M. G. J. 1953. *The Sun*, G. P. Kuiper, Ed., University of Chicago Press, Chicago, p. 88.

Müller, E. A., B. Baschek, and H. Holweger 1968. *The Structure of the Quiet Photosphere and Low Chromosphere*, C. de Jager, Ed., Reidel, Dordrecht, Holland, p. 125.

Noyes, R. W. and R. B. Leighton 1963. *Astrophys. J.* **138**, 631.

Parsons, S. B. 1967. *Astrophys. J.* **150**, 263.

Rosendahl, J. 1970. *Astrophys. J.* **160**, 627.

Rosendahl, J. D. 1973. *Astrophys. J. Lett.* **183**, L39.

Ross, J. E. 1973. *Astrophys. J.* **180**, 599.

Ross, J. E., and L. H. Aller 1972. *Sol. Phys.* **25**, 30.

Rosseland, S. 1928. *Mon. Not. Roy. Astron. Soc.* **89**, 49.

Schmalberger, D. C. 1963. *Astrophys. J.* **138**, 693.

Scholz, M. 1972. *Astrophys. Lett.* **10**, 137.

Schwartzschild, M. 1961. *Astrophys. J.* **134**, 1.

Simon, G. W. and R. B. Leighton 1964. *Astrophys. J.* **140**, 1120.

Slettebak, A. 1956. *Astrophys. J.* **124**, 173.

Smith, M. A. 1971. *Astron. Astrophys.* **11**, 325.

Smith, M. A. 1973. *Astrophys. J.* **182**, 159.

Smith, M. A. 1974. *Astrophys. J.* **190**,

Struve, O. and C. T. Elvey 1934. *Astrophys. J.* **79**, 409.

Tandberg-Hanssen, E. 1967. *Solar Activity*, Blaisdell, Waltham, Mass.

Thomas, R. B. 1961. *Nuovo Cimento Suppl.* **22**, Series 10, No. 1. Fourth Symposium on Cosmical Gas Dynamics.

van den Heuvel, E. P. J. 1963. *Bull. Astron. Inst. Neth.* **17**, 148.

van Helden, R. 1972. *Astron. Astrophys.* **21**, 209.

Vernazza, J. E., E. H. Avrett, and R. Loeser 1973. *Astrophys. J.* **184**, 605.

Voigt, H.-H. 1952. *Z. Astrophys.* **31**, 48.

Waddell, J. H. III 1956. *An Empirical Determination of the Turbulence Field in the Solar Photosphere Based on a Study of Weak Fraunhofer Lines*, thesis, University of Michigan.

Waddell, J. H. III 1958. *Astrophys. J.* **127**, 284.

Wilson, A. M. and F. J. Guidry 1974. *Mon. Not. Roy. Astron. Soc.* **166**, 219.

Wilson, O. C. 1948. *Astrophys. J.* **107**, 126.

Wilson, P. R. 1969a. *Sol. Phys.* **6**, 364.

Wilson, P. R. 1969b. *Sol. Phys.* **9**, 303.

Wolf, B. 1971. *Astron. Astrophys.* **10**, 383.

Worrall, G. and A. M. Wilson 1972. *Nature* **236**, 15.

Worrall, G. and A. H. Wilson 1973. *Vistas in Astronomy*, Vol. 15, A. Beer, Ed., Pergamon, London, p. 39.

Wright, K. O. 1946. *Publ. Roy. Astron. Soc. Can.* **40**, 183.

Wright, K. O. 1947. *Publ. Roy. Astron. Soc. Can.* **41**, 49.

Wright, K. O. 1955. *Int. Astron. Union Trans.* **IX**, 739.

Wright, K. O. and K. H. Hesse 1969. *Publ. Dom. Astrophys. Obs.* Victoria **13**, 301.

Wright, K. O. and E. van Dien 1949. *J. Roy. Astron. Soc. Can.* **43**, 15.

Zirin, H. 1966. *The Solar Atmosphere*, Blaisdell, Waltham, Mass., p. 286.

EPILOG

Thoughtful reflection of the material in the previous pages reveals the whole endeavor to be a series of successive approximations. The large number of physical variables coupled with nonlinear behavior precludes direct solution. This is illustrated in a particularly open manner in the chemical analysis procedure where an initial chemical composition is assumed for the construction of a model with which we do the chemical analysis. The same process takes place in the overall picture as well. As we progress to higher precision in measurement, the inclusion of a larger number of spectral features, and to higher spectral resolution, the data can support more complicated models. The simple plane parallel, homogeneous model photosphere is a suitable tool for mapping photospheric temperature and pressure in the HR diagram but may be inadequate for other jobs. Our understanding of the physical processes of radiation transfer, turbulence, temperature inhomogeneities, and the like will deepen and grow as the iterations progress (although not necessarily in a monotonic way). At some stage in the process it will be necessary to include the subphotosphere, especially the convection zones, and on the other end the chromosphere or other outer layers.

We certainly expect our studies of the atmospheres of stars to unify and link together a large number of astronomical areas. The progress over the last two decades has been rewarding. Disciplines ranging from stellar evolution to galactic structure, from plasma physics to astrometry have all met in the studies of the stellar photosphere.

You can judge from the literature and the references cited in these pages where the work needs to be done and where your ideas can make a contribution.

A TABLE OF USEFUL CONSTANTS

Constant	Symbol	Value*	Ref.
Speed of light	c	$2.99792458(\pm 1) \times 10^{10}$ cm/sec	1
Planck's constant	h	$6.62618(\pm 4) \times 10^{-27}$ erg/sec	1
Electron charge	e	$1.602189(\pm 5) \times 10^{-19}$ C	1
		$= 4.803237(\pm 1) \times 10^{-10}$ esu	
Electron mass	m	$9.10953(\pm 5) \times 10^{-28}$ g	1
Proton mass	m_p	$1.672648(\pm 9) \times 10^{-24}$ g	1
Atomic mass unit	amu	$1.660566(\pm 9) \times 10^{-24}$ g	1
Rydberg constant	R_∞	$1.09737318(\pm 8) \times 10^5$ cm^{-1}	1
Stefan–Boltzmann constant	σ	$5.6703(\pm 7) \times 10^{-5}$ erg/sec cm^2 °K	1
First radiation constant	c_1	$3.74183(\pm 2) \times 10^{-5}$ erg cm^2/sec	1
Second radiation constant	c_2	$1.43879(\pm 5)$ cm °K	1
Wien's law constant	a	$0.28978(\pm 1)$ cm °K	2
Gravitational constant	G	$6.672(\pm 4) \times 10^{-8}$ dyne cm^2/g^2	1
Boltzmann constant	k	$1.38066(\pm 4) \times 10^{-16}$ erg/°K	1
Gas constant	R	$8.3144(\pm 3) \times 10^7$ erg/deg mole	1
Solar mass	$M_\odot$	$1.989(\pm 1) \times 10^{33}$ g	2
Solar luminosity	$L_\odot$	$3.82(\pm 4) \times 10^{33}$ erg/sec	3
Solar radius	$R_\odot$	$6.9598(\pm 7) \times 10^{10}$ cm	2
Solar effective temperature	T_{eff}	$5770(\pm 10)$ °K	
Solar surface gravity	$g_\odot$	$2.738(\pm 3) \times 10^4$ cm/sec^2	
Astronomical unit	au	$1.495979(\pm 1) \times 10^{13}$ cm	2
Parsec	pc	$3.08568(\pm 1) \times 10^{18}$ cm	2

* The uncertainty in the last digit is indicated by the number in parentheses.

REFERENCES

1. E. R. Cohen and B. N. Taylor 1973. *J. Phys. Chem. Reference Data* **2**, 663.
2. C. W. Allen, 1973. *Astrophysical Quantities*, 3rd ed., Athlone Press, University of London.
3. H. Neckel and D. Labs, 1973. IAU Symp. No. 54. *Problems of Calibration of Absolute Magnitudes and Temperatures of Stars*, B. Hauck and B. E. Westerlund, Eds., Reidel, Dordrecht, Holland, p. 149.

TABLE OF sinc πσ

The Fourier transform of a box as given by eq. 2-5 is $W \operatorname{sinc} \pi W \sigma$. The function arises so frequently that a short table for the case of a unit width box is given here.

σ	sinc $\pi\sigma$	σ	sinc $\pi\sigma$
0.00	1.000	2.00	0.000
0.05	0.996	2.05	0.024
0.10	0.984	2.10	0.047
0.15	0.963	2.15	0.067
0.20	0.935	2.20	0.085
0.25	0.900	2.25	0.100
0.30	0.858	2.30	0.112
0.35	0.810	2.35	0.121
0.40	0.757	2.40	0.126
0.45	0.699	2.45	0.128
0.50	0.637	2.50	0.127
0.55	0.572	2.55	0.123
0.60	0.505	2.60	0.116
0.65	0.436	2.65	0.107
0.70	0.368	2.70	0.095
0.75	0.300	2.75	0.082
0.80	0.239	2.80	0.067
0.85	0.170	2.85	0.051
0.90	0.109	2.90	0.034
0.95	0.052	2.95	0.017
1.00	0.000	3.00	0.000

(*Continued*)

σ	sinc $\pi\sigma$	σ	sinc $\pi\sigma$
1.05	-0.047	3.05	-0.016
1.10	-0.089	3.10	-0.032
1.15	-0.126	3.15	-0.046
1.20	-0.156	3.20	-0.058
1.25	-0.180	3.25	-0.069
1.30	-0.198	3.30	-0.078
1.35	-0.210	3.35	-0.085
1.40	-0.216	3.40	-0.089
1.45	-0.217	3.45	-0.091
1.50	-0.212	3.50	-0.091
1.55	-0.203	3.55	-0.089
1.60	-0.189	3.60	-0.084
1.65	-0.172	3.65	-0.078
1.70	-0.151	3.70	-0.070
1.75	-0.129	3.75	-0.060
1.80	-0.104	3.80	-0.049
1.85	-0.078	3.85	-0.038
1.90	-0.052	3.90	-0.025
1.95	-0.026	3.95	-0.013

FAST FOURIER TRANSFORM
FORTRAN PROGRAM

Several versions of the fast Fourier transform can be found in the literature. Listed below is the subroutine of N. Brenner (1968, IBM Contributed Programs Library No. 360D-13.4.002).

```
      SUBROUTINE FOUR 1 (DATA, NN, ISIGN)
C   THE  COOLEY–TUKEY  FAST  FOURIER  TRANSFORM  IN  USASI
C   BASIC      FORTRAN.      TRANSFORM(K) = SUM(DATA(J)*EXP
C   (ISIGN *2PI *SQRT(–1) * (J – 1) *(K – 1)/NN)).  SUMMED  OVER
C   ALL J AND K FROM 1 TO NN. DATA IS A ONE-DIMENSIONAL
C   COMPLEX ARRAY (I.E., THE REAL AND IMAGINARY PARTS ARE
C   ADJACENT IN STORAGE, SUCH AS FORTRAN IV PLACES THEM)
C   WHOSE LENGTH  NN = 2**K, K.GE.0 (IF NECESSARY, APPEND
C   ZEROES TO THE DATA). ISIGN IS +1 OR – 1. IF A – 1 TRANSFORM
C   IS FOLLOWED BY A  + 1 ONE (OR A + 1 BY A – 1) THE ORIGINAL
C   DATA REAPPEAR, MULTIPLIED BY NN. TRANSFORM VALUES
C   ARE RETURNED IN ARRAY DATA, REPLACING THE INPUT.
      DIMENSION DATA (1)
      N = 2*NN
      J = 1
      DO 5 I = 1,N,2
      IF(I – J) 1,2,2
    1 TEMPR = DATA(J)
      TEMPI = DATA (J + 1)
      DATA(J) = DATA(I)
      DATA(J + 1) = DATA(I + 1)
```

```
      DATA(I) = TEMPR
      DATA(I + 1)= TEMPI
 2    M = N/2
 3    IF(J – M) 5,5,4
 4    J = J – M
      M = M/2
      IF(M – 2) 5,3,3
 5    J = J + M
      MMAX = 2
 6    IF(MMAX–N) 7,10,10
 7    ISTEP = 2*MMAX
      THETA = 6.28318530717959/FLOAT(ISIGN*MMAX)
      SINTH = SIN(THETA/2.)
      WSTPR = – 2. *SINTH*SINTH
      WSTPI = SIN(THETA)
      WR = 1.
      WI = 0.
      DO 9 M = 1, MMAX,2
      DO 8 I = M, N, ISTEP
      J = I + MMAX
      TEMPR = WR*DATA(J) – WI*DATA(J + 1)
      TEMPI = WR*DATA(J + 1) + WI*DATA(J)
      DATA(J) = DATA(I) – TEMPR
      DATA(J + 1) = DATA(I + 1) – TEMPI
      DATA(I) = DATA(I) + TEMPR
 8    DATA(I + 1) = DATA(I + 1) + TEMPI
      TEMPR = WR
      WR = WR*WSTPR – WI*WSTPI + WR
 9    WI = WI*WSTPR + TEMPR*WSTPI + WI
      MMAX = ISTEP
      GO TO 6
10    RETURN
      END
```

In the calling program, dimension the function (for which the transform is being computed) four times as large as the number of real components in the function.

APPENDIX D

IONIZATION POTENTIALS AND PARTITION FUNCTIONS

Table D-1 gives the first three ionization potentials in electron volts for the 92 natural elements. The data are from Moore (1949, 1952, 1958, 1970), Gray (1963), and Allen (1973). The third column in Table D-1 lists the atomic masses in atomic mass units (1 amu $= 1.6606 \times 10^{-24}$ g) based on the ^{12}C $= 12.000$ scale.

The partition functions, log $u(T)$, are tabulated in Table D-2 for an assortment of common elements. The temperature range is from 3000 to 12,000°K as indicated by the column headings. The values listed are for electron pressures typical of main sequence stars. At the hottest temperatures the pressure effects may become important. Details are given in Aller (1963). The data are from Bolton (1970) and Aller (1963), and the summary by Evans (1966).

(Table D-1 overleaf)

Table D-1 Atomic Weights and Ionization Potentials

No.	Element	Weight	I_1	I_2	I_3
1	H	1.008	13.60	—	—
2	He	4.003	24.59	54.42	—
3	Li	6.941	5.39	75.64	122.45
4	Be	9.012	9.32	18.21	153.89
5	B	10.811	8.30	25.15	37.93
6	C	12.011	11.26	24.38	47.89
7	N	14.007	14.53	29.60	47.45
8	O	15.994	13.62	35.12	54.93
9	F	18.998	17.42	34.97	62.71
10	Ne	20.179	21.56	40.96	63.45
11	Na	22.990	5.14	47.29	71.64
12	Mg	24.305	7.64	15.03	80.14
13	Al	26.982	5.99	18.83	28.45
14	Si	28.086	8.15	16.34	33.49
15	P	30.974	10.48	19.72	30.18
16	S	32.06	10.36	23.33	34.83
17	Cl	35.45	12.97	23.80	39.61
18	Ar	39.95	15.76	27.62	40.74
19	K	39.10	4.34	31.63	45.72
20	Ca	40.08	6.11	11.87	50.91
21	Sc	44.96	6.54	12.80	24.76
22	Ti	47.90	6.82	13.57	27.49
23	V	50.941	6.74	14.65	29.31
24	Cr	51.996	6.76	16.49	30.96
25	Mn	54.938	7.43	15.64	33.67
26	Fe	55.847	7.87	16.16	30.65
27	Co	58.933	7.86	17.05	33.50
28	Ni	58.71	7.64	18.17	35.17
29	Cu	63.546	7.72	20.29	36.83
30	Zn	65.37	9.39	17.96	39.72
31	Ga	69.72	6.00	20.51	30.71
32	Ge	72.59	7.90	15.93	34.22
33	As	74.922	9.81	18.63	28.35
34	Se	78.96	9.75	21.19	30.82
35	Br	79.904	11.81	21.6	36
36	Kr	83.80	14.00	24.36	36.95
37	Rb	85.468	4.18	27.28	40
38	Sr	87.62	5.69	11.03	43.6
39	Y	88.906	6.38	12.24	20.52
40	Zr	91.22	6.84	13.13	22.99

(*Continued*)

Table D-1 (*Continued*)

No.	Element	Weight	I_1	I_2	I_3
41	Nb	92.906	6.88	14.32	25.04
42	Mo	95.94	7.10	16.15	27.16
43	Tc	98.906	7.28	15.26	29.54
44	Ru	101.07	7.37	16.76	28.47
45	Rh	102.905	7.46	18.07	31.06
46	Pd	106.4	8.33	19.43	32.92
47	Ag	107.868	7.57	21.49	34.83
48	Cd	112.40	8.99	16.90	37.48
49	In	114.82	5.79	18.87	28.03
50	Sn	118.69	7.34	14.63	30.50
51	Sb	121.75	8.64	16.53	25.3
52	Te	127.60	9.01	18.6	27.96
53	I	126.904	10.45	19.13	33
54	Xe	131.30	12.13	21.21	32.1
55	Cs	132.905	3.89	25.1	35
56	Ba	137.34	5.21	10.00	—
57	La	138.906	5.58	11.06	19.18
58	Ce	140.12	5.47	10.87	20.20
59	Pr	140.908	5.42	10.55	21.62
60	Nd	144.24	5.49	10.72	—
61	Pm	146	5.55	10.90	—
62	Sm	150.4	5.63	11.07	—
63	Eu	151.96	5.67	11.25	—
64	Gd	157.25	6.14	12.1	—
65	Tb	158.925	5.85	11.52	—
66	Dy	162.50	5.93	11.67	—
67	Ho	164.930	6.02	11.80	—
68	Er	167.26	6.10	11.93	—
69	Tm	168.934	6.18	12.05	23.71
70	Yb	170.04	6.25	12.17	25.2
71	Lu	174.97	5.43	13.9	19
72	Hf	178.49	7.0	14.9	23.3
73	Ta	180.948	7.89	16	22
74	W	183.85	7.98	18	24
75	Re	186.2	7.88	17	26
76	Os	190.2	8.7	17	25
77	Ir	192.2	9.1	17	27
78	Pt	195.09	9.0	18.56	28
79	Au	196.967	9.22	20.5	30
80	Hg	200.59	10.44	18.76	34.2

(*Continued*)

Table D-1 (*Continued*)

No.	Element	Weight	I_1	I_2	I_3
81	Tl	204.37	6.11	20.43	29.83
82	Pb	207.19	7.42	15.03	31.94
83	Bi	208.981	7.29	16.68	25.56
84	Po	210	8.42	19	27
85	At	210	9.3	20	29
86	Rn	222	10.75	21	29
87	Fr	223	4	22	33
88	Ra	226.025	5.28	10.14	34
89	Ac	227	6.9	12.1	20
90	Th	232.038	6	11.5	20.0
91	Pa	230.040	—	—	—
92	U	238.029	6	—	—

Table D-2 $\log u(T)$

Element	\multicolumn{8}{c}{T, °K}								
	3000	4000	5000	6000	7000	8000	10,000	12,000
H	0.30	0.30	0.30	0.30	0.30	0.30	0.30	0.30
He I	0.00	0.00	0.00	0.00	0.00	0.00	0.00	0.00
He II	0.30	0.30	0.30	0.30	0.30	0.30	0.30	0.30
Li I	0.28	0.31	0.35	0.39	0.44	0.48	0.56	—
Be I	0.00	0.00	0.02	0.03	0.04	0.06	0.15	—
C I	0.96	0.96	0.96	0.97	0.97	0.98	1.00	1.02
C II	0.78	0.78	0.78	0.78	0.78	0.78	0.78	0.78
C III	0.00	0.00	0.00	0.00	0.00	0.00	0.00	0.00
N I	0.00	0.00	0.01	0.04	0.08	0.13	0.23	0.34
N II	0.95	0.95	0.95	0.95	0.95	0.97	0.98	0.99
O I	0.93	0.93	0.94	0.95	0.95	0.96	0.97	0.98
O II	0.60	0.60	0.60	0.60	0.60	0.60	0.61	0.65
Na I	0.30	0.30	0.31	0.34	0.38	0.45	0.66	0.88
Na II	0.00	0.00	0.00	0.00	0.00	0.00	0.00	0.00
Mg I	0.00	0.00	0.00	0.02	0.05	0.08	0.18	0.35
Mg II	0.30	0.30	0.30	0.30	0.30	0.30	0.31	0.32
Si I	0.93	0.95	0.97	0.98	1.00	1.02	1.05	0.07
Si II	0.76	0.76	0.76	0.76	0.77	0.77	0.77	0.78
K I	0.30	0.31	0.34	0.38	0.45	0.57	0.79	1.01
K II	0.00	0.00	0.00	0.00	0.00	0.00	0.00	0.00

(*Continued*)

Table D-2 (*Continued*)

Element	T, °K 3000	4000	5000	6000	7000	8000	10,000	12,000
Ca I	0.01	0.02	0.06	0.15	0.26	0.38	0.64	0.85
Ca II	0.30	0.31	0.34	0.37	0.41	0.45	0.55	0.62
Ca III	0.00	0.00	0.00	0.00	0.00	0.00	0.00	0.00
Sc I	1.01	1.02	1.08	1.16	1.24	1.32	1.49	1.64
Sc II	1.27	1.32	1.36	1.39	1.44	1.47	1.52	1.57
Ti I	1.34	1.39	1.47	1.56	1.65	1.74	1.90	2.05
Ti II	1.66	1.69	1.73	1.78	1.82	1.86	1.92	1.98
V I	1.57	1.61	1.68	1.75	1.82	1.89	2.04	2.18
V II	1.50	1.57	1.64	1.70	1.75	1.79	1.88	1.96
Cr I	0.89	0.95	1.02	1.10	1.20	1.32	1.52	1.71
Cr II	0.78	0.81	0.85	0.92	1.00	1.08	1.25	1.43
Mn I	0.78	0.78	0.80	0.85	0.91	0.98	1.17	1.36
Mn II	0.85	0.86	0.89	0.93	0.97	1.02	1.15	1.27
Fe I	1.35	1.39	1.44	1.50	1.57	1.64	1.77	1.91
Fe II	1.54	1.59	1.64	1.68	1.71	1.75	1.83	1.90
Fe III	1.31	1.33	1.35	1.36	1.38	1.40	1.45	1.50
Co I	—	1.42	1.50	1.56	1.61	1.67	1.78	—
Co II	—	1.37	1.44	1.50	1.55	1.60	1.67	—
Ni I	1.41	1.45	1.49	1.51	1.53	1.56	1.61	1.68
Ni II	0.90	0.97	1.03	1.09	1.15	1.20	1.29	1.37
Sr I	0.00	0.03	0.09	0.19	0.30	0.45	0.72	0.96
Sr II	0.30	0.31	0.33	0.36	0.40	0.45	0.53	0.61

REFERENCES

Allen, C. W. 1973. *Astrophysical Quantities*, 3rd ed., Athlone Press, University of London, London, and Oxford University Press, pp. 30, 36.

Aller, L. H. 1963. *Astrophysics The Atmospheres of the Sun and Stars*, Ronald, New York, p. 113.

Bolton, C. T. 1970. *Astrophys. J.* **161**, 1187.

Evans, J. C. 1966. *An Atmospheric Analysis of the Magnetic Star Gamma Equulei*, thesis, The University of Michigan.

Gray, D. E., Ed. 1963. *American Institute of Physics Handbook*, 2nd ed., McGraw-Hill, New York.

Moore, C. E. 1949, 1952, 1958. "Atomic Energy Levels," *U.S. Nat. Bur. Stand.*, Circ. **467**, Vols. I, II, and III.

Moore, C. E. 1970. *Ionization Potentials*, NSRDS-NBS34, Washington.

THE STRONGEST LINES
IN THE SOLAR SPECTRUM

The following summary is abstracted from *The Solar Spectrum* 2935 Å to 8770 Å by C. E. Moore, M. G. J. Minnaert, and J. Houtgast, U.S. National Bureau of Standards Monograph 61, 1966. It is useful as a guide to the spectra of solar-type stars. Between 3500 and 4000 Å, lines with equivalent widths greater than 1.50 Å are included; between 4000 and 4861 Å, lines greater than 1.00 Å, between 4861 and 6563 Å, lines greater than 0.30 Å, and between 4861 and 8750 Å, lines greater than 0.20 Å.

λ	Element	W(Å)	Name	λ	Element	W(Å)	Name
3581.2	Fe I	2.14	N	4920.3	Fe I	0.43	
3719.9	Fe I	1.66		4957.5	Fe I	0.40	
3734.9	Fe I	3.03	M	5167.3	Mg I	0.65	b4
3749.5	Fe I	1.91		5172.7	Mg I	1.26	b2
3758.2	Fe I	1.65		5183.6	Mg I	1.58	b1
3770.6	H11	1.86		5233.0	Fe I	0.35	
3797.9	H10	3.46		5269.6	Fe I	0.48	
3820.4	Fe I	1.71	L	5324.2	Fe I	0.33	
3825.9	Fe I	1.52		5328.1	Fe I	0.38	
3832.3	Mg I	1.68		5528.4	Mg I	0.29	
3835.4	H9	2.36		5890.0	Na I	0.75	D2
3838.3	Mg I	1.92		5895.9	Na I	0.56	D1
3859.9	Fe I	1.55		6122.2	Ca I	0.22	
3889.0	H8	2.35		6162.2	Ca I	0.22	
3933.7	Ca II	20.25	K	6562.8	Hα	4.02	C
3968.5	Ca II	15.47	H	6867.2	O_2 (starts)	—	B (tell.)
4045.8	Fe I	1.17		7593.7	O_2 (starts)	—	A (tell.)
4101.7	Hδ	3.13	h	8194.8	Na I	0.30	
4226.7	Ca I	1.48	g	8498.1	Ca II	1.47	
4300–4320	—	—	G	8542.1	Ca II	3.67	
4340.5	Hγ	2.86		8662.2	Ca II	2.60	
4383.6	Fe I	1.01		8688.6	Fe I	0.27	
4861.3	Hβ	3.68	F	8736.0	Mg I	0.29	
4891.5	Fe I	0.31					

APPENDIX F

COMPUTATION OF RANDOM ERRORS

It is almost always assumed that statistical errors are random and their frequency of occurrence is then a Gaussian distribution. The dispersion of the Gaussian is an inverse measure of the goodness of the observations. The best estimate of the dispersion, which we call the standard deviation, s, can be calculated from a set of N measurements of a quantity x using*

$$s \equiv \left[\frac{\sum (x_j - \bar{x})^2}{N - 1} \right]^{1/2} \tag{F-1}$$

where the summation is carried over each of the jth measurements from one to N. The mean x is given by

$$\bar{x} = \frac{\sum x_j}{N} \tag{F-2}$$

and is the best estimate of the quantity we are measuring. Should some determinations of x be more dependable than others, weights can be used in the computation of s and $\bar{x}$. Let us call the weights p_j. Then

$$s = \left[\frac{\sum p_j (x_j - \bar{x})^2}{(N - 1) \sum p_j} \right]^{1/2}$$

---

* We use the term dispersion here with its statistical meaning. Eq. F-1 gives an estimate of $\beta / \sqrt{2}$ where β is our usual notation as in eq. 2-6.

and

$$\bar{x} = \frac{\sum p_j x_j}{\sum p_j}.$$

The standard deviation, s, is also called the error of a single measurement since it is related to the width of the Gaussian distribution of the individual measurements. The uncertainty in the mean of the measured values, $\bar{x}$, is

$$\bar{s} = \frac{s}{\sqrt{N}}. \tag{F-3}$$

The percentage of measurements falling farther away from $\bar{x}$ than $\pm s$ is 32, 16% in each wing of the Gaussian. The fraction for $\pm 2s$ is 5%. When we choose to have *one-half* the measurements farther away from $\bar{x}$ than an interval h, we have defined the *probable error*. The probable error is related to the standard deviation by

$$h = 0.6745s. \tag{F-4}$$

Sometimes, especially when doing an error calculation by hand, it is convenient to use the average deviation defined by

$$A \equiv \frac{\sum |x_j - \bar{x}|}{N}. \tag{F-5}$$

For a Gaussian error distribution, the relation between A and the other width parameters is

$$h = 0.8453A$$

$$s = 1.2532A.$$

In some cases it is possible to estimate the error on a computed result from known errors on the numbers going into the computation. The first step is to express the result in terms of independent variables. When this is completed, the uncertainty in the result, $f(x_1, x_2, \ldots)$, is computed from

$$s_f = \left[\sum_k \left(\frac{\partial f}{\partial x_k} \right)^2 s_k^2 \right]^{1/2}. \tag{F-6}$$

The independent variables are denoted by x_k and their errors by s_k.

INDEX

Absolute flux, calibration, 200, 204, 213, 366, 367
  radii from, 366, 387
Absolute magnitude effects, 3, 318, 379
Abundances, *see* Chemical composition
Absorption coefficient, continuous, 140
  definition, 106
  hydrogen, 142, 147, 241
  helium, 150
  line, 221
  mass, 155
  metals, 152
  scattering, 153
  thermal, 248
  total, 249
Acoustical waves, 136, 418
Adiabatic gradient, 137, 178
Adjacency effect, 87, 93
Aerodynamic turbulence, 417
Aliasing, 36, 277–278
Angular dispersion, *see* Dispersion
Angular momentum, 392, 411
Anisotropic turbulence, 420, 426, 439
Anomalies, 63, 66
Apodization, 278
Arcturus Atlas, 16, 295
Astrophysical flux, 102
Atomic oscillator, 222
Avogado's number, 10
Axial projection, 405

Balmer jump, 142, 176–177, 201–203, 207, 373, 379
Balmer lines, absorption coefficient, 254
  electron broadening, 247
  measurement, 292
  model fits, 330
  non-LTE aspects, 304

  pressure effects, 320, 322, 380
  spectral classification, 3
  temperature effects, 313, 316, 317, 374
  Stark broadening, 241
Bar pattern, 93
Binaries, radii, 363
  rotation, 411
  surface gravities, 379, 387
Black body, continuum source function, 108, 159
  general, 111
  line source function, 302, 304, 305
  temperature distribution from, 166
Blanketing, 166
Blaze, 55, 60
Blends, 292, 350, 354
b lines, 238, 304, 319, 381, 384
Bolometric flux, defined, 215
  effective temperature from, 367
  radius from, 365
Boltzmann, constant, 10, 449
  equation, 12
Bose-Einstein statistics, 94
Bound-bound transitions, 140, 166, 221
Bound-free absorption, 140, 142, 147, 152
Brackett continuum, 142, 201
Break-up velocity, 392, 410
Broadening of lines, chapter 11, 221
  natural, 222
  pressure, 229
  rotation, 392
  thermal, 248
  turbulence, 332–333, 416

Calibration steps, 85–86
Cameras, coude, 263, 266
  scanner, 196
  spectrograph, 47

Carroll's method, 402
Cassegrain focus, 43, 193
Catalogs, 14, 393
Cathode response, 77, 81, 197, 198
Cavity radiation, see Black body
Center-to-limb variation, 421, 440
Chemical composition, analysis for, 335
  curve of growth, 323
  differential analysis, 337, 348
  line absorption, 212
  models effects, 173, 182
  solar, 351
Chromatic resolution, 47, 55, 68, 260;
    see also Instrumental profile
Chromosphere, 2, 407
Classical oscillator, 222
Clayden effect, 87
Cold box, 82
Collimator, 46, 68, 260
Color index, 9
Color terms, 203
Comparison spectra, 3
Constants, 449
Continuous absorption, 140; see also Absorption coefficient
Continuum, 140, 192, 366, 371, 379
Contrast, 86, 283, 290
Contribution functions, flux, 176
  specific intensity, 162–164
  lines, 307
Convection, conditions for, 137
  energy transport, 136
  temperature distribution, 168
  turbulence, 416, 423, 441
Convolution, defined, 29
  delta function, 31
  dispersion profile, 32
  Gaussian, 32
  occurrence, 18
Cooley-Tukey algorithm, 38, 453
Corona, 2, 360
Cosine transform, 26
Coude focus, 44, 260
Current measurement, 78
Curve of growth, abundances from, 346
  characteristics, 324, 337
  differential analysis, 348
  historical, 345
  microturbulence, 332, 342, 347, 427–428
  pressure effects, 341
  temperature measurement, 375
Curve of line widths, 428
Czerny-Turner system, 46

Damping, curve of growth, 326, 340, 428
  damping parameter, 255

gravity, 322
  microturbulence, 428
  natural, 223, 227
  pressure effects, 234, 322, 340
  radiation, 223, 227
  Stark, 237
  total, 250
  van der Waals, 238
Dark current, 82
Dark output, 82
Data sampling, see Sampling
DC measurement, 78
DDO, 7
Dead time, 91
Delta function, defined, 24
  response for, see Instrumental profile
  spectral sources, 279
Depth of formation, 307; see also Contribution functions
D lines, abundance effects, 327
  LTE, 304, 331
  pressure effects, 322, 381
  scattered light test using, 287
  Stark constant, 238
  temperature effects, 313
  van der Waals constant, 239
Detailed balance, 12, 302
Detectors, 75
Developer, 76
Differential analysis, see Curve of growth
Diffraction gratings, see Gratings
Diode arrays, 97
Dipole, 222
Discrete Fourier transform, 38
Dispersion, angular, 55
  coude, 264
  ghost grating, 65
  spectrograph, 47
Dispersion profile, 23, 225
Doppler width, 249
Double pass system, 287
Double stars, eclipsing, 363
  radii, 363
  standards, 294
  surface gravity, 387
DQE, 97
Dyes, 79
Dynode, 77

Eclipsing binaries, 363, 387
Eddington-Barbier relations, 162
Eddington solution, 133, 300
Effective temperature, defined, 3, 359
  measurement, 367
  versus B-V, 368
Einstein coefficients, 109, 301, 303
Electron broadening, 237, 247

Electron, mass and charge, 449
    pressure, 10, 171, 180
    scattering, 153
Emission coefficient, 106, 109, 301
Emulsion type, 78
Energy distribution, 192
    see also Absolute calibration; Balmer jump;
    and Paschen continuum
Energy transport, 121
Enhancement factors, 239
Entrance slit, 45, 68, 261–262
Equation of state, 9
Equivalent width, abundance analysis, 335
    curve of growth, 324
    defined, 259
    measurement, 290
Errors, 40, 462
ESW theory, 248
Excitation, equation, 11–12
    potential, 11–12
    temperature, 338, 347
Exit slit, 193, 265
Exponential integrals, 127, 129
Extinction, interstellar, 9, 213
    terrestrial, 199
    see also Absorption coefficient

Fabry lens, 267
Fabry-Perot, 69, 287
Fast Fourier transform, 38
Fatigue, 89
Fellgett's advantage, 73
Filter, Fourier, 274
    photometry, 7
    rejection band, 287
Flux, absolute, 200
    constancy, 130, 166
    continuum, 175
    defined, 102
    heliostat, 295
    integral, 102, 113
    measurement, 193, 198
    radius and effective temperature, 359
Fourier analysis, lines, 293, 402, 431, 436
    rotation, 402
    turbulence, 431, 436
Fourier filters, see Filters
Fourier transform, defined, 18
    numerical calculation, 38, 453
Fraunhofer lines, 460
Free-free absorption, helium, 151
    negative hydrogen ion, 149
    neutral hydrogen, 145
Frosting, 83
f-values, abundances, 336
    defined, 226

Einstein coefficients, 226
    hydrogenic, 226

Gain, 90
Gas laws, 9, 137
Gas pressure, behavior in models, 178
    electron pressure-temperature, 171
    hydrostatic equation, 160
    measurement, 378
Gaunt factor, 144, 146
Gaussian profile, 23, 249, 419
G band, 6, 461
Geometrical depth, 106, 174
Ghosts, 63, 268, 283
Giant, 5
Gillieson system, 193, 197
Gradient method, 177, 312
Grain noise, 76, 97
Granulation, 97, 159, 417, 441
Grating, angular dispersion, 55
    anomalies, 63
    blaze, 55, 60
    diffraction pattern, 52–53
    equation, 54
    ghosts, 63
    resolution, 52
    ruling, 48
    shadowing, 62
    transmission, 50
    use of, 193, 260
Gravity, see Surface gravity
Gravity-temperature diagram, 385
Grey case, 132
Griem theory, 247

Halides, 75
H and K lines, 6, 322, 407, 417
HD curve, 86
Helium, absorption, 150
    abundance, 335, 354
    gravity, 386
    models, 185
H jerting function, 251
Holtsmark distribution, 241, 244
Hydrogen continuous absorption, 142
Hydrogen lines, absorption coefficient, 241
    measurement, 292
    pressure, 320, 380
    temperature, 314–315, 374
Hydrostatic equalibrium, 159

Image slicers, 45, 262
Image tubes, 97
Impact approximation, 231, 240
Induced emission, 109, 115, 141, 255
Inhomogeneities, 416

Instrumental profile, definition, 267
   detector, 91–92
   measurement, 282
   removal, 269
   scanner, 196
   scattered light, 283
Interaction constants, 231, 237–238, 239
Interferogram, 70–71
Interferometers, 69, 279, 362
Intermittency effect, 87
Ionic broadening, 241
Ionization equation, 13–14, 171

K-integral, 104, 128, 132
Knife-edge noise, 266

Lasers, 280–281
Level populations, 11, 302
Lifetime, 227
Light increasing power, 45
Limb darkening, 162, 397
Line, absorption, 208
   absorption coefficient, 221
   blanketing, 166
   profile, 259
   scanner, 265
   source function, 300, 304
Linearity, Fourier transform, 32
   photoelectric, 88
   photographic, 85
Lorentz profile, 225
LTE, curve of growth, 343
   continuum, 215
   excitation, 11
   ionization, 11
   lines, 302, 304, 305, 329
   models, 159
Luminosity criteria, 3, 5, 379
Lyman continuum, 142–143
Lyman ghosts, 66

Macroturbulence, confusion with rotation,
      430, 433
   defined, 330, 416
   Fourier analysis, 431
   line profiles, 421, 423, 300
   measurement, 429, 431
   observation, 441
   separation from rotation, 433
   spectrum synthesis, 350
Magnesium b lines, non-LTE, 304, 330
   pressure effects, 319
   Stark damping, 238
Magnetic fields, 91
Magnitude, 7, 216
Mass absorption coefficient, 106, 140, 155,
      221, 255

Maxwellian velocity distribution, see
      Velocity distribution
Mean intensity, 102, 128
Mechanical waves, 136, 416, 442
Mercury, bulb, 279–280
   spectrum, 281
Metal absorption, 152
Michelson interferometer, 69
Microdensitometer, 76, 265
Microturbulence, abundances, 332, 341, 347,
      349
   anisotropic models, 330, 420, 439
   defined, 416
   depth dependence, 439
   Fourier analysis, 436
   HR diagram, 438
   line computation, 419, 423
   measurement, 332, 427
   spectrum synthesis, 350
   temperature distribution, 300, 343
   velocity distribution, 419, 443
Milne's equations, 131
Mixing length, 139
MK spectral type, 3
Model photosphere, calculation, 158
   contribution functions, 176
   pressure relations, 159, 178
   spectrum, 175, 206
   tables, 186
   temperature distribution, 162, 166
Molecular opacity, 154
Molecules, 6, 154, 350
Multiplex advantage, 73

Natural damping, 222, 227, 250
Nearest neighbor approximation, 242
Negative absorption, 109
Negative helium ion, 150
Negative hydrogen ion, 147
Neutral hydrogen absorption, 142
Noise, 40, 94, 271, 274, 462
Non-LTE, see LTE
Non-thermal velocities, see Turbulence;
      Macroturbulence; and Microturbulence
Nyquist frequency, 36, 278

Occultation, 363
Opacity, see Absorption coefficient
Optical depth, 106, 175
Optimum filter, 274
Oscillations, see Turbulence
Oscillator strength, see f-value

Parsec, 449
Partition functions, 13, 455
Paschen continuum, 142–143, 192, 371
Path difference, see Interferometer

Permeability, 222
Permitivity, 222
Phase, 18, 20–21
Photoelectric, 75, 77
Photographic, characteristics, 75
  fog, 82
  linearity, 85
  quantum efficiency, 79
Photometric standards, continuum, 200, 204
  lines, 294
Photomultipliers, dark output, 82
  electron multiplication, 77
  fatigue, 89
  gain, 90
  linearity, 88
  quantum efficiency, 77, 79
Photon noise, 94
Photon statistics, 94
Photosphere, 1–2
Physical constants, 449
Planck's law, 108, 111, 116
Plane parallel approximation, 124, 159
Plate averaging, 97
Polarization, 45, 59–60, 266
Polychromator, 197
Populations, 11, 302
Power theorem, 34
Pressure broadening, 229
Pressure indicators, Balmer jump, 213, 379
  hydrogen lines, 320, 380
  strong lines, 319, 381
  weak lines, 319, 384
Pressure relations, 105, 178
Profile, defined, 259
  measurement, 259, 290
  rotation, 394
  temperature distribution, 300
  turbulence, 423
Pulse counting, 78
Pure absorption, 108

Quantum efficiency, 75, 78, 197
Quasi-static approximation, 241

Radial-tangential turbulence, 426, 433
Radiation constants, black body, 117
  Stefan-Boltzmann, 113
  Wien, 113
Radiation standards, black body, 118, 201
  standard stars, 204
  Vega, 203
Radiation pressure, 105, 161
Radiative equilibrium, 130, 166
Radiative gradient, 179
Radiative transfer, 121
Radii, absolute flux, 366

binaries, 363
bolometric flux, 365
B-V relation, 367
interferometer, 362
occultation, 363
speckle photometry, 360
surface brightness, 364
Rate equations, 302
Rayleigh-Jean's approximation, 113
Rayleigh scattering, 154
Reciprocity failure, 87
Reconstruction, 269
Reflection grating, 48, 55
Refrigeration, 82
Replica grating, 48
Residual flux, 312
Resolution, chromatic, 47
  Fourier, 34
  grating, 52
  dispersion, 265
  instrumental profile, 267
  slit, 68
  spacial, 91
  spectrogram, 260
Resonance broadening, 230
Response, see Spectral, response
Reynold's number, 417
RGU, 7
Rossiter effect, 392
Rotation, axial projection, 405
  binaries, 411
  catalog, 15, 393–394
  clusters, 409
  Fourier analysis, 402
  measurement, 400, 402
  profile, 394
  macroturbulence, 430, 433
  main sequence, 407
  rapid, 410
Rowland ghose, see Ghost
Ruling engine, 48, 50
Rydberg, 142

Saha equation, 13, 171
Sampling, 28, 36, 276
Satellites, 63
Saturation, 323, 346, 428, 429, 436
Scale, 44
Scaling, curve of growth, 337
  temperature distribution, 159, 168, 371
Scanners, 193, 265
Scattered light, correction, 288
  defined, 283
  measurement, 286
Scattering, extinction, 107, 108, 153
  lines, 303
  spectrograph, 283

Schwarzschild criterion, 137
Seeing, 45, 193, 265
Sensitivity, *see* Quantum efficiency
Shadowing, grating, 62
  secondary mirror, 45
Shah function, 26
Side lobes, microturbulence, 436
  rotation, 403, 433
Sinc function, 22, 451
Sine transform, 26, 27
Sky corrections, 198
Slit magnification, 68, 260
Slit width, 264; *see also* Spectral, purity
Sodium D lines, chemical composition, 327–
    328, 342
  gravity, 342, 381
  impact approximation, 241
  non-LTE, 304
  pressure dependence, 322
  solar, 330
  Stark damping, 238
  temperature dependence, 313–314
  van der Waals damping, 239
Solar, abundances, 351
  atlases, 15–16
  bolometric corrections, 217
  color index, 217
  curve of growth, 342, 344–345
  limb darkening, 162
  models, 165, 188
  physical parameters, 449
  turbulence, 440, 441
Sound waves, *see* Turbulence
Source function, defined, 107
  line, 300, 303, 305
  LTE, 159
Spacial resolution, 91
Specific intensity, 100
Speckle photometry, 360
Spectra, atlas of, 4–5, 6–7
Spectral, classification, 3
  purity, 47, 68, 260
  response, 78, 81
  type, 3
Spectrographs, coude, 261
  general, 46
  low resolution, 193; *see also* Scanners
  slit magnification, 68
Spectrophotometry, 259, 294
Spectrum, calculation, 175, 206
  continuous, 192
  line, 298
  solar, 305, 460
Spontaneous emission, 109
Spread function, *see* Instrumental profile
Standard, curve of growth, 344

energy distribution, 204
  lamps, 118, 201, 280
  lines, 294
  Vega, 200, 202, 203
Stark broadening, calculation, 237, 249
  defined, 231
  linear, 241
  pressure dependence, 322, 381
Static approximation, 241
Statistical approximation, 241
Statistical weight, 12
Stefan-Boltzmann law, 113, 116
Step wedge, 85
Stimulated emission, 109, 141
Strong lines, abundance, 323, 341
  comparison of observation and theory, 330
  non-LTE, 304, 305
  pressure dependence, 319, 322, 380, 381
  solar, 460
  temperature dependence, 317, 374
Subdwarf, 367
Subphotosphere, 1–2
Supergiant, 5, 433, 439
Supergranulation, 442
Surface brightness method, 364
Surface gravity, B-V relation, 389
  defined, 2
  measurement, 378
  see also Pressure indicators
Synchronous rotation, 412
Synthesis, 306, 349

Tables, H functions, 252
  ionization potentials, 455
  model photospheres, 186
  partition functions, 455
  physical constants, 449
  sinc function, 451
  solar spectrum, 460
  stars and data, 14
Telescopes, 43
Television, 97
Telluric lines, 279, 282, 331
Temperature, black body, 111
  continuum, 206, 215, 373
  curve of growth, 343
  detectors, 81, 82
  distribution, 2, 162, 166
  effective, 3, 367
  excitation, 338
  lines, 312, 370
  measurement, 359
  minimum, 2
  photospheric, 1–2
Theorems, 32
Thermal broadening, 248, 280

Threshold wavelength, 77
Transfer equation, 121
Transition probabilities, 109, 336
Turbulence, depth dependence, 439, 440
   Fourier analysis, 431, 436
   HR diagram, 438
   isotropic, 419, 422
   measurement, 427, 431, 436
   non-isotropic, 420, 426, 439
   spectrum, 418, 423
   velocity distribution, 418, 419, 422, 426
   see also Macroturbulence; Microturbulence
Two level atom, 304

UBV system, 7
uvby system, 7

van der Waals broadening, 231, 238, 322,
   378, 381
Velocity distribution, 10, 418, 419, 422

Vertical oscillation, 442
Viscosity, 417
Vidicons, *see* Television
Visual binaries, 387
Voigt function, 250

Wave equation, 222
Weak lines, abundance, 323
   behavior, 306
   depth of formation, 307
   profile, 307
   pressure dependence, 318
   temperature dependence, 312
   turbulence, 423, 429
White dwarf, 187
Wien's law, 112
Wiggly lines, 421, 442
Windows, 28, 36, 276
Wing approximation, 244
Wood's anomaly, 63